# Defining Right and Wrong in Brain Science

# Defining Right and Wrong in Brain Science

## Essential Readings in Neuroethics

*Walter Glannon, Ph.D., Editor*

Dana Press    *New York • Washington, D.C.*

**D** The Dana Foundation, 745 Fifth Avenue, Suite 900, New York, NY 10151
900 15th Street NW, Washington, DC 20005
DANA
PRESS
DANA is a federally registered trademark.

LIBRARY OF CONGRESS CATALOGING-IN-PUBLICATION DATA
Defining right and wrong in brain science : essential readings in neuroethics / Walter Glannon, editor.
     p. ; cm.
     Includes bibliographical references.
     ISBN-13: 978-1-932594-25-6
     ISBN-10: 1-932594-25-6
     1. Brain–Research–Moral and ethical aspects.    2. Neurosciences–Research–Moral and ethical
aspects.    I. Glannon, Walter.
     [DNLM:    1. Neurology–ethics–Collected Works.    2. Neurosciences–ethics–Collected Works.   WL 21
D313 2007]
     RC343.D33 2007
     174.2'968–dc22

                                        2007016052

Copyright Information: "Visions for a New Field of 'Neuroethics'," by William Safire: from *Neuroethics: Mapping the Field*, ed. S. Marcus (2002) © Dana Press.   "Neuroethics for the New Millennium," by Adina Roskies: reprinted from *Neuron*, Vol. 35, No. 1, 21–23 © (2002).   "Emerging Ethical Issues in Neuroscience," by Martha J. Farah: reprinted from *Nature Neuroscience* 5(11) 1123–1129 © (2002). This work is protected by copyright and it is being used with permission of *Access Copyright*. Any alteration of its content or further copying in any form whatsoever is strictly prohibited.   "Monitoring and Manipulating Brain Function: New Neuroscience Technologies and Their Ethical Implications," by Martha J. Farah and Paul Root Wolpe: reprinted from *Hastings Center Report* 34 (May-June 2004) 35–45. Used by permission. "Neuroscience and Neuroethics," by Donald Kennedy: reprinted with permission from *Science* 306 (15 October 2004) 373 © (2004) AAAS.   "From the 'Public Understanding of Science' to Scientists' Understanding of the Public," by Colin Blakemore: from *Neuroethics: Mapping the Field*, ed. S. Marcus (2002) © Dana Press.   "Ethical Issues in Taking Neuroscience Research from Bench to Bedside," by Alan Leshner: from *Cerebrum* (Fall, 2004) © Dana Press.   "Models for the Neuroethical Debate in the Community," by John Timpane: from *Cerebrum* (Fall 2004) © by Dana Press   "Neuroethics in a New Era of Neuroimaging," by Judy Illes: reprinted from *American Journal of Neuroradiology*, Vol. 24 (October) 1739–1741 (2003) © by American Society of Neuroradiology.   "Ethical and Practical Considerations in Managing Incidental Findings in Functional Magnetic Resonance Imaging," by Judy Illes, John E. Desmond, Lynn F. Huang, Thomas A. Raffin, and Scott W. Atlas: reprinted from *Brain and Cognition* 50 (3) 358–365 © (2002).   "Legal and Ethical Issues in Neuroimaging Research: Human Subjects Protection, Medical Privacy, and the Public Communication of Research Results," by Jennifer Kulynych: reprinted from *Brain and Cognition* 50 (3) 345–358 © (2002).   "Incidental Findings on Research Functional MR Images: Should We Look?" by Alex Mamourian: reprinted from *American Journal of Neuroradiology* 24 (April) 520–522 (2004) © by American Society of Neuroradiology.   "Imaging or Imagining? A Neuroethics Challenge Informed by Genetics," by Judy Illes and Eric Racine: © 2005, from *American Journal of Bioethics* 5(2) 5–18. Reproduced by permission of Taylor & Francis Group, LLC., http://www.taylorandfrancis. com.   "Brains, Genes, and the Making of the Self," Lynette Reid and Francoise Baylis: © 2005 from *American Journal of Bioethics* 5(2) 21–23. Reproduced by permission of Taylor & Francis Group, LLC., http://www.taylorandfrancis.com.   "The Neural Basis of Social Behavior: Ethical Implications," by Antonio Damasio: from *Neuroethics: Mapping the Field*, ed. S. Marcus (2002) © Dana Press.   "Neuroscience: Reflections on the Neural Basis of Morality," by Patricia Smith Churchland: from *Neuroethics: Mapping the Field*, ed. S. Marcus (2002) © Dana Press   "My Brain Made Me Do It," by Michael Gazzaniga: from *The Ethical Brain* © 2005 Michael Gazzaniga. Used by permission.   "New Neuroscience, Old Problems: Legal Implications of Brain Science," by Stephen J. Morse: from *Cerebrum* (Fall, 2004) © Dana Press.   "Moral Cognition and Its Neural Constituents," by William D. Casebeer: reprinted by permission from Macmillan Publishers Ltd: *Nature Reviews Neuroscience* 4(10), 840–847 © (2003).   "From Neural 'Is' to Moral 'Ought:' What Are the Moral Implications of Neuroscientific Moral Psychology?" by J. D. Greene: reprinted by permission from Macmillan Publishers Ltd: *Nature Reviews Neuroscience* 4 (8), 847–850 © (2003).   "Psychopharmacology and Memory," by Walter Glannon: published in *Journal of Medical Ethics* (2006) © Reprinted by permission of the author.   "Shall We Enhance? A Debate," Arthur Caplan and Paul McHugh: from *Cerebrum* (Fall, 2004) © Dana Press.   "Neurocognitive Enhancement: What Can We Do and What Should We Do?" by Martha J. Farah, Judy Illes, Robert Cook-Deegan, Howard Gardner, Eric Kandel, Patricia King, Erik Parens, Barbara Sahakian, and Paul Root Wolpe: reprinted by permission from Macmillan Publishers LTD: *Nature Reviews Neuroscience* 5(5), 421–425 © (2004).   "The Promise and Predicament of Cosmetic Neurology," by Anjan Chatterjee: from *Journal of Medical Ethics* 32 (2006) 110–113. Used with permission from the BMJ Publishing Group Ltd.   "Brain Death in an Age of Heroic Medicine," by Guy M. McKhann: from *Cerebrum* (Fall, 1998) © Dana Press.   "Constructing an Ethical Stereotaxy for Severe Brain Injury: Balancing Risks, Benefits, and Access," by Joseph J. Fins: reprinted by permission from Macmillan Publishers Ltd: *Nature Reviews Neuroscience* 4(4), 323–327 © (2003).   "Hope for 'Comatose Patients'" by Nicholas D. Schiff and Joseph J. Fins: from *Cerebrum* (Fall, 2003) © Dana Press.   "Rethinking Disorders of Consciousness: New Research and Its Implications," by Joseph J. Fins: Reprinted from *Hastings Center Report* 35, March/April 2005, 22–24. Used by permission.   "Ethics in a Neurocentric World," by Steven Rose: from *The Future of the Brain: The Promise and Perils of Tomorrow's Neuroscience*, by Steven Rose; Chapter 12, 297–305 © 2005 Steven Rose; by permission of Oxford University Press, Inc.

www.dana.org

*For Yee-Wah*

# Contents

# Acknowledgments

I thank the editorial and production staff at Dana Press for their work in bringing this book from its inception to its completion. Kris Pauls, Donna Deaton, Ellen Davey, Elizabeth Rich, and Laura Rausch were all very helpful. I am most grateful to Jane Nevins, editor in chief at the Press. She enthusiastically supported this project from the time I first proposed it, gave me insightful and valuable feedback on my general introduction and section introductions, and skillfully resolved many editorial problems at every stage along the way. She is an ideal editor. The writing of this book was supported by the Canada Research Chairs program, which is gratefully acknowledged.

# Introduction

In May 2002, 150 of the best minds in the United States, Canada, the United Kingdom, and other countries met at a conference in San Francisco to discuss the most difficult emerging ethical issues in basic and clinical neuroscience. The participants reached consensus that it would be perilous to ignore these issues and that there was an urgent need to establish parameters for a new field of "neuroethics" to deal with them. The readings in this book capture the seminal thinking of the conference and the key evolving ethical concepts about cutting-edge developments in neuroscience in the five years since. They establish a strong framework for identifying and managing ethical problems arising from future applications and practices of brain science.

Long before the U.S. Congress proclaimed the 1990s the "Decade of the Brain," scientists and medical practitioners had been directly or indirectly intervening in the human brain. Egyptian, Greek, Roman, Indian, Chinese, Aztec, and Incan cultures practiced trepanning for thousands of years. This is a form of neurosurgery in which a hole is drilled or scraped into the skull. Trepanning was performed to treat epileptic seizures, headaches, and mental disorders. In the second century CE, the Greek physician Galen recommended the use of electric eels for treating headaches and facial pain. This could be described as the first example of electrical brain stimulation. Centuries later, in the 1930s, Italian psychiatrists Ugo Cerletti and Lucio Bini used electroconvulsive therapy to treat schizophrenia. More significant for our understanding of the eventual emergence of qualms about intervening in the brain and the mind was the research of English neurologist David Ferrin. In the 1880s, Ferrin and his colleagues showed that direct electrical stimulation of the brain could change behavior. This may have been the first indication that certain structures and functions of the brain correlated with certain mental states and behavior. Similar insights were obtained by Canadian neurosurgeon Wilder Penfield. Beginning in the 1930s, Penfield would carefully place an electrode on the surface of the exposed temporal lobe of some of his patients undergoing surgery for epilepsy. By eliciting memories of patients' long-forgotten experiences through electrical stimulation of the temporal lobe, Penfield provided evidence that our episodic memories are formed and stored in this region of the brain.

Not all neuroscientists and the public have embraced all interventions in the human brain. Indeed, some neurosurgical techniques have been controversial and deeply disturbing to our moral sentiments. Psychosurgery has been the most controversial of these techniques. The Portuguese neurologist Egas Moniz introduced psychosurgery in 1936 when he performed a prefrontal leucotomy, a procedure that later came to be known as frontal lobotomy. He designed this procedure believing that it could relieve symptoms of psychiatric illnesses. It consisted of destroying parts of the frontal lobes and their connections to other brain regions. Many applauded the anecdotal success stories of the operation, and Moniz received the Nobel Prize for medicine in 1949 for the "therapeutic" value of leucotomy. American neurologist Walter Freeman performed some 3,500 frontal lobotomies in the United States in the 1940s and 1950s. Although lobotomies relieved some symptoms of psychiatric illnesses, they often had severe neurological and psychological consequences. These included significant personality changes, seizures, apathy, loss of muscle control, and in some cases death.

Spanish neurophysiologist José Delgado was another mid-twentieth-century pioneer in brain science whose work was as controversial as that of his predecessors. Delgado introduced brain-implant technology with the development of the brain chip, an electronic device that could manipulate the mind by receiving signals from and transmitting them to neurons. This device was the result of Delgado's research into behavior and motor control, and his aim in implanting electrodes in the brain was to avoid lobotomies. In the 1960s and early 1970s, Delgado implanted a brain chip in 25 human subjects with schizophrenia and epilepsy. Although these individuals were desperately ill and their conditions failed to respond to all other treatments, this practice was so contentious that it was discontinued. The advent of psychotropic drugs in the 1950s obviated many of the neurosurgical interventions that I have described. Nevertheless, variations of some of these procedures are performed on some patients whose neurological and psychiatric conditions fail to respond to drug therapy. One update of surgical intervention is deep-brain stimulation, now in experimental use for Parkinson's disease, obsessive-compulsive disorder, depression, and restoring a greater degree of neurological function and consciousness for individuals in a minimally conscious state. Other forms of brain stimulation to treat different brain disorders include electroconvulsive therapy (ECT) and, although still experimental, transcranial magnetic stimulation (TMS) and vagus nerve stimulation (VNS).

The most significant advances in brain science have been in neuroimaging and psychopharmacology. These modalities have had an enormous global impact on the health and well-being of millions of people in the way they have improved the diagnosis and treatment of a wide range of neurological and psychiatric conditions. Antipsychotic and antidepressant drugs can control and relieve symptoms of schizophrenia and depression in many people who suffer from these disorders. Newer-generation dopamine agonists have improved treatment for Parkinson's by doing more to control the progression of motor difficulties associated with this disease. Drugs such as memantine and donepezil may slow memory loss in people with Alzheimer's. Computed tomography (CT), magnetic resonance imaging (MRI), positron emission tomography (PET), and functional magnetic resonance imaging (fMRI) can display the structural and functional neurobiological bases of our beliefs, emotions, memories, and other mental states. Brain imaging can enable researchers and clinicians to monitor the progression of neurological and psychiatric conditions, as well as the metabolic effects in the brain of drugs given to treat these conditions. Imaging might also reveal the first signs of neurodegeneration and thus play an important role in predicting future diseases of the brain. As you will see, though, how we should interpret these pictures is far from settled.

It is no wonder that the ability to map, monitor, and manipulate the brain raises many ethical questions. What can brain scans tell us about our capacity for moral reasoning and whether individuals are able or unable to control their impulses when they act? Will images of the brain change our understanding of free will and moral responsibility? Predictive neuroimaging showing brain anomalies may enable some people to plan their future in a more rational and prudential way. Yet the same images may cause anxiety in other people and give them a negative view of the future. Given the uncertainty of what these images mean, what should researchers or clinicians tell individuals who have anomalous brain images? Refinements in brain imaging may lead to a better way of detecting lying and our preferences for consumer goods. But will these refinements threaten the privacy of our thoughts? Will higher-resolution brain scans make neurologists more certain in diagnosing brain death? What obligations do brain researchers have to the public in explaining their findings about the brain?

Therapeutic psychopharmacology can alleviate symptoms of brain disorders. But given that the brain is by far the most complex organ in the human body, there may be unforeseen adverse effects of chronically altering brain circuits with psychotropic drugs. How do we weigh their short-term benefits against their long-term risks? Some drugs may be used to enhance normal cognitive capacities. But are the benefits of performing better on exams or having better memory worth any risk to other mental functions? Will all people have equal access to drugs that enhance cognition? Or will the limited availability of these drugs exacerbate social inequality?

How far should we go in developing and applying brain imaging, psychopharmacology, and other neurotechnologies? Should there be limits on these technologies? And who should decide? These are the main ethical questions in brain science, questions that make up the field of neuroethics.

The ethics of brain science has been a subject of great interest and debate for some time. According to William Safire, the first meeting on this subject took place in 1816 on Lake Geneva. In the 1990s French neuroscientist Jean-Pierre Changeux used the term "neuroethics" at a symposium on biology and ethics at the Pasteur Institute in Paris. One can find even earlier references to this term in the neuroscientific literature. But 2002 was the most significant and eventful year for this field. In addition to the San Francisco conference mentioned earlier, another meeting on this same topic was held in London. The *Economist* ran a lead article in May 2002 titled "The Future of Mind Control," which provocatively speculated on the possible uses and abuses of neurotechnology. It was at the San Francisco conference that Safire gave the most perspicuous definition of "neuroethics": "the examination of what is right and wrong, good and bad, about the treatment of, perfection of, or unwelcome invasion of and worrisome manipulation of the human brain."

"Right" and "wrong" are fundamental terms in normative ethics, the moral standards or norms that specify reasons to guide our behavior and tell us what we should or should not do. An action or policy is right if it is supported by a decisive or overriding reason or set of reasons for doing or implementing it. An action or policy is wrong if there is a decisive or overriding reason or set of reasons for not doing it or not implementing it. In many cases, an action or policy deemed right will be obligatory, and an action or policy deemed wrong will be prohibited. Actions and policies are permissible when there is no decisive reason for or against them. Reasons will exist for doing or implementing them, but not decisive or overriding ones. In determining whether something is right, wrong, or permissible, the reasons we cite to determine this will justify what we do or refrain from doing. This is the formal theoretical framework.

But it is important to emphasize that neuroethics involves more than just normative ethical theory in philosophy. Neuroethics lies at the intersection of the empirical brain sciences, normative ethics, the philosophy of mind, law, and the social sciences of anthropology, economics, psychology, and sociology. Accordingly, the "ethics" in "neuroethics" needs to be informed by all of these disciplines and construed broadly enough to reflect their influence on our norms of behavior.

This book puts most of its emphasis on different ethical, legal, and social dimensions of neuroimaging and psychopharmacology. To be sure, electrical and magnetic brain stimulation raises important normative questions about how to weigh its benefits and risks. Still, the number of people who have been or will be affected by brain imaging and psychotropic drugs is much greater than the number affected by other neurotechnologies, which makes these chapters timely and most worthy of discussion. Looking further into the future, we realize that neuroscience and technological medicine will likely transform cyborgs—integrated complexes that are part human and part machine—from imagination to reality. Brain-machine interfaces using neural implants will enable us to control how we translate our thoughts into actions. These neurotechnologies will force us to rethink our understanding of what makes us human. They will raise questions about the mechanization of humans and the humanization of machines. They will challenge our traditional views of personhood, personal identity, agency, and the self and in turn our ethical practices of praising and blaming, excusing, and holding individuals responsible for their behavior. Future neurotechnologies may tempt us to accept neuroscientific reductionism, the view that we are "nothing but a bunch of neurons," or mechanical replacements for them. Yet as Steven Rose points out in the Epilogue, there are persuasive reasons for rejecting this view. Instead, we should conceive of our minds as a reflection of a tripartite interaction among brain, body, and environment. We will need to ask how future neurotechnologies will affect this tripartite unity and the features of our minds that make us the persons we are.

But our existing and evolving ability to map, monitor, and manipulate the brain obliges us to confront a more current set of ethical questions. The authors of the readings included here discuss these questions in the core areas of neuroscience in an insightful and thought-provoking way. They discuss the role that neuroscience can and should play in our lives. Their observations, claims, and arguments provide a framework within which we can collectively debate with a view to defining the rightness or wrongness of different measures of and interventions in the human brain.

*Part I*

# Foundational Issues

W ILLIAM SAFIRE'S opening remarks at the 2002 neuroethics conference in San Francisco established a framework for discussion of a wide range of ethical issues in various applications of brain science. Later that same year, Adina Roskies and Martha Farah published two seminal papers in which they explored the emerging ethical issues in the field. In 2004, Farah and Paul Root Wolpe published a paper that elaborated the principal themes and concerns of the earlier articles. Around the same time, Donald Kennedy concisely expressed the main ethical concerns about neuroscience in an editorial that appeared in *Science*. In charting the ethical, social, and legal landscape in basic and clinical neuroscience, the authors of the articles in this part of the book identify and discuss existing and future challenges generated by our ability to map, monitor, and alter brain functions.

In addition to defining the nature and scope of the subject, Safire's speech, "Visions for a New Field of 'Neuroethics'," raises a set of provocative hypothetical questions to initiate exploration of some of the ethical implications of intervening in the brain. These questions include whether individuals diagnosed with mental illness can give informed consent to participate in clinical trials, positive and negative aspects of memory repression and enhancement, and the possibility of using brain imaging to detect lying. The ethical import of these questions rests on the fundamental fact that the brain "is the organ of individuality." Safire asserts that we need to consider who should set the ethical rules for these practices and whether or when such rules might be justified.

In "Neuroethics for the New Millennium," Roskies argues that although neuroethics overlaps with traditional issues in bioethics, it should not be categorized as simply a subdivision of bioethics. The connection between the brain and behavior, and between the brain and the self, involves a distinctive link between neuroscience and ethics and raises a distinctive set of ethical questions. Roskies draws a general distinction between the ethics of neuroscience and the neuroscience of ethics, which pertains to the neurobiological basis of moral reasoning. She then subdivides the ethics of neuroscience into the ethics of practice and the ethical implications of applied neuroscience. The ethics of practice pertains to traditional bioethical questions such as informed consent to treatment and research, substitute decision making, and the best interests of patients. The ethical implications of neuroscience are truly novel, since the ability to intervene in the brain through imaging, pharmacology, advanced surgical techniques, and electrical and magnetic stimulation is a fairly recent phenomenon. Roskies asks such thought-provoking questions as whether neuroscientific techniques like brain imaging and psychopharmacology for therapy and enhancement will alleviate or exacerbate social inequality. Addressing the neuroscience of ethics, she asks whether increasing knowledge about the brain will change our views about moral and legal responsibility. In addition, she raises questions about how information from brain scans might be used for lie detection, as well as the implications of these scans for privacy and confidentiality. At a deeper level, Roskies asks whether our increasing knowledge of the brain would redefine what we mean by "normal" and whether neuroscientific possibilities will alter our understanding of what makes us human. She concludes by arguing that neuroethics should not be confined to specialists but should also involve public debate with broad social participation.

Farah examines a more specific set of current and future neuroethical issues in "Emerging Ethical Issues in Neuroscience." She discusses pharmacological manipulation to enhance attention and memory as well as mood. Farah cautions that these interventions may have unforeseen adverse effects, which should make us wary of the idea that enhancement of cognition and mood could be a "free lunch." She also asks whether cognitive enhancement would undermine the value of achieving goals through one's own efforts and points out how court-ordered pharmacological interventions for violent offenders could violate individual freedom and dignity. Shifting to neuroimaging, Farah notes that the prospect of brain reading could result in a breach of privacy of a person's mind. Moreover, she points out that information from brain scans has an illusory accuracy and objectivity. Abuse of this information could lead to various forms of discrimination against individuals. Farah concludes by speculating on the future of neurotechnology and the ethical questions that will arise from it, questions that will require dialogue among neuroscientists, policy makers, and the public.

Farah and Wolpe's "Monitoring and Manipulating Brain Function" is an extended discussion of some of the issues examined by Farah. They explain the similarities and differences between neuroscience and genetics. Like some of the authors later on in Part V, Farah and Wolpe emphasize that neuroscience is distinctive because there is a more direct relation between the brain and the self than there is between genes and the self. They divide their discussion of ethical issues in neuroscience into two parts: the ability to monitor brain function through different types of neuroimaging; and the ability to alter the brain through pharmacology. In addition to the promises and pitfalls of lie detection, neuroimaging may reveal people's preferences and other sensitive information about the brain and mind. Turning to cognitive enhancement, Farah and Wolpe discuss safety issues surrounding psychopharmacology and underscore the need for professional and public awareness of the risk of unanticipated problems resulting from pharmacological alteration of the brain. They also ask whether such alteration will change our view on what it means to be a person, do meaningful work, and take pride in one's achievements.

In "Neuroscience and Neuroethics," Donald Kennedy reflects on the same generic questions addressed by the previous authors. He considers what functional magnetic resonance imaging (fMRI) might mean for our understanding of behavior, including how it has given rise to new disciplines such as "neuroeconomics." Further, he describes different uses of psychopharmacology and distinguishes between neurocognitive therapy and enhancement. In particular, he asks what bothers us ethically about altering brain states for the purpose of improving our cognitive abilities. Turning to the question of free will, Kennedy maintains that neurobiology will not threaten our belief in free will because it is unlikely that our knowledge of the brain will ever mean that it is identical with or determines the mind. He worries, however, that brain imaging may threaten the privacy of our thoughts, thus suggesting the need for debate and guidelines on this and possibly other neuroscientific techniques.

The five articles in this section provide the necessary framework for discussion of a wide range of ethical questions in brain science. They insightfully examine current problems and speculate on future challenges germane to diagnostic and predictive neuroimaging, as well as potential problems in the use of psychopharmacology for treatment and enhancement. These papers set the stage for further consideration of the relevant neuroethical issues in the remaining parts of this book.

# Chapter 1

# Visions for a New Field of "Neuroethics"*

*William Safire*

The first conference or meeting on this general subject was held back in the summer of 1816 in a cottage on Lake Geneva. Present were a couple of world-class poets, their mistresses, and their doctor. They'd been reading and discussing the disturbing works of Erasmus Darwin—who later had a grandson named Charles Darwin—about the creation of artificial life.

It was near the end of the Enlightenment, the era when, with the world in an intellectual revolt against the despotism of kings and the power of the clergy, a philosophy of rationalism and tolerance had burst upon the scene and, with it, political revolution in America and in France. Thinkers wrote about the perfectibility of man. Minds had been opened, morality reexamined, even as some of the reformers brought on their own reign of terror, and a conservative reaction was setting in.

One of the poets at that lakeside gathering, Lord Byron, had a bright idea to enliven the discussion. "Let's each of us write a ghost story," he suggested. He tried and couldn't get started. His friend, Percy Bysshe Shelley, also had a go, but quickly set it aside. Their doctor came up with a sorry tale about a vampire. And Byron's mistress only wanted Byron.

The young woman with Shelley, however, was caught up in the terror of the manipulation of life by the new science. She was the strong-minded daughter of Mary Wollstonecraft, the pioneer feminist and moral rebel. And

*From *Neuroethics: Mapping The Field* (Dana Press, 2002): 3–9.

her father was William Godwin, the social philosopher and anarchist. She wrote her ghost story and married her poet. Two years later, Mary Shelley's *Frankenstein: The Modern Prometheus* was published. Prometheus, you remember, was the god who was tortured for all eternity for bringing to man godlike powers.

In our time, two centuries later, man's Promethean presumption to create life, to interfere with what had been the exclusive domain of God or nature, is being fiercely debated all around the world. Europe is consumed with controversy about the genetic modification of foods, and, with a nod to Mary's monstrous creation, the improved—or at least manipulated — products are derided as "Franken foods." The fear of playing God, as well as the countervailing hope of creating life-saving life in the laboratory, roils the public reaction to science's breakthroughs in our own new enlightenment.

Welcome to the first symposium on one specific portion of that two-century-long growing concern: neuroethics—the examination of what is right and wrong, good and bad about the treatment of, perfection of, or unwelcome invasion of and worrisome manipulation of the human brain.

It's fitting that the Dana Foundation be the conference's sponsor. For the past decade we've been focused on the brain, not only by directly funding researchers in many fields of neuroscience—from brain imaging to neuro-immunology—but also by marshaling other support, both private and public, for brain research.

The Dana Alliance for Brain Initiatives in the United States and the European Dana Alliance for the Brain are a network of more than 300 leading neuroscientists, including more than a dozen Nobel laureates, actively reaching out to explain their work and offer help to the general public. It has been successful, and this fall in Washington, D.C., and next year in London we'll be opening centers to let more scientists, philosophers, critics, and even newspaper pundits engage in informed discussions.

One of our founding alliance members and driving forces, Zach Hall, suggested that the time was right for this conference in this place. And Dana Foundation president Ed Rover, executive vice president Francis Harper, and I were pleased to work with Barbara Koenig and Judy Illes at Stanford to help make it happen.

The only field in which I can claim some expertise in this crowd is the English language. I'm a pop grammarian and etymologist, and I regularly get asked questions like "Where does 'the whole nine yards' come from?" It comes from the capacity (measured in cubic yards) of a cement truck. I once wrote that, and I got a bunch of letters back from people saying, "It's not a cement truck, it's a *concrete* truck."

Another question I regularly get asked is, "What's the difference between ethics and morals?" The Latin *moralis* was formed by Cicero as a rendering of the Greek *ethikos,* and the words have been used interchangeably ever since. But in their usage a distinction can be drawn. To me, "moral" has to do with right and wrong, and *ethics* with good and bad.

Now, what's right is good and what's wrong is bad, so there's a lot of overlap. But I think of the difference this way: "moral" implies conformity to long-established codes of conduct set primarily by religious authorities, while "ethical" involves more subtle questions of equity. The moralist asks, "Is it right by intrinsic standards?" The ethicist asks, "Is it fair in the light of this society's customs and in these times?" Moral connotes standing firm; ethical, while still pretty stiff, can be said to swing a little.

Neuroethics, in my lexicon, is a distinct portion of bioethics, which is the consideration of good and bad consequences in medical practice and biological research. But the specific ethics of brain science hits home as research on no other organ does. It deals with our consciousness—our sense of self—and as such is central to our being. What distinguishes us from each other beyond our looks? The answer: our personalities and behavior. And these are the characteristics that brain science will soon be able to change in significant ways.

Let's face it: one person's liver is pretty much like another's. Our brains, by contrast, give us our intelligence, integrity, curiosity, compassion, and— here's the most mysterious one—conscience. The brain is the organ of individuality.

Zach Hall has made the point that when we examine and manipulate the brain—unlike the liver or, as Art Caplan would have it, the pancreas— whether for research, treatment of disease, or perhaps sinister political ends, we change people's lives in the most personal and powerful way. The misuse or abuse of this power, or the failure to make the most of it, raises ethical challenges unique to neuroscience. What's more, neuroscientists have a built-in conflict of interest that sets them apart from all other ethicists.

Everybody's brain has a personal, selfish interest in the study of the brain. It is the ultimate in self-dealing. Won't a human brain tend to do what's best for itself and take charge and take chances, plunging ahead to treat or improve the brain, as the brain might not do for the same body's liver? In possession of this power of self-improvement, of "perfectibility," how will we define and protect the integrity of our ability to judge morally and conduct ourselves ethically?

I hope the proceedings today and tomorrow will concentrate on the special challenges of neuroethics and not keep punching away at the bioethical hot buttons of embryonic stem cells and cloning so heavily debated elsewhere. Here are a few examples of the questions I hope we cover:

Remember the psychosurgeries for aggression some forty years ago? What ethical rules or legal regulations should there be for treatment to change criminal behavior?

Suppose we could develop a drug to make someone less shy, or more honest, or more intellectually attractive, with a nice sense of humor. What is there to stop us from using such a "Botox for the brain"? More seriously, if a person's brain is impaired by disease, injury, or mental illness, and he or she cannot give informed consent, who is to decide when participation in a clinical trial is humane and proper? Doctor, relative, researcher, insurer, or court?

Should we develop a drug to improve memory or to repress painful remembrances? Or to help a prosecutor elicit a professedly forgotten detail? Is it fair to implant a chip in the brain to enhance memory before an academic examination? Or is that like giving a steroid to an Olympic athlete? And here's one for the defenders of privacy: Is the imaging of suspected terrorists' brains to detect lying a form of torture, or at least a way of forcing people to incriminate themselves?

As we learn that memory is not fixed, but is constantly being reshaped as reminiscences are recalled and stored again, how do we even *define* truthful testimony and judge its reliability?

In discussions of ethics in every field there's a "but what if?" factor that fuzzes clear lines. A doctor considers it an ethical responsibility to inform a patient of the seriousness of his or her illness, but what if the patient is depressive and a suicide risk? A geneticist may consider it ethical to warn a person of the likelihood of some great vulnerability, but what if that means the patient won't be able to get insurance?

A journalist considers it unethical to reveal a source who was promised confidentiality. But what if the source turns out to be lying, or the source has evidence to save an accused from jail?

I'd like to hear about some of these "but what if 's" and other questions in neuroethics. The people in this room are better equipped than most to take them on, and they may proceed today and tomorrow to carve out new territory for an old philosophical discipline. This could well be a historic meeting that participants will look back on with great pride and that others will talk about as a seminal moment in the development of this new field.

I expect that a book of your conference papers will be published along with some of the lively and profound give-and-take. It won't have the sales of *Frankenstein*, and Boris Karloff won't star in the movie, but it might help, as you put it, map the field.

Thank you.

WILLIAM SAFIRE is chairman of the Dana Foundation and a columnist for the *New York Times*.

*Chapter 2*

# Neuroethics for the New Millennium[*]

*Adina Roskies*

The past several months have seen heightened interest in the intersection of ethics and neuroscience. In the popular press, the topic grabbed headlines in a May issue of *The Economist*[1] and was featured in a *New York Times* editorial[2]. Professional societies were a step ahead, staging several meetings devoted to ethics and neuroscience since the beginning of this year. In January 2002, *Neuron* and the AAAS sponsored a symposium entitled "Understanding the Neural Basis of Complex Behaviors: The Implications for Science and Society," which brought together a panel with expertise in the neurosciences, policy, ethics, and the law to discuss the recent advances in the neurosciences and their potential implications for science and society. The Royal Institution in London sponsored "Neuroscience Future" in March, and in May the Dana Foundation, in collaboration with Stanford University and the University of California at San Francisco, sponsored a conference boldly entitled "Neuroethics: Mapping the Field" (NMTF). The name "neuroethics" implies that such a field exists, an "unexplored continent lying between the two populated shores of ethics and of neuroscience," in the words of Al Jonsen, an organizer of the NMTF conference.

My aim here is to delineate what I see as the basic structure of this na-

*From *Neuron* 35 (July 3, 2002): 21–23.

scent field and to lay out some of the fundamental questions with which it is concerned. The views here have been informed by the presentations and discussions of numerous people who participated in these conferences, but also strongly reflect my personal perspective on what the field of neuroethics should be.

It is evident that neuroethics will overlap substantially with traditional issues in biomedical ethics. For instance, much of the recent work in ethics spurred by the Human Genome Project will be applicable, with perhaps slight modification, to some neuroethical problems. But if there is to be justification for identifying and promoting neuroethics as a new and important field, it ought not be merely a subdivision of bioethics, with issues and biomedical research. The intimate connection between our brains and our selves generates distinctive questions that beg for the interplay between ethical and neuroscientific thinking. The motivation for the newfound interest in bringing together neuroscientists, ethicists, journalists, philosophers, and policy makers arises from the intuition that our ever-increasing understanding of the brain mechanisms underlying diverse behaviors has unique and potentially dramatic implications for our perspective on ethics and for social justice. These are the issues that warrant the introduction of a new era of intellectual and social discourse.

As I see it, there are two main divisions of neuroethics: the ethics of neuroscience and the neuroscience of ethics. Each of these can be pursued independently to a large extent, but perhaps most intriguing is to contemplate how progress in each will affect the other.

## The Ethics of Neuroscience

The ethics of neuroscience can be roughly subdivided into two groups of issues: (1) the ethical issues and considerations that should be raised in the course of designing and executing neuroscientific studies and (2) evaluation of the ethical and social impact that the results of those studies might have, or ought to have, on existing social, ethical, and legal structures. Let me call, for convenience, the first the "ethics of practice" and the second the "ethical implications of neuroscience." For the most part, the ethics of practice is where traditional bioethics, as applied to neuroscience, resides. It includes familiar issues like optimal clinical trial design, guidelines for use of fetal tissues or stem cells or cloning, privacy rights to results of testing for neurological disease, and so on. However, the ethics of practice includes

some questions peculiar to neuroethics. For instance, in a liberal democratic society such as ours self-determination is highly prized, and hence the
importance of informed consent is central to medical practice and medical ethics. But neurodegenerative diseases and psychiatric disorders may impair cognition so that informed consent, as generally conceived, may be impossible. What guidelines should be in place for treatment or experimental
participation in these cases? We also take it for granted that when making
medical decisions, patients will choose what is in their best interests. Some
disorders of brain chemistry, such as depression, defy such an assumption.
Who should wield executive power when the subject cannot be counted on
to choose what is best for himself or herself?

The second subdivision of the ethics of neuroscience, the "ethical implications of neuroscience," is the area of neuroethics that is truly novel, and
perhaps the most ripe for advancement. Its aim is to investigate the implications of our mechanistic understanding of brain function for society, and
it will require integrating neuroscientific knowledge with ethical and social
thought. Advances in neuroscience have the potential to create, and to remedy, serious social inequities. How we use our knowledge will shape our
society. How, as we learn more and more about how the brain controls behavior and the causes of mental dysfunction, are we to reconcile this new
knowledge with the social structures that allow our society to run more or
less smoothly? For instance, it has been suggested that a large proportion of
inmates on death row may have damaged or injured brains. If careful epide-
miologic studies establish that this is the case, how should our views about
moral and legal responsibility change, if at all, to accommodate this surprising fact? In the future (but not currently!), it may be possible to use noninvasive imaging techniques to determine whether a person is lying. There
is some indication that such imaging technology could be used to distinguish real from false memories. Perhaps even farther in the future we could,
with some degree of certainty, diagnose behavioral dispositions, motivations,
or beliefs. In what cases can such information be used ethically? What are
the privacy issues associated with thought? What are the consequences of
reliable, but not perfect, diagnostic techniques? If someone knows that he
or she is at risk for, for example, a psychotic episode, should he or she be
held legally responsible for actions undertaken while delusional by virtue
of not having prevented the episode? How should decision making proceed
in the face of probabilistic predictions of behavior? Will the results of some
predictive studies become self-fulfilling prophecies? Once our technology
provides us access to the full spectrum of physiological states underlying

behavior, will our common practice of identifying certain behaviors as normal or abnormal have meaning? Will our knowledge prompt us to redefine what "normal" is? Such issues will undoubtedly arise as our technology and understanding of human cognition improves, and it would be well to have thought them through before we are faced with them in the flesh.

It is easy to see how consideration of the ethical implications of neuroscience will affect issues generally thought to be in the realm of the ethics of practice. Criteria for life and death are currently linked to gross generalizations about brain function. Better understanding of developmental and cognitive processes may allow us to refine these notions to better identify life and death, biological benchmarks important for a number of policy-related issues such as abortion, termination of support, etc. Other definitions are extremely relevant to the daily lives of many: what is considered a disability, what will insurance pay for? As we understand more fully the varieties of ways in which brains operate, it will be an increasing challenge to define these terms in a way that does not marginalize, but nonetheless protects, the disabled, the at-risk, and the disenfranchised. At the same time, we need to be aware of the risk of overextending these concepts and of mitigating diversity through medication. The pharmaceutical advances that will surely stem from neuroscientific ones will present the following question: when can drugs be ethically used to enhance normal capacities, rather than to just treat deficits? What effect will our policies have on existing social and economic disparities? How can regulations be enacted to promote fairness and equality? What might be the consequences for our society or another if we fail to safeguard fair and equitable access to such enhancements? We will also have to come to terms with the charge that chemical and technological enhancements make us less than human or "post-human"[3] and perhaps revisit the question of what it is to be human.

## The Neuroscience of Ethics

The second major division I highlighted is the neuroscience of ethics. Traditional ethical theory has centered on philosophical notions such as free will, self-control, personal identity, and intention. These notions can be investigated from the perspective of brain function. Although the neuroscience of ethics today is far less developed than the ethics of neuroscience, and may not progress as quickly at first, it will be the area with truly profound implications for the way ethics, writ large, is approached in the 21st century.

Already there are signs of a surge in interest in investigating the brain bases of moral cognition, and such studies are bound to burgeon in the coming years. How are decisions made in the brain? How are values represented? How are ethical decisions similar to or different from other types of decisions? Many thinkers have assumed ethical reasoning to be a variety of rational thought. But recent evidence suggests that emotions play a central role in moral cognition[4,5]. Does this undermine the view of ethics as rational or instead undermine the long-cherished division between reason and emotion? How will a better understanding of the biological basis of moral cognition and behavior modify our philosophical ethical framework? How will it affect ingrained notions of rationality and its importance to human existence?

Many of us overtly or covertly believe in a kind of "neuroessentialism," that our brains define who we are, even more than do our genes. So in investigating the brain, we investigate the self. What is the neural representation of "self" dependent upon? Is personal identity a brain-based notion? What consequences for our concepts of personal identity will alterations of the self-defining parts of us have? Some current interventions (and undoubtedly more in the future) will be such as to perhaps improve the health and functioning of the patient, but perhaps at the expense of altering the brain chemically or mechanically. Will certain medical or technological therapies change who we are? What are the ethical implications of such changes? Will we have to weigh the costs of biological death against continued life but destruction of our selves? Advances in neuroscientific research in relevant areas may change the very fabric of our philosophical outlook on life.

Although neuroscience is unlikely to answer metaphysical questions about determinism, it can certainly alter our perceptions of them. As our predictive grasp of complex behavior improves, how will the bolstered sense of the brain as a deterministic machine affect or undermine our notions of free will or of moral responsibility? Is self-determination, a driving concept in today's bioethics, merely an unscientific fiction? If not, what is the biological basis for it, and if so, what notions should replace it? What brain structures are essential to self-control? How, and to what extent, can the role for self-control in ethical and legal thought be reconciled with facts about mental illness or brain dysfunction? Even now, we have evidence from imaging studies that our brains respond selectively to race[6]. Are we seeing activity related to social, or merely perceptual, judgment? Are differences innate or learned? Will the biologizing of the moral undermine its status as moral?

It is clear here that there are a multitude of questions and few answers. We may not even have a sense of what an answer to such questions would look like. Nevertheless, it is clear that as such questions are approached scientifically, the answers we get will shape our ethical views and, thus, will affect how we approach the ethics of neuroscience. As we learn more about the neuroscientific basis of ethical reasoning, as well as what underlies self-representation and self-awareness, we may revise our ethical concepts. This will then affect how we evaluate the ethical implications of neuroscience for society. Similarly, engaging in ethical discussions about how to design and interpret neuroscientific experiments will affect what we can learn, control, and alter about the brain. The conceptual interconnections and feedback between the two main divisions of neuroethics are dense enough that it may be that distinctions between them can only be made roughly, and only in theory.

## What's in a Name?

As with all newborns, picking a name is a difficult and contentious task. One of the most animated debates at the close of the NMTF conference was about the appropriateness of the label "neuroethics." Some claimed it was an unfortunate name for this fledgling field, because ethics is the purview of philosophers, while the field clearly needs the concerted interaction of policy makers, lawyers, journalists, and the public, as well as the philosophers and neuroscientists. Others suggested that "neuroethics" was ill-chosen because ethics excluded nonethicist philosophers and other humanists. I disagree on both counts. "Neuroethics" is a name well chosen for a number of reasons. First, it is concise, catchy, and evocative. Second, it is a sad misconception of all too many that ethics is merely an academic exercise of philosophers. Rather, our ability to think and act ethically is arguably one of the defining things of what it is to be human: it is an inclusive rather than exclusive term. Part of what it is to be a scientist, a doctor, a lawyer, a politician, or a journalist is to execute one's office in accordance with the values of one's profession and the society at large. Witness the Hippocratic oath, the courtroom oath, the swearing in before taking office, and the injunction not to fabricate stories or data. Ethics should therefore not be a domain foreign to the nonethicist professional. Moreover, in the time of Plato and Aristotle, it was considered imperative for every citizen to have a moral education and to take part in the ethical deliberations of society. It is perhaps reflective of some of the ills in our society that ethics is thought to be a philosopher's concern and not the common man's. But this is not a misconception we should yield to—it is an invitation to reeducate the public that ethics is a forum that needs the participation of everyone. Rather than capitulate to a narrow view of what ethics is and whom it concerns, we should embrace the dialectical model of the NMTF meeting and demonstrate that ethics is as broad and inclusive a category as any.

We should not merely pay lip service to this inclusiveness. Neuroethics has the potential to be an interdisciplinary field with wide-ranging effects. However, because it ultimately impinges on the well-being of the individual and our society, it is not a study that can or should be undertaken in the ivory tower. It is imperative that neuroethics take part in a dialogue with the public. To make this possible, however, it is important in the short term to strive for "neuroliteracy" of the public and the media. We must make a concerted effort to make the subtleties of neuroscientific research accessible to the lay public via the media and refrain from the current practice of feeding it sound bites. For it is only with a nuanced understanding of the sciences, and a renewed trust in the goals of neuroscientists, that real progress will be made on these difficult issues. In the last few months, we have heard just the first noises of such a dialogue. As Dana Foundation executive director Francis Harper aptly noted at the close of NMTF, "You can call it what you want, but the neuroethics train has left the station."

REFERENCES

1. Open your mind, *The Economist*. (May 23, 2002).

2. W. Safire, The But-What-if Factor. *The New York Times*. (May 16, 2002).

3. F. Fukuyama, *Our Posthuman Future: Consequences of the Biotechnology Revolution*, (New York: Farrar Straus & Giroux, 2002).

4. A.R. Damasio, *Descartes Error: Emotion, Reason, and the Human Brain*, New York: Avon Books, (2005).

5. J.D. Green, R.B. Sommerville, L.E. Nystrom, J.M. Darley, and J.D. Cohen, An fMRI investigation of emotional engagement in moral judgment. *Science* 293, (2001) 2105–2108.

6. E.A. Phelps, Faces and races in the brain. *Nature Neuroscience*. 4, (2001) 775–776.

ADINA ROSKIES, PH.D., is assistant professor of Philosophy and director of the MALS program in Mind and Brain Studies at Dartmouth College.

*Chaper 3*

# Emerging Ethical Issues in Neuroscience*

*Martha J. Farah*

In less than a year, "neuroethics" has joined the vocabulary of most neuroscientists. Exactly what the word signifies may not be clear to most of us, however. Both the world and the field to which it refers come largely from individuals outside neuroscience. Newspaper columnist William Safire gave the field its name, and defining statements of the issues are found in such sources as *Brain Policy*[1] by bioethicist Robert Blank, *Our Posthuman Future*,[2] by historian Francis Fukuyama, and a cover story in the *Economist* magazine (May 23, 2002). Neuroscientists themselves have been relatively scarce in public discourse on neuroethics, perhaps because many of the issues under discussion seem far-fetched. Need we devote serious attention now to the needs and rights of cyborg human beings with computer-augmented brains? Probably not, given the current state of technology. Yet neuroscientists are just the people to guide the discussion toward issues of current and near-term priority. How does neuroethics, as presented to us in the literature, relate to the current state of neuroscience and its foreseeable future? Here I attempt to triage the issues that have been raised, separating those that are both new and immediate from those that are not new or are likely to arise only in the distant future. Although all three categories deserve our continued attention, the first poses the most immediate intellectual and social challenges.

*From *Nature Neuroscience* 5 (2002): 1123–1129.

Three broad issues survive the triage for novelty and imminence: enhancement of normal function, court-ordered CNS intervention, and "brain reading." Each emerges from work in multiple areas of neuroscience, from molecular to cognitive neuroscience. The nature of the ethical issues raised are similarly varied, and include the rights to equal opportunity, privacy, and freedom.

## Enhancement of Normal Function

If drugs and other forms of central nervous system intervention can be used to improve our mood, cognition, or behavior of people with problems in these areas, what might they do for normal individuals? Some treatments can be viewed as "normalizers," which have little or no effect on systems that are already normal (for instance, the mood stabilizer lithium)[3] and will therefore not figure in debates over enhancement. Other treatments can indeed make normal people "better than normal." Pharmacological enhancement is arguably being practiced now in several psychological domains: enhancement of mood, cognition, and vegetative functions, including sleep, appetite, and sex.

The enhancement potential of some psychiatric treatments is, in itself, nothing new. Until recently, however, psychotropic medications had significant risks and side effects that made them attractive only as an alternative to illness. With our growing understanding of neurotransmission at a molecular level, it has been possible to design more selective drugs with better side-effect profiles. In addition, adjuvant therapy with other drugs is increasingly used to counteract the remaining side effects. For example, the most troublesome side effect for users of selective serotonin reuptake inhibitors (SSRIs) is sexual dysfunction, which responds well to the drug sildenafil (Viagra). Other drugs specifically developed to counteract the sexual side effects of SSRIs are in development and clinical trials (Vernalis press release, May 22, 2002). The result of both new designer drugs and adjuvant drugs is the same: increasingly selective neurochemical alteration of our mental states and abilities.

Peter Kramer's book *Listening to Prozac*[4] first focused society's attention on the possibility of safe mood enhancement. The growth in sales of SSRIs clearly indicates that more people, with less severe depression, are using them. Has the threshold for SSRI use dropped below the line separating the healthy from the sick? This question is hard to answer for several reasons. First, the line between healthy and sick is a fuzzy and perhaps arbitrary one. There is no simple discontinuity between the characteristic moods of patients with diagnosable mood disorders and the range of moods found in the general population.[5] Second, diagnostic thresholds are clearly moving downward as a result of these very changes in treatment. For a given severity of illness, the better tolerated the treatment, the more likely patients are to present for diagnosis and the more likely physicians are to diagnose and treat. As a related point, other more common and less debilitating conditions are also being treated with SSRIs, such as cyclic changes in women's moods before menstruation.[6] Third, although depression is usually a remitting-relapsing disease with typically years between episodes, patients today are likely to be treated prophylactically with antidepressant medication for periods of one to three years, even when symptom free.[7] Thus there are many people now on antidepressant medication who are healthy, with only a vulnerability to depression as opposed to depression. These changes in psychiatric practice have resulted in many people using SSRIs and other antidepressants who would not have been prescribed these drugs ten years ago. There is no reason to predict that their ranks will not continue to swell, and to include healthier and higher-functioning people.

What changes might healthy individuals hope to experience through the use of antidepressant medication? Mood enhancement belongs on the docket of new and imminent bioethical issues in neuroscience only if current and foreseeable medications can deliver pleasing results to happy people. A handful of studies have assessed the effects of SSRIs on mood and personality in normal subjects over short periods of a few months or less (for example, refs. 8, 9). The effects are relatively selective, reducing self-reported negative affect (such as fear, hostility) while leaving positive affect (happiness, excitement) the same. The drugs can also increase affiliative behavior in laboratory social interactions and cooperative/competitive games played with confederates, for example decreasing the number of spoken commands and increasing the number of suggestions. In one double-blind crossover design, subjects not only were more cooperative in a game but showed real-world changes in behavior as well: roommates found them less submissive on citalopam, though no more dominant or hostile.[9] Much more research is needed to clarify the effects of SSRIs and other antidepressant agents on mood and behavior of normal subjects, but the evidence so far suggests subtle salutary effects.

Pharmacological manipulations of other neurotransmitter systems can alter cognitive abilities, including attention and memory. Attention, in the sense of sustained effort and resistance to distraction, is primarily modulated by dopamine and norepinephrine. Stimulant medication, such as methylphenidate (Ritalin) and amphetamines (Adderall) affect both systems and are effective in treating attention deficit hyperactivity disorder (ADHD). In normal individuals, these drugs induce reliable changes in vigilance, response time, and higher cognitive functions, such as novel problem-solving and planning.[10] As it turns out, thousands of normal, healthy children and adults have discovered similar effects on their own.

The question of whether and when to treat ADHD medically is a complex and contentious one for many reasons, most of which are not related to enhancement. However, as with affective disorders, it is difficult to locate a discontinuity between normal attentional functioning and ADHD (NIH consensus statement, 1998). To the extent that we intervene too "high up" the continuum, we are practicing enhancement. According to most experts, pharmacological enhancement of children's attention is routine in some communities.[11] Parents who are eager to give their children every edge in school may press their pediatricians for medication, and teachers often welcome the greater orderliness in a classroom of attentive children. Because ADHD in children is diagnosed primarily on the basis of parent and teacher questionnaire responses, it can be difficult to free the diagnostic process from the values and standards of the respondents.

Whereas the diagnostic "over-reach" is a reason that some arguably normal children receive stimulants, many young adults with no pretense at all to a diagnosis are using stimulants to enhance their performance in college. Methylphenidate is considered by some to be the most widely used recreational drug on American campuses.[12] Students have often approached me after talks on the topic to relate their own stories about Ritalin use among their non-ADHD peers, for example recalling a hockey coach who always reminded her team to take their Ritalin before playing another school.

Loss of cholinergic neurons is responsible for many of the cognitive changes in Alzheimer's disease, including the pronounced impairment of memory. Drug therapies such as donepezil (Aricept) that increase acetylcholine can slow or reverse the loss of memory ability in the early stages of the disease. Can this or other treatments improve the memory of healthy individuals? Discussions of memory enhancement must take age into account. Although certain specialized pursuits could conceivably benefit from super-memory, the forgetting rates of normal young humans seem to be optimal for most purposes.[13] Empirically, prodigious memory is linked to difficulties with thinking and problem solving,[14] and computationally, boosting the durability of individual memories decreases the ability to generalize.[15] Memory enhancement is of more interest in middle age and beyond, when the normal process of memory loss is first noticeable in healthy individuals.[16] Rejuvenation of memory function in healthy older people is a form of memory enhancement with broad appeal. Indeed, memory-enhancing nutritional supplements are a billion-dollar industry (*Nutrition Business Journal*, 1998), despite little evidence of efficacy. Ginkgo biloba, the most popular of the memory-enhancing supplements, was recently found to be equivalent to placebo.[17]

How close are we to more specific and effective memory enhancement for healthy older adults? Many drug companies are not directing enormous research efforts to the development of memory-boosting drugs (*Neuroinvestment*, September 2001). The candidate drugs target various stages in the molecular cascade that underlies memory formation, including the presynaptic neurotransmitter release (for example, existing cholinesterase inhibitors such as donepezil) and postsynaptic effects (such as the class of drugs known as ampakines). These drugs are currently considered treatments for dementia and so-called "mild cognitive impairment," which is more severe than normal age-related cognitive decline. No drug companies have yet targeted normal memory for enhancement, but there is reason to believe that some of the products under development would work for that purpose as well. For example, treatment of healthy human subjects with an ampakine improved performance on several memory tests.[18]

Advances in the neurochemistry of sleep, appetite, and sex are paving the way for better pharmacological control of these functions as well, with results that will be of interest to normal people. The drug modafinil (Provigil), approved for the treatment of narcolepsy, can prolong alert wakefulness for days.[19] Its use by healthy people is currently being explored by the military.[20] The appeal of such a drug to average people who would like more time in their lives is obvious, and media coverage of modafinil has been extensive. Weight control is a societal preoccupation, and Wallace Simpson's quip that "a woman cannot be too rich or too thin" sums up the likely attitude of most people to a safe, long-term appetite suppressant. There is currently a very limited choice of medication for weight loss, and what is available is less effective than the Fenfluramine-Phenylpropanolamine combination, withdrawn from the market in 1997 due to severe adverse effects.[21] However, findings that hormones such as leptin, ghrelin, and melanocortin are involved in appetite control have given pharmaceutical researchers new avenues to explore for drug development. Men without erectile dysfunction have discovered sildenafil (Viagra) and created a new market for the drug as an enhancer of sexual performance. Although a prescription medication, sildenafil is easily obtained for such purposes after completing a short diagnostic questionnaire on the Internet.[22] Pharmaceutical companies are pursuing drugs that more selectively target the neural bases of sexual function, which would have fewer cardiovascular side effects than sildenafil.

In sum, enhancement is not just a theoretical possibility. Enhancement of mood, cognition, and vegetative functions in healthy people is now a fact of life, and the only uncertainties concern the speed with which new and more appealing enhancement methods will become available and attract more users.

## Ethical Issues in Enhancement

Most of us would love to go through life cheerful and svelte, focusing like a laser beam at work and enjoying rapturous sex each night. Yet most of us also feel uneasy about the idea of achieving these things through drugs. With the necessary technology at or near hand, it is important to examine the reasons for this unease (for a more detailed discussion of enhancement in other domains, see ref. 23). Objections to enhancement can be divided into two broad categories: problems for the individual user and problems for society if use becomes widespread.

The first problem that springs to mind for many people is the possibility of serious side effects for the individual, including long-term or delayed effects that might evade current FDA safeguards. Perhaps a youth spent scaling the heights of academic and job success thanks to enhancement by Ritalin will be followed by a middle age of premature memory loss and cognitive decline. By and large, a concern with long-term or hidden side effects is not unique to enhancement but applies to therapeutic treatments as well. Its special salience in the case of enhancement may reflect an underlying wariness of "free lunches." There is one respect in which enhancement might deserve extra scrutiny for hidden costs, which is suggested by evolutionary considerations. We understand little about the design constraints that were being satisfied in the process of creating a modern human brain. Therefore we do not know which "limitations" are there for a good reason.

As already mentioned, normal forgetting rates seem to be optimal for informational retrieval.

A concern unique to enhancement is the moral objection to in effect, gain without pain. Most people in our society feel there is value to earning one's happiness, success, and so on. When wealthy parents make their teenage children take summer jobs to earn their spending money, they are applying this principle in a way that most of us would find reasonable. However, our judgments often deviate from this principle. Although we recognize the value of earning life's rewards, our lives are full of shortcuts to looking and feeling better. We do not disapprove of people who dislike vegetables improving their health by taking vitamin pills. Nor do we begrudge college applicants their SAT prep books or Stanley Kaplan classes. Psychopharmacological enhancement can therefore be seen as fitting in with an array of practices that are already accepted and widespread.

One variant of the "no pain, no gain" objection is specific to our emotional lives. Many people hold the belief that one cannot experience the beauty and joy in life unless one is also acquainted with life's pain. In the words of Nietzsche, "If you take away my devils, you will take away my angels too." As an empirical claim, supporting evidence is so far lacking. Anecdotal reports of generalized emotional blunting notwithstanding, the small literature on short-term SSRI effects in normal subjects suggests no change in either direction on positive affect, only a selective decrease in negative affect. In any case, even if emotional blunting were a side effect of current mood enhancers, it is not a basis for rejecting mood enhancement in general. There is no *a priori* reason that newer medications would have the same effect.

Other objections stem from potential harm to society. One worry is that enhancement will not be fairly distributed. It is likely that wealthy and privileged will have the choice of self-enhancement and the less privileged will not. Is this what lies at the root of our unease with enhancement? Probably not, given that our society is already full of such inequities. No one would seek to prohibit private schools, personal trainers, or cosmetic surgery on the grounds that they are inequitably distributed. Besides, consider a scenario in which the entire populace is given full and equal access to Ritalin, Prozac, and other enhancers. If our qualms about enhancement were linked to equal opportunity, then this should set our minds at ease, but more than likely it does not.

Another social problem with enhancement is that widespread enhancement will raise our standards of normalcy. This in turn would put individuals who choose not to enhance at a disadvantage, in effect a form of indirect coercion. Even the enhancement of mood, which at first glance lacks a competitive function, seems to be associated with increased social ability, which does not confer an advantage in many walks of life. Such coercion may already be felt by parents whose children attend schools with high rates of Ritalin use. Clearly coercion is not a good thing. Yet it would seem at least as much of an infringement on personal freedom to restrict access to safe enhancements for the sake of avoiding the indirect coercion of individuals who do not wish to partake.

The idea of self-enhancement through manipulations of brain function feels wrong or dangerous to many people. Yet the root cause of that feeling is difficult to find. Perhaps it is a misleading feeling, which we will get over once we have discussed the issue of enhancement thoroughly and rationally. Or perhaps further discussion will reveal the cause of our reflexive worry.

## Court-Ordered CNS Intervention

Another controversial use of our current psychopharmacopia is to improve the behavior of others when that behavior is medically unremarkable but socially undesirable. Rehabilitation has long been intertwined with punishment in our criminal justice system. Successful rehabilitation benefits both the offender and society, insofar as it reduces repeat offenses. It may be offered as an option or as a mandatory component of a sentence. Furthermore, court-ordered therapy or rehabilitation is not confined to medically diagnosed illnesses. Judges may require healthy individuals to undergo such interventions as parenting classes or anger management therapy.

Addiction, aggression, impulse control, and even parenting behavior have been studied for several decades, and we are increasingly able to manipulate the relevant neural systems in animals by drugs and other interventions. Some of this work has been successfully generalized to humans. For example, impulsive violence has been linked to serotonergic abnormalities in patient,[24] criminal,[25] and healthy community populations.[26] Accordingly, SSRIs have been tried as a treatment for aggressive behavior and found to be helpful.[27] For example, in three double-blind studies, fluoxetine (compared against placebo) reduced aggression in patients with personality disorder.[25,28,29]

How close do our current practices come to directly altering brain function under the rubric of court-ordered rehabilitation? For any person deemed a threat to self or others, including criminal offenders, judges routinely order compliance with medication. Although the ethical issues raised by involuntary treatment are far from trivial, there is nevertheless broad consensus in favor of applying recognized treatments in such cases. A more controversial use is sentencing sexual offenders to pharmacological treatments aimed at reducing their sex drive. Several states in the United States have enacted laws that either allow or require sex offenders to take the synthetic hormone medroxy-progesterone acetate, which lowers serum testosterone and significantly reduces recidivism.[30] Other pharmacological approaches involving serotonin are being explored in research studies.[30]

The issue of diagnostic creep is also relevant here. Many behavioral tendencies that the layman would consider "bad" but not medical illnesses have acquired diagnostic codes in the *Diagnostic and Statistical Manual of Mental Disorders* of the American Psychiatric Association.[31] These diagnoses include drug abuse, compulsive shoplifting, and sexual attraction to children. Psychiatrist Alvin Poussaint has even suggested that racism is a psychiatric illness and should be treated by therapy (*New York Times*, August 26, 1999). The "medical model" of condemnable behavior has been criticized when used to excuse, not simply explain, behavior.[32] In the future, the model's impact may be less friendly to offenders by subjecting more of them to involuntary regimens of psychotropic medication.

Court-ordered CNS intervention has not been highlighted in recent discussions of neuroethics but deserves greater attention for three reasons. First, some of the relevant technologies are already available, for example SSRIs to reduce violent behavior. Second, the practice of requiring nonpharmacological treatment aimed at changing the behavior of healthy offenders is well established. And if this, in itself, does not put us on the slippery slope toward court-ordered CNS modification of healthy offenders, then the third fact surely does, namely the use of antiandrogen treatment with convicted sex offenders.

## Ethics of Court-Ordered Intervention

Court-ordered CNS intervention need not simply subjugate an individual's interests to those of society, in the style of Soviet psychiatry or *A Clockwork Orange*. Such uses do not challenge our moral intuitions or social policies; they are clear violations of an individual's freedom and human dignity. The harder questions arise when we consider uses of neuroscience in the criminal justice system for genuinely therapeutic purposes. For example, a judge's order to attend anger management class or a parenting support group is intended to help the offender, in addition to whatever society gains from having fewer hotheads and abusive parents among us. Substituting medications that improve anger management or parenting skills renders the effect no less therapeutic. Yet many people's intuitions raise a flag here. And if not here, then at the thought of more permanent interventions such as implanted stimulators or neurosurgery to achieve these same goals.

What moral obligation triggers this flag? Primarily an intuition about individual freedom, of a kind that we have not previously denied even to prisoners—the freedom to think one's own thoughts and have one's own personality. In anger management class, a person is free to think, "This is stupid. No way I am going to use these methods." In contrast, the mechanism by which Prozac curbs impulsive violence cannot be accepted or resisted in the same way. Offering CNS interventions in the context of a choice, with conventional therapies and incarceration as alternatives, mitigates this worry but does not eliminate it. Sentencing alternatives are rarely appealing options, introducing implicit coercion.

## "Brain Reading"

Mind reading is the stuff of science fiction, and the current capabilities of neuroscience fall far short of such a feat. Even a major leap in the signal-to-noise ratio of functional brain imaging would simply leave us with gigabytes of more accurate physiological data whose psychological meaning would be obscure. Nevertheless, the accomplishments of the field to date include neural correlates of many psychological traits and states. Furthermore, the demand for "scientific" measures of personality, veracity, attitudes, and behavioral dispositions in our society ensures that, ready or not, these measures will have an increasing role in our lives.

Most of our knowledge of individual variation in mental and neural function comes from biological psychiatry and concerns patterns of brain activity in mental disorders. This work has important future clinical implications, especially in a field where the major diagnostic categories remain syndromal, that is, defined in terms of clusters of signs and symptoms. The current state of the art in functional neuroimaging does not earn it a place in psychiatric diagnosis. In general, abnormalities that characterize particular illnesses can be demonstrated when small groups of patients are compared to control subjects but are not diagnostic at the individual patient level. Nevertheless, diagnostic imaging is currently the goal of many research groups, with encouraging results for some disorders, such as ADHD.[33]

Although current imaging methods cannot reliably place most patients in a diagnostic category, this limitation does not rule out occasional revelations about an individual. Even though most patients' scans will be impossible to classify with certainty, other individual scans will deviate enough from the normal pattern to constitute a "positive" finding. One such example comes from studies of drug craving. Drug-free cocaine addicts experience a craving state when shown pictures of drug paraphernalia, which results in reliable group differences in PET activation of the amygdala, anterior cingulate and orbitofrontal cortex.[34] Although some of the individual scans in the patient group are indistinguishable from normal, others clearly differ from normal. In one laboratory, at least half of recently detoxified cocaine users could be identified by differential amygdala response to drug-related versus non-drug-related pictures (A. R. Childress, personal communication). Drug use is not unique in this respect; other stimuli to which individuals are strongly attracted evoke activity in the same circuits. For instance, subjects aroused by sexually explicit videos activate many of the same limbic system areas.[35] Furthermore, the conscious attempt to suppress arousal may also engender a distinct pattern of brain activation,[36] suggesting an advantage of such scans over more peripheral measures capable of revealing sexual preferences.

The significance of such results for individuals is not in their use for classification or diagnosis, because of the ambiguity of most people's scans, but in the information they reveal about some fraction of the subjects (the size of which varies from study to study) whose scans fall clearly outside the normal range. Although subject cooperation is required for such scans, because of the need to remain still and focus on the visually presented stimuli, the subject need not know the scan's purpose.

Many recent studies have sought neuroimaging correlates of the dimensions of personality found in classic theories of normal personality, such as extraversion and neuroticism (see ref. 37 for a review of the social and ethical issues). These studies use small groups of subjects, but at least a small fraction of the subjects can be classified by visual inspection of the scans (T. Canli, personal communication). Other socially relevant characteristics such as racial group identity and unconscious racial attitudes also have neural correlates that can be measured in small groups of subjects. For example, a study in which four black and four white subjects viewed photographs of black and white faces found significant differences in response to in-group and out-group faces.[38] A correlational study of unconscious attitudes found that white subjects with more negative evaluations of black faces had more of an increase in amygdala activity to pictures of unfamiliar black than white faces.[39]

One of the most sought-after uses of "brain reading" is the detection of deception. In the wake of the 9/11 tragedy, there is renewed interest in lie detection for security purposes, to screen individuals for their attitudes and allegiances, as well as for traditional forensic purposes. The company Brain Fingerprinting Laboratories is already marketing a system that uses scalp-recorded ERPs (event-related potentials) to detect so-called "guilty knowledge," such as familiarity with certain people, objects, or scenes. Research seeking more neuroanatomically specific measures of deception using fMRI is under way.[40]

## Ethical Issues in Brain Reading

One problem posed by these developments concerns privacy. As with any testing method that reveals information about an individual (such as genetic testing for breast cancer risk), it may not always be in the person's best interest to have that information available to others. However, there is an added dimension of ethical significance when the information concerns the kinds of personal traits and states that neuroimaging may reveal. The goal, in some cases already partially realized, involves breaching the privacy of a person's own mind.

Another, more immediate problem concerns the way that brain scans are interpreted outside the neuroimaging community. Physiological measures, especially brain-based measures, possess an illusory accuracy and objectivity as perceived by the general public. One commentator, in proposing the use of Brain Fingerprinting as a screening tool at airports, wrote, "Although people lie . . . brainwaves do not" (www.skirsh.com). Brain-based measures do, in principle, have an advantage as indices of psychological traits and states. Measures of brain function are one causal step closer to these traits and states than the behavioral or even peripheral autonomic signs that form the basis of more familiar measures, from responses on personality questionnaires to polygraph tracings. Imaging may therefore, one day, provide the most sensitive and specific measures available of psychological processes. For now, however, this is not the case, and there is a risk that juries, judges, parole boards, the immigration service, and so on will weigh such measures too heavily in their decision making.

## Long–standing Issues in Neuroethics

The emerging field of neuroethics is concerned with a broad array of issues beyond the three just discussed. Some are familiar, though by no means settled. Others remain hypothetical, pending future developments in neuroscience, but are fairly certain to materialize within many readers' lifetimes. In both cases, bioethicists, policy makers, and society in general will benefit from having the perspective of informed neuroscientists included in their discussions.

The familiar issues can themselves be divided into those that relate to neuroscience and to other biomedical sciences as well and those uniquely related to our growing understanding of brain function. Common biomedical issues are exemplified by questions such as the following: How safe are the new methods of neuroscience, such as transcranial magnetic stimulation or high-field MRI, and who should decide? What is the appropriate course of action when an incidental neurological abnormality is found in the course of research data collection? What considerations should guide the development of therapies for diseases such as Parkinson's based on fetal tissues or embryonic stem cells? How should promising new therapies be rationed? When and why should predictive testing be offered for future neurological or neuropsychiatric illness when no cure is available, as with Alzheimer's and Huntington's diseases? These are difficult questions on which reasonable people can disagree. They are also questions with a history in bioethics, which offers helpful general principles and precedents.

Other ethical issues arise exclusively in neuroscience because of the particular subject matter of the field. The brain is the organ of the mind, consciousness, and selfhood. Although the issues in this category are not new, they are evolving as the field evolves and in some cases developing new wrinkles.

The definition of death is one such issue. Until the 1960s, the generally accepted criterion for death was permanent cessation of respiration and circulation. The Harvard criteria for death, published in 1968, shifted the focus to brain function. This definition was refined by a presidential commission in 1981, which defined brain death as "the irreversible cessation of all functions of the entire brain, including the brain stem." This definition, in turn, has been found wanting.[41] With our growing understanding of mind-brain relationships, and our ability to assess them with functional neuroimaging, a narrower focus on the status of higher brain functions seems indicated.[42] However, any such move will raise profound questions about personhood and the brain.

Informed consent for research participation or for treatment[43] is another issue that is special in neuroscience, because in many cases the subjects or patients in question have brain disorders that affect their decision-making ability. The ethics of psychosurgery is a related issue, not least because thousands of patients ostensibly consented to the destructive and unproven method of prefrontal leucotomy.[44]

Although relatively rare today, psychosurgery continues to be practiced as a last resort for patients suffering from refractory depression, obsessive-compulsive disorder (OCD), and anxiety disorders. The most common procedures are cingulotomy, stereotactic subcaudate tractotomy, anterior capsulotomy, and limbic leucotomy, all of which disrupt the interconnecting pathways of the limbic system and the prefrontal cortex.[45,46] According to one recent review, at least one third of depressed patients experience improvement as a result of these operations, with just under one third of OCD and anxiety patients improving.[45] This could be considered a favorable record with patients who have failed to benefit from multiple other treatments. Should we therefore approve of psychosurgery as a less-than-last resort?

Our notions of responsibility and blame, which guide our legal as well as personal ethics, seem at odds with deterministic views of human behavior. Whether we are moved by the "Twinkie defense" (the apocryphal defense of a murderer based on his loss of control caused by junk-food consumption) or the "abuse excuse" depends on how we reconcile commonsense notions of free will with mechanistic views of the causation of behavior. Although the perceived conflict between free will and determinism does not hinge on the particulars of any specific deterministic account, progress in cognitive and behavioral neuroscience certainly increases the salience of the deterministic view. The abstraction that all human behavior is explainable in terms of the laws of physics does not encroach much on our intuitions about a defendant's responsibility for his actions. In contrast, a detailed account of the mechanisms linking childhood abuse to diminished impulse control seems much more likely to temper our intuitions about responsibility and blame. As the neuroscience of intentional behavior continues to develop, it will challenge our ways of thinking about responsibility and blame.

## Neuroethical Questions on the Horizon

The future will bring new ways of enhancing, controlling, and "reading" the brain. The current ability of TMS (transcranial magnetic stimulation) to improve cognition and mood[47] by the activation or inhibition of specific brain areas may be refined in the service of enhancement or control. In the more distant future, similar extensions of deep brain stimulation techniques can be envisioned, and genetic manipulations of targeted neural systems and neurosurgery could permanently modify brain function. Nanotechnology and neural prostheses might eventually create a breed of enhanced human cyborgs. Such possibilities may sound like science fiction in 2002, but consider that space travel and test tube babies were once just science fiction and seemed every bit as far-fetched in the decades before they became reality.

In addition to altering brain function, our ability to monitor and interpret it could one day achieve equally fantastic results. After all, twenty years ago it would have seemed implausible that neuroscientists would have even candidate brain indices of truth versus lie,[40] veridical versus false memory,[48] the likelihood of future violent crime[49], styles of moral reasoning,[50] the intention to cooperate,[51] and even the specific content of thoughts (visualizing houses versus faces).[52] What might we have in another twenty years, or fifty? Our track record for predicting the rate of scientific progress has not been impressive. Gene therapy has yet to achieve the promise that seemed imminent ten or fifteen years ago, whereas the cloning of animals took the world by surprise.

One need not project very far into the future to see the increasing role of neuroscience in our lives, and the social and ethical concerns it will bring. Like the field of genetics, neuroscience concerns the biological foundations of who we are, of our "essence." The relationship of self to brain is, if anything, more direct than that of self to genome, and neural interventions are more easily accomplished than genetic interventions. Yet compared to molecular geneticists, who instigated public discussion in the early days of recombinant DNA research, neuroscientists have paid relatively little attention to the social implications of their field. The time is now ripe for examination of these implications, among scientists themselves and in dialogue with policy makers and the public.

REFERENCES

1. R.H. Blank, *Brain Policy: How the New Neuroscience Will Change Our Lives and Our Politics* (Washington, DC: Georgetown University Press, 1999).

2. F. Fukuyama, *Our Posthuman Future* (New York: Farrar, Straus & Giroux, 2002).

3. B. Shastry, On the Functions of Lithium: the Mood Stabilizer, *Bioessays* 19 (1997): 199–200.

4. P.D. Kramer, *Listening to Prozac* (New York: Penguin, 1993).

5. G. L. Flett, K. Vrendenburg, and L. Krames, The Continuity of Depression in Clinical and Nonclinical Samples, *Psychology Bulletin* 121 (1997): 395–416.

6. G. Redmond, Mood Disorders in the female patient, *International Journal of Fertility Women's Medicine* 42 (1997): 67–72.

7. R. M. Berman, J. K. Belanoff, D.S. Charney, and A.F. Schatzberg, *Neurobiology of Mental Illness*, eds. D.S. Charney, E.J. Nestler, and B.S. Bunney, 419–432 (New York: Oxford University Press, 1999).

8. B. Knutson et al., Selective alteration of personality and social behavior by serotonergic intervention, *American Journal of Psychiatry* 155 (1998): 373–379.

9. W.S. Tse and A.J. Bond, Serotonergic intervention affects both social dominance and affiliative behavior, *Psychopharmacology* 161 (2002): 324–330.

10. R. Elliot et al., Effects of methylphenidate on spatial working memory and planning in healthy young adults, *Psychopharmacology* 131 (1997):196–206.

11. L.H. Diller, The run on Ritalin: Attention deficit disorder and stimulant treatment in the 1990's, *Hastings Center Report* 26 (1996): 12–14.

12. Q. Babcock and T. Byrne, Student perceptions of methylphenidate abuse at a public liberal arts college, *Journal of American College Health* 49 (2000): 143–150.

13. J. Anderson, The adaptive characteristics of thought (Hillsdale, NJ: Erlbaum, 1990).

14. A.R. Luria, *The mind of a mnemonist* (Cambridge, MA: Harvard University Press, 1968).

15. J.L. McClelland, B.L. McNaughton, and R.C. O'Reilly, Why there are complementary learning systems in the hippocampus and neocortex: Insights from the successes and failures of connectionist models of learning and memory, *Psychology Review* 102 (1995): 419–457.

16. F.I.M. Craik and T.A. Salthouse, *The handbook of aging and cognition* (Hillsdale, NJ: Erlbaum, 1992).

17. P. Solomon, F. Adams, A. Silver, J. Zimmer and R. DeVeaux, Gingko for memory enhancement : a randomized controlled trial, *JAMA* 288 (2002): 835–840.

18. M. Ingvar et al., Enhancement by an ampakine of memory encoding in humans, *Experimental Neurology* 146 (1997): 553–559.

19. D. Lagarde, D. Batejat, P. Van Beers, D. Sarafian, and S. Pradella, Interest of modafinil, a new psychostimulant, during a sixty-hour sleep deprivation experiement, *Fundamental Clinical Pharmacology* 9 (1995): 1–9.

20. J. Caldwell, N. Smythe and K. Hall, A double-blind, placebo-controlled investigation of the efficacy of modanifil for sustaining the alertness and performance of aviators: a helicopter simulator study, *Psychopharmacology* 150 (2000): 272–282.

21. M.L. Campbell and M.L. Mathys, Pharmacologic options for the treatment of obesity, *American Journal of Health System Pharmacology* 58 (2001): 1301–1308).

22. K. Armstrong, J. Schwartz and D. Asch, Direct sale of sildenafil (Viagra) to consumers over the internet, *New England Journal of Medicine* 341 (1999): 1389–1392).

23. E. Parens, ed. *Enhancing human traits: ethical and social implications* (Washington, DC: Georgetown University Press, 1998).

24. E.F. Coccaro, R.J. Kavoussi and R.L. Hauger, Serotonin function and antiaggressive responsive to fluoexetine: a pilot study, *Biological Psychiatry* 42 (1997): 546–552.

25. D.R. Cherek, S.D. Lane, C.J. Pietras and J.L. Steinberg, Effects of chronic paroxetine administration on measures of aggressive and impulsive responses of adult males with a history of conduct disorder, *Psychopharmacology* 159 (2002): 266–274.

26. S.B. Manuck et al., Aggression, impulsivity, and central nervous system serotonergic responsivity in a nonpatient sample, *Neuropsychopharmacology* 19 (1998): 287–299.

27. M.T. Walsh and T.G. Dinan, Selective serotonin reuptake inhibitors and violence: a review of the available evidence, *Acta Psychiatrica Scandinavica* 104 (2001): 84–91.

28. E.F. Coccaro and R.J. Kavoussi, Fluoxetine and impulsive aggressive behavior in personality-disordered subjects, *Archives of General Psychiatry*, 54 (1997): 1081–1088.

29. C. Salzman et al., Effect of fluoxetine on anger in symptomatic volunteers with borderline personality disorder, *Journal of Clinical Pharmacology* 15 (1995): 23–29.

30. L.S Grossman, B. Martis and C.G. Fichtner, Are sex offenders treatable? A research overview, *Psychiatric Services*, 50 (1999): 349–361.

31. *Diagnostic and Statistical Manual of Mental Disorders IV-R* (Washington, DC: American Psychiatric Association, 1994).

32. S.J. Morse, Hooked on hype: addiction and responsibility, *Law and Philosophy* 19 (2000): 3–49.

33. D.D. Dougherty et al., Dopamine transporter density in patients with attention deficit hyperactivity disorder, *Lancet* 354 (1999): 2132–2133.

34. A.R. Childress et al., Limbic activation during cue-induced cocaine craving, *American Journal of Psychiatry* 156 (1999): 11–18.

35. H. Garavan et al., Cue-induced cocaine craving: neuroanatomical specificity for drug users and drug stimuli, *American Journal of Psychiatry* 157 (2000): 1789–1798.

36. M. Beauregard, J. Levesque and P. Bourgouin, Neural correlates of conscious self-regulation of emotion, *Journal of Neuroscience* 21 (2001): 1–6.

37. T. Canli and Z. Amin, Neuroimaging of emotion and personality: Scientific evidence and ethical considerations, *Brain and Cognition* 50 (2002): 414–431.

38. A. Hart, P. Whalen, S. McInerney, H. Fischer and S. Rausch, Differential response in the human amygdala to racial outgroup versus ingroup face stimuli, *Neuroreport* 11 (2000): 2351–2355.

39. E. Phelps et al., Performance on indirect measures of race evaluation predicts amygdala activation, *Journal of Cognitive Neuroscience* 12 (2000): 729–738.

40. D.D. Langleben et al., Brain activity during stimulated deception: an event-related functional magnetic resonance study, *Neuroimage* 15 (2002): 727–732.

41. R.D. Truog, Is it time to abandon brain death? *Hastings Center Report* 27 (1997): 29–37.

42. P. Churchland, *The Engine of Reason, the Seat of the Soul: A Philosophical Journey into the Brain* (Cambridge, MA: MIT University Press, 1995).

43. A.M. Capron, Ethical and human rights issues in research on mental disorders that may affect decision-making capacity, *New England Journal of Medicine* 340 (1999): 1430–1434.

44. E.S. Valenstein, *Great and Desperate Cures: The Rise and Decline of Psychosurgery and Other Radical Treatments for Mental Illness* (New York: Basic, 1986).

45. G.S. Mahli and P. Sachdev, Novel physical treatments for the management of neuropsychiatric disorders, *Journal of Psychosomatic Research* 53 (2002): 709–719.

46. F. Ovsview and D.M. Frim, Neurosurgery for psychiatric disorders, *Journal of Neurology, Neurosurgery, and Psychiatry* 63 (1997): 701–705.

47. M. George, E. Wasserman and R. Post, Transcraial magnetic stimulation: A

neuropsychiatric tool for the 21$^{st}$ century, *Neuropsychiatry and Clinical Neuroscience* 8 (1996): 373–382.

48. R. Cabeza, S.M. Rao, A.D. Wagner, A.R. Mayer and D.L. Schacter, Can medial temporal regions distinguish true from false? *Proceedings of the National Academy of Sciences, USA* 98 (2001): 4805–4810.

49. A. Raine et al., Reduced prefrontal and increased subcortical brain functioning assessed using positron emission tomography in predatory and affective murderers, *Behaviorial Science and the Law* 16 (1998): 319–322.

50. J. Greene et al., An fMRI investigation of emotional engagement in moral judgement, *Science* 293 (2001): 2105–2108.

51. K. McCabe, D. Houser, L. Ryan, V. Smith and T. Trouard, A functional imaging study of cooperation in two-person reciprocal exchange, *Proceedings of the National Academy of Sciences, USA* 98, (2001): 11832–11835.

52. K.M. O'Craven and N. Kanwisher, Mental imagery of faces and places activates corresponding stimulus-specific brain regions, *Journal of Cognitive Neuroscience* 12 (2000): 1013–1023.

MARTHA J. FARAH, PH.D., is director of the Center for Cognitive Neuroscience and Walter H. Annenberg Professor in the Natural Sciences at the University of Pennsylvania.

*Chapter 4*

# Monitoring and Manipulating Brain Function: *New Neuroscience Technologies and Their Ethical Implications*[*]

## *Martha J. Farah* and *Paul Root Wolpe*

Congress christened the 1990s the "Decade of the Brain," and this was apt from the vantage point of the early 21st century. Great strides were made in both basic and clinical neuroscience. What the current decade may, in retrospect, be remembered for is the growth of neuroscience beyond those two categories, "basic" and "clinical," into a host of new applications. From the measurement of mental processes with functional neuroimaging to their manipulation with ever more selective drugs, the new capabilities of neuroscience raise unprecedented ethical and social issues. These issues must be identified and addressed if society is to benefit from the neuroscience revolution now in progress.

Like the field of genetics, cognitive neuroscience raises questions about the biological foundations of who we are. Indeed, the relation of self and personal identity to the brain is, if anything, more direct than that of self to the genome. In addition, the ethical questions of neuroscience are more urgent, as neural interventions are currently more easily accomplished than genetic interventions. Yet compared to the field of molecular genetics, in which ethical issues have been at the forefront since the days of the 1975 Asilomar meeting on recombinant DNA, relatively little attention has been paid to the ethics of neuroscience.

[*]From *Hastings Center Report* 34, no. 3 (May–June 2004): 35–45.

This situation is changing, as bioethicists and neuroscientists are beginning to explore the emerging social and ethical issues raised by progress in neuroscience. In the Society for Neuroscience's recently formulated mission statement, bioethical issues figure prominently.[1] Numerous articles, meetings, and symposia have appeared on the subject.[2] The term "neuroethics," which originally referred to bioethical issues in clinical neurology, has now been adopted to refer to ethical issues in the technological advances of neuroscience more generally.[3] (Unfortunately, the term is also used to refer to the neural bases of ethical thinking, a different topic.[4])

Neuroethics encompasses a broad and varied set of bioethical issues. Some are similar to those that have arisen previously in biomedicine, such as the safety of new research and treatment methods, the rationing of promising new therapies, and predictive testing for future illnesses when no cure is available (as with Alzheimer's or Huntington's disease). Other neuroethical issues, however, are unique to neuroscience because of the particular subject matter of that field. The brain is the organ of the mind and consciousness, the locus of our sense of selfhood. Interventions in the brain therefore have different ethical implications than interventions in other organs. In addition, our growing knowledge of mind-brain relations is likely to affect our definitions of competence, mental health and illness, and death. Our moral and legal conceptions of responsibility are likewise susceptible to change as our understanding of the physical mechanisms of behavior evolves. Our sense of the privacy and confidentiality of our own thought processes may also be threatened by technologies that can reveal the neural correlates of our innermost thoughts.

Many of the new social and ethical issues in neuroscience result from one of two developments. The first is the ability to monitor brain function in living humans with a spatial and temporal resolution sufficient to capture psychologically meaningful fluctuations of activity. The second is the ability to alter the brain with chemical or anatomical selectivity that is sufficient to induce specific functional changes. For each of these developments, we will review advances in the enabling technology and provide examples of ethically challenging uses of the technology and an analysis of the ethical issues they raise.

## Neuroimaging

The history of modern brain imaging began in the 1970s with computed axial tomography or CAT scans and proceeded at a rapid and accelerating rate for the remaining decades of the twentieth century. The idea of passing X-rays through the head from multiple directions and reconstructing a three-dimensional structural image, revolutionary at the time, was quickly adapted to radiological signals other than X-rays. These included radiation from exogenous tracers to enable imaging of brain function, as in positron emission tomography (PET) and single photon emission computed tomography (SPECT), and endogenously generated magnetic fields to image either structure or function, as in magnetic resonance imaging (MRI). Pioneering research on cognition and emotion was undertaken with PET and SPECT in the 1980s, and by the 1990s MRI, the noninvasive alternative to PET, became commonplace in research.[5]

In an MRI, atoms are first aligned by a strong static magnetic field, then knocked out of alignment by a radio frequency pulse, and then allowed to realign. The fluctuating field created as the atoms "relax" to the aligned state is the signal that is measured. Although early functional MRI used an injected contrast agent, current methods use the magnetic properties of the blood itself as a tracer, and are therefore entirely noninvasive. In blood oxygen level dependent (BOLD) MRI, the different magnetic susceptibility of oxygenated and deoxygenated hemoglobin provides a measure of regional brain activity.[6] In arterial spin labeling (ASL) MRI, the atoms are aligned by a magnetic field at the neck and relax as they circulate through the brain, indicating regional perfusion.[7] The spatial and temporal resolution of functional MRI (fMRI) is limited by haemodynamics rather than by the physics of the method; blood flow changes over seconds in response to neural activity, and these changes extend into nearby tissue. In practice, fMRI has a spatial resolution of one millimeter and a temporal resolution of about one second, which is adequate to distinguish among at least some psychologically meaningful differences in brain activity.[8] A few additional methods figure in the cognitive neuroimaging revolution. One is structural MRI, from which precise measurements of brain size and shape can be made. Combined with reliable methods for delineating and measuring particular brain structures, this has opened up the field of brain morphometry, in which slight anatomical variations are correlated with psychological traits.[9] The venerable techniques of electroencephalography (EEG) and event-related potentials (ERPs) have acquired new capabilities by the application of signal processing techniques that allow better localization of brain activity and analysis of temporal patterns of activity.[10] Optical methods, such as near infrared spectroscopy (NIRS), provide another noninvasive measure of regional brain activity based on the absorption of different wavelengths of light as it passes through the head.[11]

By and large, these methods have been developed for long-standing clinical and scientific goals, from localizing seizure foci to studying the neurochemical abnormalities in psychiatric illness. These uses are associated with ethical issues of a familiar nature: for example, the risks of radiation, obtaining adequate informed consent (especially from the mentally ill), and the possibility of discovering incidental brain anomalies. However, neuroimaging also yields information that can be used for different purposes, raising new ethical issues. In principle, and increasingly in practice, imaging can be used to infer people's psychological traits and states, in many cases without the person's cooperation or consent. It can be used, in effect, as a crude form of mind reading.

## Imaging of Personal Information

Our society's attitude toward mental illness has come a long way since 1972, when Senator Thomas Eagleton was forced to withdraw his vice presidential candidacy after his history of depression became known. Nevertheless, psychiatric illness continues to carry a stigma, and a currently healthy individual might well wish to avoid disclosing a psychiatric history. The finding that depression, schizophrenia, and other illnesses leave their marks on the brain raises the possibility that psychiatric history and risk could be inferred from a brain scan without an individual's knowledge or consent. For the most part, the currently available markers are morphometric, relying on structural rather than functional imaging.[12]

Although the abnormalities that characterize particular illnesses can be demonstrated when small groups of patients are compared to control subjects, they are not currently diagnostic at the individual patient level. Nevertheless, diagnostic imaging is currently the goal of many research groups, with encouraging results for some disorders, for example ADHD.[13] Should diagnostic imaging become reliable, the possibility of inferring current or prior psychiatric illness from images taken for other purposes will also become a concern.

## Imaging of Personality

A number of recent studies have sought neuroimaging correlates of personality found in classic theories of normal personality, including extraversion/introversion, neuroticism, novelty seeking, harm avoidance, and reward dependence.[14] Most of the studies employed resting scans (that is, scans that were obtained while subjects were simply resting rather than performing any particular task) of groups of twelve to thirty healthy subjects, not selected for being especially extreme on any dimension, and they performed correlations between personality scale scores and brain activation in regions of *a priori* interest throughout the brain. Despite the seemingly low power of such designs, a number of positive results have been reported, with both converging and diverging results among the studies. The areas that distinguish normal people with differing personality at rest include a large number of cortical and subcortical areas, particularly paralimbic cortical areas such as the insula, the orbitofrontal cortex, and the anterior cingulate, as well as subcortical structures, such as the amygdala and putamen.

Canli and colleagues have sought correlates of personality in the brain's response to emotionally evocative stimuli. Given that many aspects of personality are most apparent in the context of frightening, happy, sad, or tempting stimuli, such an approach has the potential to identify important differences not apparent in resting scans. In one study, Canli focused on two personality traits: extraversion, which is the tendency to seek out and enjoy social contact and maintain an upbeat outlook, and neuroticism, which is the tendency to worry and focus on negative information.[15] They found that extraversion was correlated with brain response in several areas to pictures with positive emotional valence such as puppies, ice cream, and sunsets. The effect was specific to positive and not negative stimuli, and this was confirmed in a later study with pictures of happy and fearful faces.[16] Neuroticism, in contrast, is associated with differences in response to negative but not positive stimuli. Photographs of spiders, cemeteries, crying people, and other negatively valenced images evoked more response in certain brain areas the more neurotic the subject. Positive pictures did not show such an effect.

## Imaging of Social and Moral Attitudes

In a now well-known study, Phelps and colleagues studied white subjects' attitudes toward unfamiliar black faces, using both behavioral measures and fMRI.[17] Using previously developed behavioral measures, they were able to estimate the degree of unconscious negative evaluation of unfamiliar black as opposed to white faces. They then measured brain response to unfamiliar black and white faces and found a moderately strong correlation between individuals' amygdala activation and the degree of negative evaluation of black faces.

Racial group identity also has neural correlates that are roughly measurable with current brain imaging methods. In a study of black and white subjects viewing photographs of black and white faces, significant differences in response to in-group and out-group faces were found.[18]

Differences in the way people view particular actions as right or wrong, across specific moral dilemmas and across individuals, have measurable neural correlates. In particular, Greene and colleagues used fMRI to demonstrate different patterns of brain activation associated with the logical weighing of rights and wrongs. For example, they found that the emotional centers of the brain were more active when subjects made moral decisions based more on their visceral reactions than on a rational weighing of costs and benefits.[19]

## Imaging of Preferences

The objects of a person's desires may also be discernable in some cases with functional neuroimaging. The first experiments to demonstrate this concerned drug craving. Drug-free cocaine addicts experience a craving state when shown pictures of drug paraphernalia, which results in reliable group differences in PET activation of limbic and paralimbic areas, including the amygdala, anterior cingulate cortex, and orbitofrontal cortex.[20]

Drug use is not unique in this respect; other stimuli to which individuals are strongly attracted have been found to evoke activity in these neural circuits. Subjects aroused by sexually explicit videos activate many of the same limbic system areas as drug craving does.[21] Furthermore, the conscious attempt to suppress arousal may also engender a distinct pattern of brain activation.[22] For this reason neuroimaging may be more informative than peripheral measures that are capable of revealing sexual preferences.

Objects that are feared or disliked may also be discerned by brain imaging. Amygdala responses to photographs of upsetting scenes and unpleasant facial expressions are among the most reliable findings in the imaging literature on emotion.[23] Indeed, the amygdala response to such stimuli is detectable even when the photographs have been presented at subliminal exposure durations and subjects are not aware of having seen them.[24]

## Forensic Imaging

The ability to know a person's attitudes and thoughts and to predict their actions would be particularly useful within the criminal justice, intelligence, and immigration enforcement communities, where interviewees are often motivated to lie or to withhold desired information. Several different applications of functional neuroimaging are being explored with support from these communities.

Lie detection is one of the most sought-after applications. The work of Langleben and colleagues attracted tremendous media attention when it showed differences in subjects' brain activation when bluffing versus telling the truth about symbols on playing cards.[25] Lee et al. mapped the differences in brain activation in a memory task between honest test performance and simulated malingering.[26] Such research has a long way to go before it can be used to detect spontaneous, genuine deception. The forms of deception being detected in these studies involve highly constrained questions and may reflect nothing more than the additional cognitive effort required to deceive.

The "guilty knowledge test," used for decades with peripheral measures of autonomic response, has been adapted for use with scalp-recorded event-related potentials (ERPs) and marketed by ERP researcher Lawrence Farwell. The method is based on the difference in the P300 ERP evoked by familiar and unfamiliar stimuli. In Farwell's "brain fingerprinting," people, objects, or scenes associated with a crime are presented to an individual to determine whether the brain recognizes the image as familiar (such as whether a crime scene appears "familiar" to the brain despite the subject's claim that he has never been there). The Brain Fingerprinting company's Web site describes the method as "a new paradigm in criminal investigations and counterterrorism,"[27] and indeed it has been admitted as evidence in court[28] and is being promoted as a means of screening for terrorists, despite the reservations of leading ERP researchers such as Emmanual Donchin.[29]

In addition to the problem of discriminating intentional lies from truth, brain imaging is potentially applicable to a related problem of great legal significance: the problem of discriminating false memory from veridical memory. A false memory is a kind of memory error that occurs when a person mistakenly believes that he or she remembers an event that did not actually take place. When false memories are induced in the laboratory, they evoke patterns of activity in memory-related areas of the brain that are distinctive from both veridical memories and correct judgments that an event did not happen. Whereas both veridical and false memories activate the hippocampus, the parahippocampal region is activated more strongly by veridical memories.[30]

Finally, the effort to predict future violent crime may eventually be aided by functional neuroimaging. Some offenders commit one violent crime and live the rest of their lives without harming anyone, whereas others continue to be violent. Personality factors correlate to some degree with these tendencies, but more recently PET and fMRI have been used on an experimental basis to distinguish these two populations.[31]

## Imaging Specific Thoughts

Perhaps the most science-fictionesque example of brain imaging as mind reading comes from studies of high-level vision. Although visual processing does not have the obvious personal and social relevance that we associate with social attitudes, emotions, or tendencies to violence, the striking thing about work in this area is the specificity of the mental content that can be recovered by analyzing a brain image. Haxby and colleagues scanned subjects while they viewed numerous pictures each of faces, cats, houses, chairs, scissors, shoes, and bottles.[32] They found that the overall pattern of activation in the ventral extrastriate cortex enabled them to classify the stimulus category being viewed by the subject with 96 percent accuracy.

Working with a reduced set of stimulus categories, O'Craven and Kanwisher accomplished a similar feat with subjects' purely mental images, formed from memory in the absence of a visual stimulus.[33] After first showing subjects pictures of faces and houses and noting the locations of maximum activation to each type of stimulus, they instructed the same subjects to imagine faces and houses. For a majority of the scans, the researchers were able to tell whether a subject was thinking about a face and a house just by explaining the scan.

## Ethical Issues in Neuroimaging

The main ethical problem that the scientific trends just reviewed pose concerns for privacy. As with any testing method that reveals new kinds of information about an individual (genetic testing for breast cancer risk, for example), it may not always be in the individual's best interest to have that information available to others. There is an added dimension of ethical significance when the information concerns the kinds of personal traits and states that neuroimaging may reveal. The current technology can, in some cases, breach the privacy of a person's own mind, for example laying bare a disavowed attitude toward particular races. It may eventually be possible for employers, juries, parole boards, or law enforcement to examine your brain in order to answer: Are you prone to depression? How neurotic are you? To whom are you sexually attracted? How do you feel about other races? What scares you? Have you abused illegal drugs?

An individual need not know when images are used to obtain personal information. Images used for one purpose—for example, medical diagnosis—may nevertheless reveal unrelated private information to whatever party evaluates (or subpoenas) the image. The experimental paradigm used by Phelps and colleagues to correlate amygdala activation with racial attitudes simply required subjects to view pictures of faces, and it could be administered in the guise of a face perception study. Brain activation can reveal attitudes and feelings that the subject may not be aware of having. For example, although subjects in Whalen's study were not aware of having seen fearful facial expressions when the expressions were presented subliminally, and cortical brain regions did not react to them, the amygdala nevertheless responded.

What obstacles lay between the present state of imaging technology and the ability reliably to read personality, psychiatric history, truthfulness, and so on from an individual's brain scan? One important limitation of the current technology is the need to aggregate data over multiple observations. When the individual subject is the unit of analysis, the need for multiple trials of data collection may be impractical. Although fears or cravings can be evoked repeatedly if necessary, the recall of a specific memory cannot be repeated without changing the nature of the memory itself. For most of the examples cited here, subject groups must be compared in order to obtain reliable differences between groups (between formerly depressed and never depressed individuals, for example), or to detect a relation to a trait (such as extraversion).

Nevertheless, even a scanning protocol that is incapable of reliably classifying all individuals may be able to classify individuals with relatively extreme patterns of brain activity, and so may be seen as a useful screening tool in certain circumstances. In one lab, for example, at least half of recently detoxified cocaine users could be identified by differential amygdala response to drug-related versus non-drug-related pictures.[34] In another, simple visual examination of whole brain activity patterns allowed at least a fraction of the subjects to be sorted by personality trait.[35] Even when patterns of brain activation are not extreme, they provide information sufficient to narrow the range of an individual's likely values on psychological traits of interest. Using only the published data in reports of imaging correlates of personality traits, a new individual's trait level could be bracketed within a range of 2.0 to 3.5 standard deviations (depending on the study), compared to the 4.0 standard deviation range of the population.[36]

## Illusory Accuracy

In addition to privacy concerns, neuroimaging is liable to overreliance on, or misapplication of, information from brain scans. The ability to assess personality, attitudes, and desires would be of interest in screening for employment, school tracking, or military service. The ability to distinguish between truth and falsehood, or veridical and false memory, would find wide use in the legal system. The demand for these abilities, coupled with the inevitable misunderstandings of brain imaging among the lay public, sets the stage for misuse. Physiological measures, especially brain-based measures, possess an illusory accuracy and objectivity as perceived by the general public. In proposing the use of brain fingerprinting as a screening tool at airports, one commentator wrote, "Although people lie . . . brainwaves do not."[37]

Although brainwaves do not lie, neither do they tell the truth; they are simply measures of brain activity. Whether based on regional cerebral blood flow or electrical activity, brain images must be interpreted like any other correlate of mental activity, behavioral or physiological. Brain images and waveforms give an impression of concreteness and directness compared to behavioral measures of psychological traits and states, and high-tech instrumentation lends an aura of accuracy and objectivity. Nevertheless, the psychological interpretations of these measures are far from direct or intrinsically objective. As the foregoing review suggests, progress has been made in the use of such measures, and some inferences to socially relevant traits and states can now be made with a degree of certainty under specific and highly controlled conditions. However, the current state of the art does not allow reliable screening, profiling, or lie detection.

There is no reason to doubt that the state of the art will improve in the coming years. Brain-based measures do, in principle, have an advantage as indices of psychological traits and states over more-familiar behavioral or autonomic measures. They are one causal step closer to these traits and states than responses on personality questionnaires or polygraph tracings. Imaging may therefore one day provide the most sensitive and specific measures available of psychological processes. For now, however, this is not the case, and there is a risk that juries, judges, parole boards, the immigration service, and so on will use these technologies prematurely.

## Brain Enhancement

The psychopharmacology of the mid-20th century depended entirely on serendipity. The antihistamine chlorpromazine was accidentally found to calm agitated schizophrenic patients and reduce their psychosis. Another early drug investigated for its antipsychotic properties, imipramine, turned out to be ineffective for that purpose but was observed to lift the mood of some of the patients taking it. When a small number of patients with major depression tried it, the therapeutic effect was dramatic, and imipramine continues to be used as an antidepressant today. The second antidepressant to be discovered, iproniazid, was hitherto used as an antibiotic for treating patients with tuberculosis when its mood-elevating properties were observed. Similar accidental discoveries led to the identification of amphetamine as a stimulant in the course of refining a treatment for asthma, and meprobamate as an anti-anxiety treatment in the course of testing an antibiotic.[38]

Such lucky accidents were then augmented by trial-and-error tests with other molecules of similar structure. Parallel to this development, researchers began to understand the effects of these drugs on brain function, identifying the specific neurotransmitter systems affected by the drugs and the mechanisms by which the drugs interacted with these systems. The advent of direct-binding assays in the 1960s provided the first direct approach to testing and comparing the affinity of a drug for different neurotransmitter receptors, and the tools of the molecular biology revolution, including the cloning of rare subtypes of receptors, allowed for the design of highly selective agonists, antagonists, and other molecules to influence selectively the process of neurotransmission.

The continual improvement in side-effect profile of modern psychotropic medications is due to the increasing selectivity of drug action made possible by the methods of molecular neuroscience. "Selective" is the first S in "SSRI," the class of drugs to which fluoxetine (Prozac) belongs. New drugs with ever more selective actions on the neurochemistry of mood, anxiety, attention, and memory are under development. Although intended for therapy, many of these drugs affect brain function in healthy people, raising the possibility of their use for enhancement of normal function rather than remediation of dysfunction.

The enhancement potential of some medications is, in itself, nothing new, and the attempts of human beings to use chemical substances to alter normal affective and cognitive traits is as old as the drinking of alcohol. Until recently, however, psychotropic drugs had significant risks and side effects that limited their attractiveness. This situation is changing as side-effect profiles become more tolerable. In addition, therapy in conjunction with other drugs is an increasingly common strategy for counteracting the remaining side effects. For example, the most troublesome side effect for users of SSRIs is sexual dysfunction, which responds well to the drug sildenafil (Viagra). Other drugs specifically developed to counteract the sexual side effects of SSRIs are in development and clinical trials. The result of both new designer drugs and adjuvant drugs is the same: increasingly selective alteration of our mental states and abilities through neurochemical intervention, with correspondingly less downside to their use by anyone, sick or well.

Technical advances in non-pharmaceutical methods for altering brain function are also creating potential enhancement tools. Transcranial magnetic stimulation (TMS) and, more rarely, vagus nerve stimulation and deep-brain stimulation have already been used to improve mental function or mood in patients with medically intractable neuropsychiatric illnesses.[39] Research on the effects of non-pharmaceutical methods on brain function in normal individuals has been limited to the relatively less invasive TMS. Mood effects on normal healthy subjects have been investigated in the context of basic research on mood and brain function,[40] and at least one laboratory is devoted to the development of TMS methods for enhancing normal cognition.[41] Finally, there is growing research interest in computer augmentation of brains. Most research on brain-machine interfaces currently focuses on capturing and using movement command signals from the brain and carrying sensory inputs to the brain, for example from a video camera.[42] One research program is tackling memory augmentation by developing a prosthetic hippocampus that can be interfaced with a rodent brain.[43] The motivation for this research is partly scientific, to better understand neural coding of sensory, motor, and memory information, and partly clinical, to help patients with paralysis and peripheral sensory impairments.

Nevertheless, the military's substantial support for this research suggests that some think normal healthy individuals might someday be enhanced by neural prostheses.[44]

## Enhancement of Normal Mood

Amphetamines, barbiturates, benzodiazepines, and other "mother's little helpers" have long been used to improve the moods of healthy people. However, the high potential for addiction and tolerance with these drugs dissuades most people from using them. Pre-SSRI antidepressants, while presenting no such risks, have unpleasant side effects that limit their appeal to only those faced with clinical depression as the alternative. The SSRIs, in contrast, have relatively narrower neurochemical effects and consequently fewer side effects. The result, as Peter Kramer described in *Listening to Prozac*, is that many people who would never have taken a tricyclic antidepressant are taking SSRIs.[45]

Of course, most people using SSRIs meet *DSM-IV* criteria for some psychiatric disorder, although not necessarily major depression: dysthymia (a mild depression), social phobia (an extreme form of shyness and self-consciousness), premenstrual dysphoric disorder (a recurrent negative mood associated with PMS), and various eating disorders respond well to SSRIs. It nevertheless remains controversial whether some of these diagnostic categories are medicalized labels for normal variants of human personality, which do not necessarily require pharmaceutical treatment. In addition, some people using SSRIs have no recognized illness. These include people who have suffered from depression in the past and choose to continue medication prophylactically, as well as people who, in Peter Kramer's words, feel "better than well" when taking an antidepressant.

What is the effect of SSRIs on normal, healthy individuals? While no systematic studies have examined individuals who choose to take these medications, a handful of studies have assessed the effects of SSRIs on mood and personality in randomly selected healthy subjects over short periods of a few months or less.[46] Effects on mood are relatively selective, reducing self-reported negative affect while leaving positive affect neither increased nor decreased. The drugs also increased affiliative behavior in laboratory social interactions and cooperative/competitive games played with confederates. For example, subjects on the drug spoke fewer commands and instead made more suggestions. In one double-blind crossover design, subjects not only were more cooperative in a game but showed real-world changes in behavior as well: flatmates found the subjects less submissive on citalopram, though no more dominant or hostile.[47] Although more research is needed to clarify the effects of SSRIs and other antidepressant agents on mood and behavior of normal healthy subjects, and long-term studies are needed on those who choose to take SSRIs in real-life settings, the evidence so far suggests subtle but salutary effects without significant short-term side effects.

## Enhancement of Cognition

Our current ability to enhance cognition through the direct alteration of brain function involves two types of cognitive function: attention and memory. "Attention" is used here in its broadest sense, including active use of working memory, executive function, and other forms of cognitive self-control. These are the cognitive abilities most obviously deficient in the syndrome of attention deficit hyperactivity disorder (ADHD). These same abilities vary in their strength within the normal population. Indeed it seems likely that ADHD represents the lower tail of the whole population distribution rather than a qualitatively different state of functioning, discontinuous with the normal population.[48]

Drugs targeting the neurotransmitter systems dopamine and norepinephrine are effective in treating ADHD, and have been shown to improve normal attentional function as well. Methylphenidate (Ritalin) and amphetamine (Adderall), as well as modafinil (Provigil, a newer drug approved for regulating sleep) have been shown to enhance attention across a variety of different tests in healthy young volunteers.[49]

Do these laboratory-measured improvements translate into a noticeable improvement of real-world cognitive performance? No experimental evidence is available, but the growing illicit use of ADHD medications on college campuses suggests that many young adults believe their cognition is enhanced by the drugs.[50] Parents also appear to find real-world benefits for their normal children with ADHD medication: in certain school districts the proportion of boys taking methylphenidate exceeds the most generous estimates of ADHD prevalence.[51]

Memory is the other cognitive ability that can, at present, be manipulated to some degree by drugs. Interest in memory enhancement has so far been confined to the middle-aged and elderly, whose memory ability undergoes a gradual decline even in the absence of dementia. The most commonly used method involves manipulation not of memory circuits per se but of cerebrovascular function. Herbal supplements such as ginkgo biloba affect memory mainly by increasing blood flow within the brain. However, the effectiveness of this treatment is questionable.[52] How close are we to more specific and effective memory enhancement for healthy older adults?

As the molecular biology of memory progresses, it presents drug designers with a variety of entry points through which to influence the specific processes of memory formation. A huge research effort is now being directed to the development of memory-boosting drugs.[53] The candidate drugs target various stages in the molecular cascade that underlies memory formation, including the initial induction of long-term potentiation (LTP) and the later stages of memory consolidation. There is reason to believe that some of the products under development would work for enhancement as well as therapy. For example, treatment of healthy human subjects with an ampakine, which enhances LTP, improved performance in a dose-dependent manner.[54]

Few consider memory enhancement for the young to be a goal. Although some specialized pursuits, such as certain competitive card games, could conceivably benefit from super-memory, evidence suggests that the forgetting rates of normal young humans are optimal for most purposes.[55] Empirically, prodigious memory has been linked to difficulties with thinking and problem solving,[56] and computationally, the effect of boosting the durability of individual memories is to decrease the ability to generalize.[57]

Indeed, in some circumstances reduced learning would confer benefit. Memories of traumatic events can cause lifelong suffering in the form of post-traumatic stress disorder (PTSD), and methods are being sought to prevent the consolidation of such memories by intervening pharmacologically immediately following the trauma.[58] Drugs that interfere with the consolidation of memories in general, such as benzodiazepines, are well known.[59] Extending these methods beyond the victims of trauma, to anyone wishing to avoid remembering an unpleasant event, is yet another way in which the neural bases of memory could be altered to enhance normal function.

## Ethical Issues in Enhancement

Although the promise of enhancement is easy to identify—smarter, more cheerful, and more capable people—the risk is harder to articulate. Most people feel at least some ambivalence about neuropsychological enhancement, but distinguishing realistic or compelling arguments from generalized fear is often difficult.

Many of the ethical issues raised by neuropsychological enhancement also arise with other types of enhancement.[60] Cosmetic surgery and the use of human growth hormone for healthy children who are naturally short, for example, are medical enhancements that do not affect brain function, and though both are controversial, both are generally accepted. Enhancement techniques that affect brain function through more familiar and non-neuroscience-based interventions such as biofeedback, meditation, tutoring, or psychotherapy are not seen as objectionable, and, in fact, are often seen as laudable. What, then, are the objections to using pharmaceutical or other neurotechnological means to achieve the same ends as behavioral techniques? Much recent discussion has focused on this question.[61] Although few if any ethical concerns arise uniquely in connection with neuroscience-based methods, two concerns seem particularly salient in the context of neural interventions for enhancement compared with other biomedical interventions whose targets are not psychological, on the one hand, and behavioral interventions for psychological enhancement, on the other.[62]

The first of these concerns is safety. Safety is a concern with all medications and procedures, but in comparison to other comparably elective treatments such as cosmetic surgery or growth hormone treatment, neuroscience-based enhancement involves intervening in a far more complex system. We are therefore at greater risk of unanticipated problems when we tinker. Would endowing learners with super-memory interfere with their ability to understand what they have learned and relate it to other knowledge? Might today's Ritalin users face an old age of premature cognitive decline? These are empirical questions, of course, which can be answered only in time. So far, medications such as SSRIs and stimulants have good safety records, and their long-term effects may even be positive. For example, SSRIs have been shown to be neuroprotective over the long term.[63] A recent study of the effects of Ritalin on rat brain development showed both desirable and undesirable effects on later adult behavior.[64] Nevertheless, drug safety testing does not routinely address long-term use, and relatively little evidence is available on long-term use by healthy subjects. It remains an open empirical issue whether the net effects of these or other yet-to-be developed drugs are positive or negative.

The second concern about neuroscience-based enhancement is more complex and difficult to state succinctly. This is actually a group of related concerns resulting from the many ways in which neuroscience-based enhancement intersects with our understanding of what it means to be a person, to be healthy and whole, to do meaningful work, and to value human life in all its imperfection. The recent report of the President's Council on Bioethics emphasized these issues in its discussion of enhancement. At the heart of this group of concerns is the problem of reconciling our understanding of persons and brains.[65]

Among the widely shared intuitions about persons are the following: Persons have a kind of value that is independent of any commodity or capability they bring to the world. Persons are responsible for their actions and deserve blame or respect depending on those actions. Persons lead lives that have meaning, and although it is difficult to say exactly what is meant by "meaning" in this context, most of us would agree that accomplishments in life are made meaningful partly by the effort they require. Finally, persons endure over time; although some of their characteristics may change, there is a self that remains constant for as long as the person can be said to exist.

Brains are physical systems and as such do not share any of the foregoing qualities. Of course, neuroscience-based enhancements work because changes to the brain result in changes to the person. To use such enhancements, without infringing on our personhood, can seem a contradiction, or at least perplexing, and raises a number of concerns. Maximizing the performance capabilities of an already healthy, functional person can be viewed as commodifying human abilities. Improving behavior pharmacologically seems to detract from the responsibility of the person for his or her own actions. Reducing the effort needed for personal accomplishments by neurochemical means may reduce their meaning as well. And the changing of abilities, memories, and moods at will by swallowing a pill may undermine the idea of a constant "self."

## Pending Challenges

Technologies for monitoring and manipulating the brain have developed rapidly over the last few decades and are poised for continued growth. Some of the ethical problems posed by these developments have immediate practical consequences. Examples of such problems include the illusory accuracy of brain images in forensic contexts and the unknown safety of long-term stimulant use by healthy adults and children. Other ethical problems are on the horizon, pending further technological progress. For example, brain imaging will not pose a serious threat to privacy until scanning methods can reliably deliver useful information about individual subjects. Although this is not the case at present, the development is foreseeable and could have enormous practical consequences.

Another way in which developments in neuroscience will influence society is less tangible than those just mentioned, but no less consequential. Both brain imaging and brain-based enhancement are forcing us to confront the fact that we are physical systems. If specific abilities, personality traits, and dispositions are manifest in characteristic patterns of brain activation and can be manipulated by specific neurochemical interventions, then they must be part of the physical world. Our intuitions about personhood do not mesh easily with this realization. At the very least, the realization calls for a considerably more nuanced idea of personal responsibility in law and morality.[66] More generally, it will prove challenging to traditional ideas regarding the soul, or the nonmaterial component of the human mind.

## Acknowledgments

The authors thank the editor and reviewers for extremely helpful comments on an earlier draft of this article, and Arthur Caplan for discussion and encouragement. The writing of this article was supported by NSF grants 0226060 and 0342108, and NIH grants R21- DA01586, R01-DA14129, and R01-HD043078.

REFERENCES

1. Available at www.sfn.org.

2. M. J. Farah, Emerging ethical issues in neuroscience, *Nature Neuroscience* 5 (2002): 1123–29; J. Illes, M.P Kirschen, and J. Gabrieli, From neuroimaging to neuroethics, *Nature Neuroscience* 6 (2003): 205; S.J. Marcus, ed., *Neuroethics Mapping the field*, conference proceedings (New York: Dana Foundation, 2002); A. Roskies, Neuroethics for the new millennium, *Neuron* 35 (2002): 21–23.

3. P.R. Wolpe, Neuroethics, *Encyclopedia of Bioethics*, 3rd ed. (Farmington Hills, MI: Macmillan Reference, 2004).

4. See Roskies, Neuroethics for the new millenium.

5. M.I. Posner and M.E. Raichle, *Images of the mind* (New York: Scientific American Books, 1994).

6. S. Ogawa et al., Brain magnetic resonance imaging with contrast dependent on blood oxygenation, *Proceedings of the National Academy of Sciences* 87 (1990): 9868–9872.

7. D.C. Alsop and J.A. Detre, Multisectional cerebral flow MR imaging with continuous arterial spin labeling, *Radiology* 208 (1998): 410–16.

8. G.K. Aguirre, Functional neuroimaging, in *Behavioral neurology and neuropsychology*, 2nd ed. (New York: McGraw Hill, 2003).

9. J. Ashburner and K.J. Friston, Voxel-based morphometry: The methods, *Neuroimage* 11 (2000): 805–21.

10. Z.J. Koles, Trends in EEG source localization, *Electroencephalography and Clinical Neurophysiology* 106 (1998): 127–37.

11. A. Villringer and B. Chance, Noninvasive optical spectroscopy and imaging of human brain function, *Trends in Neuroscience* 20 (1997): 435–42.

12. For example, K.N. Botteron, et al., Volumetric reduction in left subgenual/prefrontal cortex in early onset depression, *Biological Psychiatry* 15 (2002): 342–344; P. Milev et al., Initial magnetic resonance imaging volumetric brain measurements and outcome in

schizophrenia: A prospective longitudinal study with 5-year follow-up, *Biological Psychiatry* 54 (2003): 608–15.

13. D.D. Dougherty et al., Dopamine transporter density in patients with attention deficit hyperacivity disorder, *Lancet* 354 (1999): 2132–2133.

14. H. Fischer, G. Wik, and M. Fredrikson, Extraverion, neuroticism, and brain function: A PET study of personality, *Personality and Individual Differences* 23 (1997): 345–352; D.L. Johnson et al., Cerebral blood flow and personality: A positron emission tomography study, *American Journal of Psychiatry* 156 (1999): 252–257; M. Sugiura et al., Correlation between human personality and neural activity in cerebral cortex, *NeuroImage* 11 (2000): 541–546; T. Youn et al., Relationship between personality trait and regional cerebral glucose metabolism assessed with positron emission tomography," *Biological Psychology* 60 (2002): 109–120; T. Canli and Z. Amin, Neuroimaging of emotion and personality: Scientific evidence and ethical considerations, *Brain and Cognition* 50 (2002): 414–431.

15. T. Canli et al., An fMRI study of personality influences on brain reactivity to emotional stimuli, *Behavioral Neuroscience* 115 (2001): 33–42.

16. T. Canli et al., Amygdala response to happy faces as a function of extraversion, *Science* 296 (2002): 2191.

17. E.A. Phelps et al., Performance on indirect measures of race evaluation predicts amygdala activation, *Journal of Cognitive Neuroscience* 12 (2000): 729–738.

18. A.J. Hart et al., Differential response in the human amygdala to racial outgroup vs. ingroup face stimuli, *Neuroreport* 11 (2000): 2351–2355.

19. J. Greene et al., An fMRI investigation of emotional engagement in moral judgement, *Science* 293 (2001): 2105–2108.

20. A.R. Childress et al., Limbic activation during cue-induced cocaine craving, *American Journal of Psychiatry* 156 (1999): 11–18.

21. H. Garavan et al., Cue-induced cocaine craving: Neuroanatomical specificity for drug users and drug stimuli, *American of Journal of Psychiatry* 157 (2000): 1789–1798.

22. M. Beauregard, J. Levesque, and P. Bourgouin, Neural correlates of conscious self-regulation of emotion, *Journal of Neuroscience* 21 (2002).

23. K.L. Phan, T. Wager, S.F. Taylor, and I. Liberzon, Functional neuroanatomy of emotion: A meta-analysis of emotion activation studies in PET and fMRI, *Neuroimage* 16 (2002): 331–48.

24. P.J. Whalen et al., "Masked presentations of emotional facial expressions modulate amygdala activity without explicit knowledge, *Journal of Neuroscience* 18 (1998): 411–18.

25. D. Langleben et al., Brain activity during simulated deception: An event-related functional magnetic resonance study, *Neuroimage* (2002): 727–732.

26. T.M. Lee et al., Lie detection by functional magnetic resonance imaging, *Human Brain Mapping* 15 (2002): 157–164.

27. Available at www.brainwaves.com.

28. *Terry J. Harrington v. Supreme Court of Iowa*, February 23, 2003 decision.

29. GAO, *Investigative contacts: Federal agency views on the potential application of "brain fingerprinting"* (Washington, D.C., U.S. General Accounting Office, 2001): 1–24.

30. R. Cabeza et al., Can medial temporal regions distinguish true from false? *Proceedings of the National Academy of Sciences* 98 (2001): 4805–4810.

31. A. Raine et al., Reduced prefrontal and increased subcortical brain functioning assessed using positron emission tomography in predatory and affective murderers, *Behavioral Science and Law* 16 (1998): 319–332.

32. J.V. Haxby et al., Distributed and overlapping representations of faces and objects in ventral temporal cortex, *Science* 293 (2001): 2425–2430.

33. K.M. O'Craven and N. Kanwisher, Mental imagery of faces and places activates corresponding stimulus-specific brain regions, *Journal of Cognitive Neuroscience* 12 (2000): 1013–1023.

34. A.R. Childress, personal communication.

35. T. Canli, personal communication.

36. M.J. Farah, D. Foster, and C. Gawuga, "Reading personal information from functional brain scans, or oops your personality is showing (paper to be presented at the 11th annual meeting of the Cognitive Neuroscience Society, San Francisco).

37. Available at www.skirmish.com.

38. S.H. Barondes, *Better than Prozac: Creating the next generation of psychiatric drugs* (London: Oxford University Press, 2003).

39. G.S. Mahli and P. Sachdev, Novel physical treatments for the management of neuropsychiatric disorders, *Journal of Psychosomatic Research* 53 (2002): 709–719.

40. M. George, E. Wasserman, and R. Post, Transcranial magnetic stimulation: A neuropsychiatric tool for the 21st century, *Journal of Neuropsychiatry Clinical Neuroscience* 8 (1996): 373–382.

41. A.W. Snyder et al., Savant-like skills exposed in normal people by suppressing the left fronto-temporal lobe, *Journal of Integrative Neuroscience* 2 (2003):149–158.

42. J. Donoghue, Connecting cortex to machines: Recent advances in brain interfaces," *Nature Neuroscience Supplement* 5 (2002): 1085–1088.

43. Available at www.usc.edu/programs/ pibbs/site/faculty/berger_t.

44. H. Hoag, Neuroengineering: Remote control, *Nature* 423 (2003): 796–798.

45. P.D. Kramer, *Listening to Prozac* (New York: Penguin, 1993).

46. B. Knutson et al., Selective alteration of personality and social behavior by serotonergic intervention, *American Journal of Psychiatry* 155 (1998): 373–379; W.S. Tse and A.J. Bond, Serotonergic intervention affects both social dominance and affiliative behaviour, *Psychopharmacology* 161 (2002): 324–330.

47. W.S. Tse and A.J. Bond, Serotonergic intervention affects both social dominance and affiliative behaviour, *Psychopharmacology* 161 (2002): 324–330.

48. NIH, *Diagnosis and treatment of attention deficit hyperactivity disorder*, NIH Consensus Statement 16, no. 2 (1998): 1–37.

49. R. Elliott et al., Effects of methylphenidate on spatial working memory and planning in healthy young adults, *Psychopharmacology* 131 (1997): 196–206; M.A. Mehta et al., Methylphenidate enhances working memory by modulating discrete frontal and parietal lobe regions in the human brain, *Journal of Neuroscience* 20 (2000): RC65; D.C. Turner et al., Cognitive enhancing effects of modafinil in healthy volunteers, *Psychopharmacology* 165 (2003): 260–69.

50. Q. Babcock and T. Byrne, Student perceptions of methylphenidate abuse at a public liberal arts college, *Journal of American College Health* 49 (2000): 143–45.

51. L.H. Diller, Running on Ritalin: Attention deficit disorder and stimulant treatment in the 1990s, *Hastings Center Report* 26 (1996): 12–14.

52. P.E. Gold, L. Cahill, and G.L. Wenk, Ginkgo Biloba: a cognitive enhancer?, *Psychological Science in the Public Interest* 3 (2002): 2–11.

53. G. Lynch, Memory enhancement: The search for mechanism-based drugs, *Nature Neuroscience* 5 (2002): 1035–1038.

54. M. Ingvar et al., Enhancement by an ampakine of memory encoding in humans," *Experimental Neurology* 146 (1997): 553–559.

55. J. Anderson, *The adaptive characteristics of thought* (Hillsdale, NJ: Erlbaum, 1990).

56. A.R. Luria, *The mind of a mnemonist* (Cambridge, MA: Harvard University Press, 1968).

57. J.L. McClelland, B.L. McNaughton, and R.C. O'Reilly, Why there are complementary learning systems in the hippocampus and neocortex: Insights from the successes and failures of connectionist models of learning and memory, *Psychology Review* 102 (1995): 419–457.

58. R.K. Pitman, K.M. Sanders, R.M. Zusman, A.R. Healy, F. Cheema, and N.B. Lasko, Pilot study of secondary prevention of posttraumatic stress disorder with propranolol, *Biological Psychiatry* 15 (2002): 189–192.

59. S.E. Buffett-Jerrott and S.H. Stewart, Cognitive and sedative effects of benzodiazepine use, *Current Pharmaceutical Design* 8 (2002): 45–58.

60. See C. Elliot, *Better than well: American medicine meets the American dream* (New York: Norton, 2003); E. Parens, (Ed.) *Enhancing human traits: Social and ethical implications* (Washington, DC: Georgetown University Press, 2000).

61. M.J. Farah et al., Neurocognitive enhancement: What can we do? What should we do? *Nature Reviews Neuroscience* 5 (2004): 421–425; P.R. Wolpe, Treatment, enhancement, and the ethics of neurotherapeutics, *Brain and Cognition* 50 (2003): 387–395.

62. J.D. Moreno, Neuroethics: An agenda for neuroscience and society, *Nature Reviews Neuroscience* 4 (2003): 149–153.

63. V. Sanchez, J. Camarero, B. Esteban, M.J. Peter, A.R. Green, and M.I. Colado, The mechanisms involved in the longlasting neuroprotective effect of fluoxetine against MDMA ("Ecstasy")-induced degeneration of 5-HT nerve endings in rat brain, *British Journal of Pharmacology* 134 (2001): 46–57.

64. W.A. Carlezon Jr., S.D. Mague, and S.L. Andersen, Enduring behavioral effects of early exposure to methylphenidate in rats, *Biological Psychiatry* (2003): 1330–1337.

65. L. Kass, *Beyond therapy: Biotechnology and the pursuit of happiness* (New York: HarperCollins 2003).

66. S. Morse, Brain and blame, *Georgetown Law Journal* 84 (1996).

MARTHA J. FARAH, PH.D., is director of the Center for Cognitive Neuroscience and Walter H. Annenberg Professor in the Natural Sciences at the University of Pennsylvania.

PAUL ROOT WOLPE, PH.D., is professor of psychiatry and senior fellow in the Center for Bioethics at the University of Pennsylvania. He is president of the American Society for Bioethics and Humanities.

*Chapter 5*

# Neuroscience and Neuroethics[*]

*Donald Kennedy*

Neuroethics, it appears, is a subject that has "arrived." The Dana Foundation is, for the second time since 2002, sponsoring a special lecture on this topic at this year's annual meeting of the Society for Neuroscience. AAAS, publisher of *Science*, also joined with Dana to produce a conference, "Neuroscience and the Law," earlier this year. The U.S. President's Council on Bioethics is now devoting serious attention to the topic. Companies are deploying functional magnetic resonance imaging (fMRI) to map brain activity as they assess the product preferences of prospective consumers (Coke or Pepsi?). There's even a new discipline called neuroeconomics. So something is going on here.

What got it started, and where is it headed? I think it emerged as new techniques and insights into human brain function gave us a dramatically revised notion of what might be possible. The first microelectrode recordings in active, behaving, nonhuman primates made it possible to look seriously at how valuation, choice, and expectation are encoded by single cells in particular parts of the brain. It further evolved with the development of fMRI and other noninvasive techniques for tracing neural activity in people. These studies are beginning to explain how particular brain structures are involved in higher functions (making difficult moral choices, for example) or in predisposing the individual to a particular kind of behavior.

In a different area, the successes of psychopharmacology in altering brain

[*]From *Science* 306, October 15, 2004.

states and behavior have raised new problems of their own, not least in terms of how we may feel about the chemical manipulation of innate capacities. The list is long and ever growing: antidepressants, methylphenidate (Ritalin) for attention deficit hyperactivity disorder (ADHD), compounds that enhance alertness, and a new wave of drugs that may enhance memory formation and heighten cognitive ability.

Some of the questions now being raised by our expanded neuroscientific capacity are not exactly new. Consider, for example, the old issue of treatment versus enhancement. A child deficient in growth hormone could benefit from replacement therapy, and few would object to that, but its use by an aspiring teenage basketball player of normal height would raise questions. Now to the nervous system: Children with ADHD are often given methylphenidate after a physician considers their need. High school and college students without benefit of evaluation are using the same drug in the hope of improving their exam performance. Aside from the health risks associated with such drugs, what is it that bothers us here?

Perhaps it is our belief that the playing field should be level—we worry about the students who can't access the drug. Well, what about the kids who can't afford a preparatory course for taking a standardized test? Don't they raise the same questions about distributive justice? And suppose that we make the playing field level: All kids get the drugs, and all the sprinters get the steroids. Risks aside, are we comfortable with competition run in this way? Will the winners examine their enhanced selves and wonder "Was that really me?"

The ability to peer into brain processes also intensifies old privacy questions. Suppose that fMRI records become individually diagnostic with respect to some behavioral anomaly or predictive of some future tendency. Surely we would worry if they were used in insurance or employment contexts or in criminal litigation. Privacy protection would be guaranteed if the record were obtained as part of a medical procedure, but of course there are other possible sources. In the future, brain imaging techniques could conceivably be employed in the context of a court procedure as a test of truth-telling or subpoenaed in a case involving violence.

Finally, special issues arise when we penetrate into the philosophical territory where dualists and determinists debate over free will. As we learn more about the neurobiology of choice and decision, will we reach a point at which we feel less free? Perhaps more important for society, will we eventually know enough to change our view about individual responsibility for antisocial acts? There are those who worry about this. I am not among them, only because it seems so unlikely to me that our knowledge of the brain will deepen enough to fuse it with the mind. So, remaining convinced that my will is free, I am left to worry about the privacy of my inclinations and my thoughts.

DONALD KENNEDY, PH.D., is editor in chief of *Science*.

# Part II

# Professional Obligation and Public Understanding

A MAJOR CHALLENGE for neuroscientists is informing the public of actual and possible uses of neurotechnology in mapping, monitoring, and manipulating the brain. Brain imaging raises concern about privacy and confidentiality of sensitive information related to our thoughts. Neurosurgery and neurostimulation raise additional concerns about potential adverse effects of these interventions on cognition, mood, personality, and motor functions. Given the magnitude of the potential benefits and risks of intervening in the brain, neuroscientists have a professional and social obligation to explain the science of the brain to nonprofessionals directly in public forums or indirectly through the media.

Public misconception about science has consequences for both the public and science. For example, one common misconception is the belief that antibiotics designed to treat bacterial infections can treat influenza and other viral infections. The result has been overuse of these drugs, giving rise to antibiotic-resistant germs and a costly scientific scramble to find new drugs to kill them. A different problem is mistrust in medical science among the general public in the wake of high-profile clinical trials that have had disastrous outcomes. One recent shock to public confidence in science was a 2006 clinical trial in the United Kingdom for a monoclonal antibody to control autoimmune disease, which resulted in life-threatening side effects in six healthy volunteers. What needs to be different in the process of doing and applying research so that responsibility can be assigned and changes made to restore public trust in science?

Promoting understanding of science and retaining or restoring public trust in science is not just a "top-down" process of scientists lecturing to the public. It should also be "bottom-up," with the public playing an active role. The public needs to express its concerns about the potential uses and abuses of science. Policy makers also play an important role in this process, since they formulate and implement guidelines and policies for the application of science. The media arguably play the most critical role in disseminating knowledge about science in general and neuroscience in particular, providing channels of communication between and among researchers, policy makers, and the public. All of these stakeholders need to be part of a deliberative democratic process in coming to a common understanding of what neuroscience can do and in reaching consensus on how we should use it.

In "From the 'Public Understanding of Science' to Scientists' Understanding of the Public," Colin Blakemore points out that, in Britain at least, discussion of the costs and benefits of science has involved both scientists and the public for some time. This has occurred thanks to the recognition of the importance of informed debate of science consisting of politicians, commerce, and the public in discussing how science should be applied. Noting the erosion of public trust in science in the wake of AIDS, bovine spongiform encephalopathy (BSE), and Creutzfeldt-Jakob disease (CJD), Blakemore stresses that scientists have a duty to explain their work to nonscientists. Scientists need to engage the public in a two-way interaction with a view to restoring public confidence in the scientific process.

Alan Leshner further emphasizes the need for dialogue between scientists and the public in deciding how neuroscience should move from basic research to clinical practice in "Ethical Issues in Taking Neuroscience Research from Bench to Bedside." He underscores the distinctive nature of neuroscience among the sciences, given that the brain enables and mediates cognition, emotion, and personality. Altering the brain may entail potentially great benefits and costs. Leshner asserts that neuroscientists have an obligation to apply their research findings for the benefit of humanity. As part of this obligation, neuroscientists must engage other professionals and the public in discussing how we should use neuroscientific techniques. This type of debate could minimize the likelihood of discrimination using data from brain imaging studies in drawing unwarranted inferences from groups to individuals, or vice versa. Debate should focus on the need to protect vulnerable individuals by upholding the principle of informed consent in different applications of neuroscience. The obligation of neuroscientists extends beyond their research to dissemination of knowledge about the brain to all who may be affected by the research.

In "Models for the Neuroethical Debate in the Community," John Timpane strengthens the case for public deliberation about neuroscientific interventions and therapies. He insists that deliberation should start at the grassroots level. Scientists and policy makers need to be guided by the opinions of the people they serve. Timpane uses his model to explain how ethical debate on stem cell research should proceed and extends the model to frame and deliberate ethical questions in two areas of neurotechnology—personality enhancement and brain fingerprinting. He emphasizes the need to anticipate developments in neurotechnology, noting that it is "best to start now before the wave is upon us."

This last comment underscores an important aspect of public debate on existing and emerging neurotechnologies. By thinking "upstream" and anticipating developments before they occur, neuroscientists, policy makers, the media, and the general public will be able to constructively debate the benefits and risks of interventions in the brain and develop guidelines or polices on how these interventions can and should be used. In this way, we avoid unreflectively reacting to findings about the brain, or effects of intervention in the brain, when they occur. Now is the time for all stakeholders to collectively engage in thoughtful, deliberative discussion of the possibilities and limits of the science of the brain.

# Chapter 6

# From the "Public Understanding of Science" to Scientists' Understanding of the Public*

*Colin Blakemore*

Britain is ahead of the United States in very few areas of science, but in generating problems, controversies, and public confrontations involving science, we're really in the lead. I guess that's why I've been invited to talk in this session on the role of scientists in public communication of ethical problems.

Regarding my credentials in this area, I'm director of the Oxford Centre for Cognitive Neuroscience but am also chairman of the British Association for the Advancement of Science, the major national organization in Britain devoted to public communication. And as it happens, I'm also much involved with the Dana Alliance in Europe.

I did my first radio broadcast in 1976, when I gave a series of lectures, called the Reith Lectures, on BBC radio—six straight half hour, no-illustrations presentations—which, amazingly, had a considerable audience. And since then I've been involved in what's now approaching 500 radio and TV programs, including a thirteen-part television series on the brain and mind. I say this not to impress you about me, just to let you know that I have a foot in both camps.

*From *Neuroethics: Mapping the Field* (Dana Press, 2002): 212–221.

The agenda for the "public understanding of science"—that phrase is very much recognized in Britain—actually began about seventeen years ago. The two main issues that drove it were the recognition of the importance of having a scientifically informed public, given the speed of progress of science, and the need, in a democratic society, to involve the public in the decisions of politicians and commerce about how science should be applied.

What are the advantages of a scientifically informed public? A very considerable one is giving people a better capacity to assess risks in their own lives. If risks are demonstrably so large that people should be protected from their own inclinations, then legislation tends to do that. But there is a wide range of public activity with an element of risk for which legislation isn't appropriate, and here is where people need to somehow be informed by scientists in order to make proper decisions about how to run their lives. And of course, some sources of risk come directly from technology and science.

In addition, it's important that people be able to assess the potential benefits of new developments in technology and therefore be better equipped to perform cost-benefit analysis in their heads; that way, they may make sensible decisions about what they want to do with technology.

Equally important is the empowerment of people to participate in public discussion and debate about where science should go and how technology should be applied. A broader, more metaphysical advantage is involving people who are not themselves specialists in science in the culture of science; this recognizes and endeavors to change the fact that while science makes a very important contribution to the culture of human society, only about 5 percent of the population is in any sense professionally involved.

And finally, from a political perspective, an educated public is more likely to be supportive of science-based policy, which, of course, is a basic principle for all developed countries.

In 1985 the Royal Society—the Academy of Sciences in Britain—commissioned a report, chaired by Sir Walter Bodmer, on the public's knowledge of science. It followed a paper in *Nature* that was a survey of the public's knowledge—or rather ignorance—of absolutely basic facts of science. The paper reported on answers to such *Who Wants to Be a Millionaire*–type questions as "Does the Earth go around the sun, or the sun around the Earth?" and "Do antibiotics kill viruses?" It turned out that people were abysmally bad at those things. I'm not sure it really matters very much that most of them don't know whether the Earth goes around the sun—it actually hinges rather little on their everyday lives—but it's an indication of the depth of the problem.

One of the Bodmer Report's major conclusions was a message to scientists: Learn to communicate with the public, be willing to do so, and consider it your duty to do so. The notion of scientists' *duty* was central.

In 1986 the Royal Society followed up on this report by establishing the Committee on the Public Understanding of Science (COPUS), at that time a very powerful organization. It was jointly administered by the three major scientific organizations that interface with the public and its leaders: the Royal Society itself, the British Association, and the Royal Institution.

During the following ten years the public-understanding-of-science agenda became deeply embedded in the scientific ethos of Britain. By 1995 another government-commissioned report, the Wolfendale Report, concluded that scientists and engineers in receipt of public funds have a duty—there's that word "duty" again—to explain their work to the general public. This report actually recommended that every holder of a publicly funded grant be required to participate in public activities. They would have to specify what public activities they had been involved in—whether it was local radio and newspapers, or national television, or whatever—before they would be eligible to apply for renewal. It was really quite draconian.

The research councils—the government funding agencies that disperse government funds—have all become involved in public-understanding-of-science activities. For example, the corporate plan of PPARC—the Particle Physics and Astronomy Research Council—states: "We believe that those engaged in publicly funded research have a duty to explain their work to the general public." On average, a research council spends about 0.25 percent of its total annual budget on such activities.

However, against that background of a decade of increasing recognition that we have a duty to go out and explain our work to the public, a series of problems have confronted Britain and shaken to its roots the public's confidence in the scientific process. This public disillusionment applies particularly to the scientific advisory process regarding AIDS, mad cow disease (BSE) [bovine spongiform encephalopathy] and the associated variant form of CJD [Creutzfeldt-Jakob disease] in humans, embryo research, and animal rights (a topic in which Britain seems to specialize, unfortunately). GM [genetically modified] foods, of course, was immensely controversial, with that controversy even spreading back to the United States, where the technology had been much more happily accepted than in Europe.

Then too, there has been controversy over cloning and stem cell technology, cellular telephones and the new police version of telecommunications, the MMR vaccine (the triple vaccine for mumps, measles, and rubella, which, it's been claimed, might be associated with increased incidence of autism), and the foot-and-mouth disease epidemic in Britain just last year.

In this series of events, none of which can be directly attributed to the activity of scientists, a link was made by the media and by many members of the public that if things somehow go wrong with technology, it must be the generators of technology who are at fault. So the blame for many of these things was laid at the door of science—quite unreasonably, but that was a problem we faced.

Even more, these events shook the public's trust in the scientific advisory process, especially for government. Why did government ministers continue until late 1995 to say that there was literally no possibility of a risk of transmission of mad cow disease to human beings? They shamefacedly had to admit only a few months later (in 1996) that there was indeed very good evidence for such a shift. This series of events implied either that the scientists advising government were incompetent or that the process of transforming their advice into public statements was distorted. I'm absolutely sure that the latter is true, but the public got the impression that the former was the problem.

In the last few years, trust in scientists in general has—at least at times—been rather low. A poll by MORI [Market and Opinion Research International, a United Kingdom firm] in 1996 showed that 75 percent of the people had great or fair faith in scientists who were associated with environmental groups, while 45 percent had faith in scientists working in industry, and 32 percent had faith in government scientists. The public is more willing to trust scientists who don't, as it were, have vested interests—scientists associated with nongovernmental organizations that are committed to charitable acts and generally to fighting the establishment.

Trust ratings consistently show that if you simply ask people to rank in descending order their trust of different professions, priests are always very high, doctors are high, teachers are high. Close to the bottom of the scale are scientists. Then, I'm relieved to say, even lower than scientists are journalists, and right at the bottom—almost off the scale—are politicians.

Scientists' rankings in such surveys is very disappointing to those of us who've spent a good fraction of our lives trying to communicate with the public—in the hope that this would improve relations and understanding—and then discover that trust and confidence, if anything, has decreased during that decade. Interestingly, surveys around Europe show an inverse correlation between the level of public knowledge about science (at least with respect to those *Who Wants to Be a Millionaire*–type questions) and public trust in science. The more people know, apparently, the less they trust the scientists.

The more confident of those scientists who've been involved in the public-understanding-of-science process tend to say: "Well, this must be a transition period. It's a good thing; it's a sign of healthy understanding and inquiry. We'll move through it to the point where we can carry the public with us eventually." I hope so, but the results are nevertheless a reason for concern.

Actually, there has been a shift in attitude, post-BSE, with a 2000 report on science and society by the House of Lords' Science and Technology Committee. Headed by former Tory minister Patrick Jenkin, the study surveyed the successes and failures of the public understanding of science's agenda over the last ten years, and it came to the conclusion that despite all that effort, society's relationship with science is in a critical phase — there is a crisis of confidence in science among the public — but also that this crisis of trust has produced a new mood for dialogue.

In effect, the terminology has been turned around completely. "Public understanding of science" has become an even dirtier word, it seems, than its acronym (PUS), and now everybody talks instead about dialogue and debate — two-way processes of interaction between scientists and the public, rather than a one-way didactic presentation of the truth from scientists to ordinary people.

Why should scientists themselves be involved in this process of two-way communication? There are obvious advantages, the first being the authenticity of the evidence that people receive about scientific fact.

Second, more subtly, is that involving scientists in presenting the results and evidence of science reveals the process of science. There's general agreement that members of the public know very little about how science really works. They tend to see the headlined breakthroughs and achievements. They expect science to come in irrefutable, uncontested truths, rather than in the turmoil of conflict and controversy that we know underlies the generation of, in the end, accepted ideas.

Showing scientists as people, showing they actually are human beings without two heads and without antennae sticking out of them, that they're the kinds of people who might live next door and have mortgages and kids and so on, is considered very important too. And so is policing the media.

The disadvantages are obvious as well: the opportunities given scientists to become showmen, to abuse the privilege of an audience to which statements can be made without peer review, to distort that evidence for whatever reason, political or professional.

And finally, there's the conflict of an involvement in public communication with the normal career path. It's generally recognized that activity in the public domain is still insufficiently incorporated into the career assessment of scientists, despite the fact that there's been a lot of lip service to its importance.

In any case, the emphasis in Britain now is on public engagement—not expecting the public to tell scientists what to do or expecting the public to judge the scientific facts and the worth of science, but encouraging the public to be involved in deciding where science should go, where its limits should be, and how it should be applied. The result, it is hoped, will be an increase in public confidence in the applications of science.

A variety of events in this spirit have been introduced over the last year or so by communications organizations in Britain. I'll just mention one of them: SciBars (or *cafés scientifiques*), which have spread very quickly around the country. These are free events in pubs and wine bars—anyone can walk in off the street, buy their own drinks, and participate in a public discussion about some scientific issue, usually led by a scientist. They've been fantastically popular and successful.

Finally, just to mention the Dana Foundation's involvement, it has very generously helped sponsor a new building associated with the Science Museum in London, three floors of which will be office space in which the European Dana Alliance will be accommodated along with the British Association. But another three floors are public space devoted to the science-public dialogue. Virtually any organization that claims it has science at its heart and wants to communicate—science and nongovernmental organizations, lobbying groups, and so on—will be able to have access to this space for genuine public dialogue.

Thank you.

## Question and Answer

*Marilyn Albert:* Colin, aside from the issue of BSE and hoof-and-mouth disease, the list of crises that you mentioned is similar in the United States and Great Britain. Yet I think it's fair to say that there's less suspicion of scientists here. Maybe one of the reasons is that the involvement of scientists with advocacy organizations that have to do with disease—specifically with the translation of basic science to treatment— is particularly clear to the American public. Could you could say something about whether or not you think there's a difference between Great Britain and the United States in that aspect of public activity?

*Blakemore:* I think that's quite right—it's one of the obvious differences between the U.S. and Britain. Another, and greater, cause of the difference in trust is that there is less of a tradition of openness in Britain and in Europe in general. We don't yet have a Freedom of Information Act, for example, though we're just about to introduce one. There's a far less open presentation by government of the process of government. Agency Web sites and so on still include rather little of the information on which government bases its decisions. This lack of openness and transparency is a major problem, though one that has been very much recognized now and is changing rapidly. European leaders are coming to understand that when there are problems and members of the public can't get easy access to the reasons for the problems, they become paranoid and mistrustful of what they see are the responsible organizations.

*Michael Williams:* Last year I was the chair of the AMA's Council on Scientific Affairs, and we faced these kinds of issues. But despite our openness, I think that large segments of the public remain skeptical. For example, the African American community generally distrusts science, medicine, and researchers, and I think that clinical research as an enterprise has faced a number of challenges in the last several years as a result. We got burned very badly at Johns Hopkins last year in that arena. My question for you is, Despite some differences between the American public and the British public, do you have any insights on things that are better or worse for us to consider in how to enhance this dialogue?

*Blakemore:* I don't know to what extent the public in the States has had the opportunity to engage face-to-face with practicing scientists (except maybe during Brain Awareness Week, as far as brain research is concerned). This is a relatively new development in Britain, which has been extremely successful. Every year we have a National Science Week, with hundreds of events and thousands of participants. Meanwhile, the rate of science broadcasting on radio and TV has increased even further; it was high already. And the involvement of scientists themselves in broadcasting has increased tremendously. These are a few areas, I think, where there have been real achievements. I don't know to what extent they're mirrored in this country, but certainly some lessons could be learned from them.

COLIN BLAKEMORE, FMEDSCI, FI BIOL, HON FRCP, FRS, is chief executive of the British Medical Research Council and Waynflete Professor of Physiology at Oxford University. He has been president and chairman of the British Association for the Advancement of Science and president of the British Neuroscience Association, the Physiological Society and the Biosciences Federation. His books for general readers include *Mechanics of the Mind, Images and Understanding, Mindwaves, The Mind Machine, Gender and Society* and *The Oxford Companion to the Body.*

*Chapter 7*

# Ethical Issues in Taking Neuroscience Research from Bench to Bedside*

*Alan I. Leshner*

The dramatic progress made by neuroscience has generated both rich promise and quickly rising expectations for improvements in the health and well-being of people throughout the world. However, moving those advances from the research bench to actual clinical practice will require more than just scientific and technological progress. The unique attributes of the brain as an organ system and its centrality to our concept of our own humanity raise an array of ethical issues that must be resolved in an open dialogue involving both the scientific community and the wider public before we will see widespread application of the fruits of neuroscientific progress.

## What's So Special About Brain Research?

At a minimum, the same overarching ethical standards should be applied to brain research as to any other area of clinical work: prevent harm, protect the vulnerable, and ensure fairness and equity of access to the benefits of the research. Moreover, of course, the brain is not the only organ system whose integrity is essential to a healthy life.

*From *Cerebrum* 6 (2004): 66–73.

However, the brain is the most complex organ in the body. No other system has so many roles and consists of so many interoperating parts—the brain's millions of interconnected cells and circuits. No part of the brain is an "island"; individual parts of the brain neither act alone nor appear to be involved in only one function. This complexity makes studying and eventually intervening effectively in the operations of the human brain among the most difficult challenges facing the scientific enterprise. The interconnectedness of its parts and the multitasking nature of its individual structures mean that any intervention, however small or precise we try to make it, is unlikely to have a single consequence. Therefore, the decision to in any way alter brain structure or activity involves potentially great cost-benefit trade-offs.

What makes the brain so special is that it is the seat of the mind. At least, most people think *mind* when they hear or think about the brain. Although the brain does much in addition to generating what we think of as the mind, mental activity is so central to our very humanness that the relationship between brain and mind always haunts any thoughts of normal or abnormal brain function. The brain is the essence of the "self" and, therefore, doing anything to the brain is potentially altering one's essential being. This close association among brain, mind, and self colors any discussion of real or imagined brain interventions, whether to enhance a normal brain or to correct neural malfunction.

In addition, although behavior is determined by an interaction among one's genes, one's personal life history, the environmental context in which the behavior will occur, and other aspects of an individual's biological state, the brain is the final common path for the experience and expression of all mental activity. For that reason, any intervention in our brains raises the specter of not only causing potential physical disability but also changing our cognition, emotion, or even our personalities.

## Do We Really Want to Know?

The idea that research discoveries are possible that might best be left unmade is a concept almost uniquely relevant to the issues surrounding research on human biology and behavior. Donald Kennedy, editor of the journal *Science*, took up this question in a workshop on neuroethics supported by the Dana Foundation and organized by Stanford University and the University of California–San Francisco in 2002 (see reading list). His remarks caught my attention because it is such an unusual question to be asked by scientists.

In almost every other area of science, we would immediately answer, "Of course we want to know. We want to know everything!" However, society at large has at times seemed at best ambivalent about what it wants to know about human behavior and its relationship to biology. The best-known cases surround issues such as the neurogenetics of intelligence or of violence. Earlier attempts by scientists to tackle these issues have been met with a great hue and cry from many quarters, with people concerned primarily about how the information might be misused to stereotype or stigmatize individuals or groups.

I believe these negative reactions reflect scientists' failure to accurately communicate the studies and their potential implications as much as they do misuse of or overgeneralization from scientific findings. For example, as I will discuss further, the fact of a possible genetic predisposition to greater or lesser intelligence does not automatically imply that members of one or another racial or ethnic group will be more or less intelligent. Nevertheless, findings on the genetics of intelligence have too frequently been interpreted that way. The same has been true for studies of genetic contributions to levels of aggressiveness or violence.

Another form of the same question was posed in a recent report of the President's Council on Bioethics titled *Beyond Therapy*. The council members grouped ethical questions around behavioral biology and similar domains into two categories. One set relates to interpersonal issues of preventing harm and protecting vulnerable people (also discussed more below). The second set is of a higher order, having to do with our sense of our own humanity. The council raised for discussion, without coming to a clear conclusion, the issue that, as we learn much more about genetics and about the brain and how to use our findings to intervene, we may be at risk of "fooling with Mother Nature" or "playing God." For example, the council suggested we ponder whether we may be at risk of doing "unnatural things" when we think about brain-based behavioral enhancements.

This issue, of course, can play out at the individual level, the group level, or even at societal levels. Thus, we also need to think about whether we would also be at risk of changing our entire society.

Whatever the origins of past problems or future concerns, I share the belief of the scientific community that we have an obligation to apply the full power of science to solving the toughest problems facing humanity, even if they are potentially contentious. And as scientists and clinicians, we must do whatever we can to relieve pain and suffering. However, I also believe that, when entering these kinds of domains, scientists have a duty to be extremely sensitive to the potential implications and uses of the results of their work, and that they need to engage fully with other members of the public, including philosophers, clergy, and ordinary citizens in developing a moral consensus and guidelines about how we will proceed.

## Moving from Animal Models to Humans

Discussions such as this always point out that humans are animals, and therefore much of what we have learned from studies of animal model systems ought to be generalizable to humans. In very many cases, that is true. But many animal models of complex human conditions, such as mental illnesses or addictions, are actually quite weak; they only very roughly approximate the human condition. Similarly, many animal models of complex human behaviors yield findings that, at best, generalize only minimally to humans, either because the human behavior the animals are supposed to model is really much more complex or the models are only superficial approximations of the human condition thought to underlie the behavior. The same is true for in vitro models, which are not whole organisms but cells in a petri dish or test tube.

This inability to generalize readily from animal models to humans raises a set of issues concerning whether and how to proceed from basic knowledge derived in animal or in vitro studies to tests of its relevance to humans. What criteria should we use to decide when basic-science findings are strong enough to be tested in humans? How do we decide the appropriateness or relevance of the models used to prepare for studies in humans? In seeking to replicate animal findings through noninvasive, observational studies of humans—using functional magnetic resonance imaging (fMRI), for example—we might set the bar relatively low, but when modifying brain function might be a part of the study, the decision becomes much more difficult.

An interesting recent article by Vivienne Parry in the British *Guardian* newspaper illustrates well the concern by pointing out some examples where basic science-derived approaches to neurotherapeutics—mostly developed either in animal or in in vitro models—were prematurely tested in human subjects, with catastrophic results. Parry argues that the great promise of stem cell research—for example, for alleviating the disability of diseases such as Parkinson's—could lead us to premature testing in humans. But at some point, if we are going to reap the benefits from basic research, we will need to take the leap. I believe Parry's point is a good one and that clear standards need to be set now, because the basic science advances are coming at a rapid pace.

## Pitfalls of Generalizing from the Many to One

One aim of neuroscience research is to improve our ability to predict future health conditions and behavior and our knowledge of how to modify both. However, scientific conclusions are often drawn from averages characterizing relatively large groups of subjects; they may not hold for individual cases within the group. It thus becomes important to avoid prejudging or stigmatizing a particular person merely because he or she belongs to a distinctive group; the individual may or may not have the characteristic in question.

This is especially true in interpreting studies of the genetics of behavior. Most behavioral traits such as intelligence, emotionality, or aggressiveness vary in intensity along a continuum, and genes predispose people to behave more or less intensely in response to stimuli in their environment. It is not the case that an individual is aggressive or not or that he or she is emotional or not; rather, genes predispose people to be more or less aggressive or more or less emotional. Moreover, genes do not doom an individual to behave in a particular way. They are only one of many things that determine what a person is like, including the individual's personal history and the environmental context and triggering stimuli.

The same could be said of studies of other kinds of so-called predisposing or risk factors, such as early nutrition, exposure to drugs or alcohol, infections, or parental behaviors. In each case, when one looks at the whole group under study, one can see an overall effect. However, not all individuals within the group will necessarily share the outcome that characterizes the average of the subjects. The effects of prenatal cocaine exposure on later cognitive development are a powerful example that has received much public attention. The latest data show that, when the subjects are averaged together, the effects are not very great, although some individuals are affected very dramatically and others appear to be unaffected. Caution in generalizing from group data to the individual is especially critical in studies of brain and behavior, because any improved ability to predict behavior will likely be of great interest to law enforcement, employers, insurers, and schools. Such entities may use that knowledge in ways not always in the best interests of individuals.

## Vulnerable People

Clinical brain research is subject to the same ethical guidelines and regulations as any other field of research with humans. These include regulations governing conflicts of interest, confidentiality, data and safety monitoring of clinical trials, institutional review boards, and informed consent. These issues have been deeply considered and extensively written about, and a good source of current thinking and information on regulations and guidelines can be found at the Web site of the Office of Extramural Research of the National Institutes of Health (http://www.hhs.gov/ohrp).

But many human subjects in clinical brain research are in one way or another particularly vulnerable, and protecting them therefore requires special consideration. The general rules are not adequate.

Perhaps the most widely discussed issues have to do with informed consent, since many clinical brain research subjects are either cognitively or emotionally impaired. They might not really understand what they are consenting to, or they might be particularly susceptible to inducement or coercion.

It can be extremely difficult to get genuinely informed consent from patients with dementias or other mental illnesses that compromise their intellect and emotions. The same can be true for children too young to really understand the issues at hand. In these cases, researchers often secure consent through a responsible family member or a specially appointed surrogate or proxy for the impaired patient.*

In my own field of addiction research, we have had substantial discussion about the ethics of administering drugs of abuse to addicted individuals, whether they are currently abstaining or not. The problems are manifold: First, by definition, addicted individuals have severely compromised abilities to control their cravings and thus might be especially susceptible to improper inducements. It also is well known that in abstinent addicts even a very small "taste of the drug" can induce phenomenal cravings and relapse to drug use. Is it right for us to ask the "clean" addict to take that risk? Final-

---

*For more information, see also http://grants.nih.gov/grants/policy/questionablecapacity.htm.

ly, every clinician's first priority ought to be to get addicted people into treatment, since addiction is a serious, life-compromising illness. In this case, current best practice is first to work hard to get potential subjects into drug treatment, and only if that fails to include them in experiments where they might be exposed to abusable substances.

A second issue arises from the need to know what drugs of abuse do to naïve subjects—people never before exposed to the substance under study. Given the powerful addictive quality of many drugs of abuse, is it ever ethical to give drugs of abuse to naïve subjects? In this case, both government guidelines and consensus in the field suggest that drugs of abuse should be given to naïve subjects only under the most exceptional circumstances and with the strongest justification.*

Some newly emergent issues affect biomedical research generally but seem of particular concern when dealing with vulnerable populations. A particularly thorny area is the testing in children or adolescents of brain-targeting medications that have been approved for use in adults but are previously untested in children. Historically, the general trend was to consider children and adolescents as simply little adults and assume they would respond the same way adults do. However, in the past few years it has become clear that the brain undergoes very rapid developmental changes all through adolescence; therefore, the effects and risks of a substance on the brain of a preadolescent might be quite different from those for an older adolescent. We need clearer guidelines than we have now about when and how medications—even if approved for similar uses in adults—can be tested in younger people. A similar argument can be made for research on elderly patients whose brains are in a different state of flux.

## Beyond the Bench and Bedside

All scientists wrestling with the ethical issues embedded in translating basic brain research into life-saving therapies must also be mindful of the non-therapeutic uses to which their work might be applied. I have not spent any time here on the very important questions about how society will use in other settings the dramatic new understanding of the brain that science is providing. However, I believe it is important to make explicit that the roles and obligations of scientists do not end with them conducting their studies ac-

---

*For more information, see also http://www.drugabuse.gov/Funding/HSGuide.html.

cording to the highest ethical standards nor with simply publishing their results and communicating them to their colleagues.

Most new discoveries about the brain are both subtle and complex. Interpreting them appropriately for practical and policy use requires a deep
understanding of the strengths and weaknesses of the relevant experiments
and of the nuances and caveats that surround the scientific findings and
theories derived from them. For that reason, many neuroscientists must be
willing to go beyond their traditional "bench-based" roles. Any scientist who
chooses to work on subjects with such significant ethical and legal ramifications must be willing also to play a central role in the important dialogues
that will ensue, involving ethicists, the clergy, policy makers, and the interested public.

## Some Interesting Reading

Baker, C. 2004. *Behavioral Genetics*. Washington, DC: American Association for the
Advancement of Science.

Gray, J. R., and Thompson, P.M. 2004. Neurobiology of intelligence: Science and
ethics. *Nature Reviews Neuroscience* 5: 471–482.

Marcus, S. J. 2002. *Neuroethics: Mapping the field*. New York: Dana Press.

Parry, V. 2004. A matter of life and death: Scientists hope that by injecting human
embryo stem cells into the brain, they will be able to cure Parkinson's disease. But is such
radical treatment worth the risk? *Guardian*. July 6.

President's Council on Bioethics. 2003. *Beyond therapy: Biotechnology and the pursuit of
happiness*. New York. Dana Press.

ALAN I. LESHNER, PH.D., is chief executive officer of the American
Association for the Advancement of Science and executive publisher of the
journal *Science*. He was director of the National Institute on Drug Abuse
from 1994 to 2001, following his service as deputy director and acting director of the National Institute of Mental Health.

# Chapter 8

# Models for the Neuroethical Debate in the Community*

## John Timpane

In the fourth chapter of the Hindu text, *Bhagavad-Gita* (*The Song of the Blessed One*), you can find what may be the most consoling sentence ever written: "In truth, there is nothing in the world so purifying as knowledge."

I believe that, but not everyone does. Some behold the onward rush of neuroscience and wish to go slow in territory so intimate, so long forbidden. Our minds; the patterns of our choices; our cherishment of memories; our soulprints—who would not say, "Go carefully, if at all"?

In truth, however, nothing in the world is as purifying as knowledge. Not only does it often improve life (even nuclear technology, the great counter-example, has benefited billions more than it has harmed), but also it purges us of delusions and prejudices born of ignorance. Knowledge, so far, has created more than it has destroyed.

But many are the uninformed or the mal-informed; many are scared. That creates an ideal arena for the opportunist politician or leader (of any party) who would create public policies that tie scientists' hands and prevent the very advances that may most purify. We've seen such short-circuiting before. In the reproductive revolution, state legislatures obstructed as many reproductive choices as possible with labyrinthine legal hedgeworks, often bypassing a fully democratic debate. Today, with early manifestations of the genom-

*From *Cerebrum* 6 (2004): 100–107.

ic revolution, we are seeing devious (and hilarious) government policies on stem cell research and government "peer review" of federally funded science. Here again, leaders have sidled past the obligation to hold a fully inclusive social debate. Not that the damage can't be healed; of course it can, but healing will take years and cost so much that many will decide not to try. I would hate to see the same thing happen with the neurobiological revolution.

The great project facing neurobiologists and neuroethicists is to mediate an informed public discussion of coming inventions and therapies. This has to be, I think, a local, roots-up process, not top down. To short-circuit the short-circuiters, Americans must deliberate among themselves on the great issues raised by neuroscience and the values citizens wish to see expressed in public policies regarding our coming power to map, read, and manipulate the human mind. In this essay, I'd like to suggest some forms such a deliberative process might take and give some examples of how the issues might be framed.

It's important to acknowledge the efforts already under way to get just such a deliberation started. In its offices in Washington, London, and elsewhere, the Dana Foundation has hosted discussions ranging from formal debates to pub lunches. And I was thrilled to be present when the David A. Mahoney Society at the University of Pennsylvania announced the inception of the Eliot Stellar Program in Neuroscience and Society, which will seek to encourage discussion of neuroethics in the community.

I speak best, however, from my experience as the commentary page editor of the *Philadelphia Inquirer*. For years, editorial page editor Chris Satullo, along with Harris Sokolow of the University of Pennsylvania, has been conducting what we call Citizen Voices forums. We often do a round of Citizen Voices when there is an election or a public issue—in the Philadelphia area, two such issues have been health care and urban revitalization—of especial moment. This is a deliberative process, not a debate, a crucial distinction I will explain. Our model is the one created by the National Issues Forums Institute (www.nifi.com) with help from the Kettering Foundation.

We're a newspaper. It's just such local, grassroots organizations—churches, schools, local media outlets, and social groups—that must initiate these discussions. We announce the forums to our readers and invite them to participate. If, in the makeup of the participant pool, we see an imbalance in gender, age, income level, educational or professional background, or ethnicity, we try to redress it, throwing out the call throughout the community. Once a pool of respondents is established, we hold introductory meetings at which we explain the subject, the process, and the goals.

Imagine, then, a pool of participants, say, 200 people, in eight community-based groups of 25 apiece. They will come together for some parts of this process, but most of the work they will do in their own groups. This is not just an exercise; it is the very work of citizenship.

## Getting Started

Our citizens will have a concrete goal: to communicate the results of their deliberations, in the form of specific policy recommendations to guide scientists in their research and to guide political leaders in writing laws concerning research. Neither scientists nor leaders can be self-regulating; both need the advice of those they serve.

Experts in neuroscience are important in this process, however: they will educate the participants and act as consultants or referees during discussions. Neuroscience is one of the most complex realms of human understanding, befitting the human brain, the most complex thing ever beheld. Thus, any citizens' forum on neuroscience would begin with an introduction to the current state of knowledge and research. The object is not to turn the participants into experts, but rather to give them enough shared knowledge that they know what they want to argue about. We also would need something like the "issue books" distributed by the National Issues Forum Institute, presenting nonpartisan overviews of issues and possible public responses. In the expert's second role, as consultant and referee, he or she will consult with the moderator to extract ideas or direct discussion and adjudicate matters of fact.

## Reasoning Together

Our citizens meetings tend to resemble town meetings, but they could take other forms. A moderator directs and regulates discussion; he or she notes important points on the board, elicits discussion of alternatives, risks, and benefits, and keeps the discussion fair.

What our now-informed citizens are doing is deliberating the issues. In the words of the National Issues Forum Institute guidelines, "participants use discussion to discover, not to persuade or advocate." Having acquired a common fund of knowledge, they now have to identify the salient issues, discuss as many viewpoints and options as they can, and set out these deliberations in some formal fashion.

The aim is not to see which side "wins"; that's what happens in a debate, but it is not what's happening here. The object is to create a chart of the range of policy options in each case, considering all sides and the consequences of all options. In our actual meetings, people do argue and debate, but as different opinions arise, the moderator notes them on a blackboard or chart and urges the group to follow those opinions together to their logical conclusions.

Here is what I imagine our citizen deliberators would care most about and the kinds of valuable contributions they would make to major neuroethical issues.

## Deliberation #1: Stem Cell Research

The moderator might begin deliberation by presenting the group with a forthright position statement:

> We should give scientists unlimited freedom to use embryonic stem cells to explore ways to treat Alzheimer's disease, Parkinson's disease, and other neurodegenerative disorders.

The moderator then asks the group to brainstorm benefits and risks of the proposal. In that discussion, the participants identify more than half a dozen credible arguments on both sides of the question. On the plus side of the proposal are:

- Finding treatments for these diseases will be of incalculable value to the human race. That in itself justifies the use of embryos.
- It's efficient. Embryonic stem cells are so far most promising for this research, and using them will allow us to achieve results more quickly.
- Using embryonic stem cells now will allow us to create new stem cell lines, meaning that someday we won't have to rely on embryos left over from fertility treatments, as is the case today. Using embryos now will allow us to stop using them sooner rather than later.
- It better honors these embryos to use them in such research than to throw them away.

Looking at the proposal from the standpoint of risks, however, they find an equal number of arguments the other way. These are:

- In treating human life (the embryo) as a means to an end, such research erodes respect for human life.

- This research starts us down a slippery slope. If we can use the embryo for research, what about the fetus? What, eventually, of the adult? Will we have a world in which women sell their embryos to the highest bidder? And at the end of the slope lies the frightening specter of human cloning.
- We would sacrifice values for "efficiency." Use of embryonic tissues may get us to "pay dirt" fast, but at the cost of trampling on essential human values.
- We don't know whether this or any other research will find any cures at all. The end is too unsure to justify such radical means.

The moderator now asks the group to look for common ground as a way to make a policy recommendation. It might seem that little common ground exists here, but it does emerge. Here is what the group recommends:

> A closely monitored pilot project of some predetermined length — 5 years? 10 years? — might be tolerable. But it must take place under the strictest supervision, with tight controls on the sources and numbers of viable cell lines (and they must be truly viable), with frequent review and reconsideration. If this pilot program operates for a reasonable time without success, we should have the option to shut it down.

## Deliberation #2: The Personality Makeover

In another approach, our citizen-deliberators read a fictional dialogue. They are asked to determine the values that each side treasures most and formulate a common ground. Instead of a proposed policy, this discussion might result in ethical guidelines that could underlie policies.

> Smith and Jones learn that drugs or surgery can now cure people of various disorders of behavior: sociopathic violence, depression, compulsive gambling, overeating, uncontrollable temper. Jones thinks this is a great idea. "This is a miracle that improves human life," he says. "For thousands of years, humankind has been burdened by these afflictions, and now there is a way to dispel them. This will allow us more time to seek even higher levels of perfection for ourselves. It's an incredible contribution to the history of our species."

Smith is less enthusiastic. "What kind of world are we creating?" he asks. "These people are not learning self-control; they're having it implanted. They're not better people through their own efforts; they took a pill to make their limitations or issues go away. They avoided their problems rather than solved them. I'd rather live in a world in which individuals, taking full responsibility and through their own authentic agency, lived through their problems and worked to solve them. Pain, risk, and failure are essential to the formation of the human person, and if we deny free play to bad luck, difference, and reasonable suffering, we deny ourselves some of our greatest opportunities to grow."

Now the moderator directs the group to brainstorm a list of values for each speaker. The ideals they credit each with championing are:

*Jones*: improvement of our species, efficiency, technology, knowledge, easing of pain, usefulness, new horizons, discoveries and breakthroughs.

*Smith*: self-control, self-determination, active solution vs. technological enhancement, responsibility, authentic agency, value of lived experience, constructive value of pain, risk, and failure, nobility, free will.

The group must now look for common ground between the two sets of values. Smith might agree with Jones that on-demand relief of all pain and risk would not necessarily be good. They might agree that needless, intractable pain and incurable disease—misfortunes that prevent any authentic choices or learning—do merit the full application of new technologies. And both might agree that society must prevent or mitigate unfair advantages for the wealthy.

## Deliberation #3: Felons and the Brainprint

It's 2075. Congress just approved a federal law under which felons automatically forfeit the privacy of their brainprints—maps of the characteristic cognitive and affective operations of a person's mind. If you are convicted of any felony, your brainprint goes into a database closely guarded by the government. Your print is compared with those of other felons to determine patterns of behavior and probability of recidivism. Some members of Congress want to publish the brainprints of felons on the Internet. What benefits and risks has our future Congress created?

Our deliberators will see some obvious benefits to the legislation. It can help law enforcement track behavior of criminals to predict and prevent crime, and distribution of the brainprint could be a fitting punishment for severe crime, equal to the permanent forfeiture of the right to vote in some states.

But the risks are just as clear. Although the Constitution contains no language defining the right to privacy, a very strong traditional presumption of it exists in American life. That assumption exists in the Third, Fourth, Fifth, and Ninth Amendments of the Constitution, as well as in Supreme Court decisions such as *Roe v. Wade.* A person's brainprint would be tantamount to his or her soul. Only extremely rarely should such intimate information become anyone else's property; others should have access only with the fully informed consent of the owner.

Some would see grave risks in allowing forfeiture of brainprint to become a means of punishment. It might give too much power to the state. Unscrupulous officials might manipulate people with this powerful information. And what of unauthorized sharing of the information? And would not brainprints create prejudice, encouraging the fallacy that the existence of a brain pattern always predicts and determines the conduct of the person? Some would argue that distribution of a brainprint would unfairly limit the felon's access to health care and other social goods. The predictive power of brainprints might create pressure to institute preemptive punishments à la *Minority Report.*

Taken all by itself, the brainprint question is pretty simple, and the deliberators will have a quick and clear recommendation:

> Forfeiture of brainprint should never become a means of punishment. And punishment should never be preemptive; such a notion, even if technology were to make it possible, is unconstitutional and unjust. Actual deeds, not potential deeds, are the only fitting subjects of correction. The future must always be allowed to play itself out. As tempting as it may be to "head off crime," we should avoid injuring human rights simply for the sake of efficiency.

The brainprint will drive the discussion into other realms. Participants will consider commercial ramifications, as in this scenario: Congress has approved a law that allows health insurance companies to charge higher premiums to, or refuse to cover, people who will not allow access to their brainprints. What benefits and risks has our future Congress created? Answering that question will force discussants to balance the rights of entities in a free-market system against the need for privacy and health care.

But the discussants now have pulled the thread that unravels the sweater. Inevitably, they will trend toward the largest philosophical issue of all: What if science eventually "explains" consciousness? The potential of that Great Explanation raises quandaries familiar to neuroscientists but not yet familiar enough to citizens. If we want to arrive democratically at good policy for neuroscience, we must encourage discussions at this level. Time and again in our Citizen Voices projects, we've seen that supposedly "lay" people are good at turning philosophy into policy.

Some will feel that the Great Explanation will detract from the magic and dignity of human life, that being explained somehow diminishes the human being. Some may feel, with understandable dismay, that it does away with free will and replaces God with prejudice and totalitarian control. On the other hand, the Great Explanation may lead to tremendous advances: We can lay hands directly on the sick spirit. Explaining my soul does not change the fact that my life is singularly mine, and gives me more, not less, reason for wonder, more, not less, reason to believe if I wish.

And what is free will? The mind's reflexive assumption that its choices are not directly constrained. Will the Great Explanation really impinge on my mind's assumption of its self-sufficiency? A question to ask, because, far from being conceptual or abstruse, someday our answers will be embedded in policy. Discussants likely will require that the Great Explanation never be allowed to interfere with the progress of democratic institutions or with the individual's right to self-determination. The more we understand consciousness, the clearer it will be that such interference is evil.

The great desideratum here is a set of sensible overarching recommendations:

- *Explanation must never become legislated into compulsion.* The potential for abuse is real; act now to head it off. Leaders must not be allowed to argue that, because consciousness has been explained, the sacredness or dignity of the human person, or that person's rights and obligations to others, now have been qualified or diminished. Democracies can have no place for systems based on categories of consciousness. Generations before these innovations actually exist, we should stipulate that the Great Explanation should not affect the basis of democracy.
- *Preserve the social contract.* Safeguard human rights, especially for the less privileged, the elderly, the old, the sick, the less talented, and anyone who can't keep up with the pace of innovation. Government must

always ensure a "sacred minimum": a stipulation of common equality for everyone no matter his or her circumstances. Granted, the social contract is based on a pretense, but that's just an interesting irony much overborne by the necessity of social organization. We must continue to pretend that all people are created equal, even as evidence accrues that they are not.

- *Legislate vigorously against prejudice.* No one's bad luck (a genome that codes for a less excellent or less adaptable consciousness) should be held against him or her.
- *Ensure equal access and opportunity.* That includes equal access to education and social services, and an equal chance to succeed or fail within reasonable tolerances.

## Time to Face the Neuroscience Revolution

In applying the deliberative model to neuroethics, little may go the way I have imagined it. But the model holds great promise in helping citizens discover and discuss the ethical issues and policy ramifications of the coming neuroscience revolution. In an article I wrote earlier in 2004 in the *Inquirer*, titled "Making Up Our Minds," I argued that the time couldn't be better to begin the process. Best to start now before the wave is upon us.

I look forward to being part of that discussion in my community. My newspaper is looking for partners in the effort; we're lucky we have so many universities and medical schools nearby. We may pick a nearby year—say, 2006—and make it the Year of Neuroscience. May the same happen in many other places. Scientists need the guidance, and nonscientists need the awareness. An exciting, challenging, possibly hazardous time is coming. Like the warrior-archer, Arjuna, in the *Bhagavad-Gita*, we are nearing a knowledge through which "you can behold the entire universe." But, like Arjuna, who could not face the battle, unless we have prepared for it, we may never enjoy the fullness of that purifying force.

JOHN TIMPANE, PH.D., is commentary page editor of the *Philadelphia Inquirer*. He earned his doctorate in English and humanities from Stanford University and has taught at Lafayette College, Rutgers University, the University of Southampton, and Stanford. He is author or coauthor of four books, including *Usonia, New York: Building a Community with Frank*

*Lloyd Wright* (Princeton Architectural Press, 2001) and *It Could be Verse: Anybody's Guide to Poetry* (Ten Speed Press, 1995). His scientific essays and articles have appeared in periodicals such as *Scientific American* and *Science,* and his general-audience essays and reviews in newspapers such as the *Philadelphia Inquirer,* the *Boston Globe,* and the *Miami Herald.*

*Part III*

# Neuroimaging

THE USE OF COMPUTED TOMOGRAPHY (CT), magnetic resonance imaging (MRI), positron emission tomography (PET), functional magnetic resonance imaging (fMRI), and other techniques to produce images of the structural and functional landscape of the brain has been a major achievement of neuroscience. Brain imaging can enable researchers and clinicians to confirm a diagnosis of a neurological or psychiatric disease. In addition, imaging may enable medical professionals to monitor the effects of drugs used to treat these diseases. It may also detect subtle brain abnormalities long before the appearance of symptoms associated with diseases of the brain and mind. These features of the brain might predict who would develop these diseases. Neuroimaging may one day be used as a form of lie detection in the criminal law as well.

Yet scans of the brain are fraught with uncertainty. The fundamental problem is how to interpret what can be ambiguous information about the brain, which is based on indirect measures of brain structure and function. Neuroimaging cannot establish with certainty that an individual was lying in denying that he or she committed a criminal act. Nor can it determine that an individual with a brain anomaly will in fact develop a neurological or psychiatric disease. Informing an individual that he or she has a brain abnormality may harm that person by causing anxiety about a condition that may never develop. This can be especially harmful among control subjects in imaging clinical trials who are healthy and asymptomatic. Imaging can also cause anxiety by revealing early signs of brain diseases that have no effective treatment. On the other hand, early detection of an abnormality may lead to interventions that could prevent a serious disease or fatal event, such as a hemorrhage from an aneurysm. Images of the brain might also be accessible to third parties such as employers and insurers, which could lead to discrimination in access to employment and health care. Even more disturbing, disclosure of information about the brain may violate the privacy of individual thought, which is the core of personal identity and the self.

What are the responsibilities of researchers and clinicians in disclosing the results of brain imaging to research subjects and patients? What should radiologists, neurologists, and psychiatrists tell people about the medical significance of incidental findings in what were supposed to be "normal" brains? Given the uncertainty surrounding information about the brain, can a patient or participant in research give fully informed consent to undergo a brain scan? Can we protect those undergoing scans from discrimination by third parties? And can we protect the privacy of individual thought? The authors in this section address these and related ethical questions raised by brain imaging.

Judy Illes presents the main ethical challenges that brain imaging poses for researchers and clinicians in "Neuroethics in a New Era of Neuroimaging." She distinguishes the more specific neuroimaging of ethics to reveal the neurobiological basis of moral reasoning (to be discussed in Part IV) from the more general ethics of neuroimaging, which has a broader range of legal and social implications. Illes points out that information about the brain from neuroimaging is not purely scientific but value-laden and culturally determined to some extent. She also notes the potential misuses and abuses of this information and how they may contribute to stigma and discrimination regarding access to employment and health insurance. Incidental findings of structural and functional brain scans are especially challenging because of difficulties in interpreting these findings and communicating them to patients and volunteers in imaging studies. These and other problems surrounding ambiguous information about the brain make it imperative for all relevant parties to carefully weigh the potential benefits against the risks of diagnostic and predictive neuroimaging.

Illes and colleagues explore the ethical implications of unexpected anomalous findings from brain scans in "Ethical and Practical Considerations in Managing Incidental Findings in Functional Magnetic Resonance Imaging." Noting that some asymptomatic individuals serving as healthy controls in imaging studies display brain abnormalities, Illes et al. raise a series of questions with a view to establishing guidelines or policies that would protect these individuals from harm. These questions include whether prospective subjects in imaging trials are fully informed of the risks and benefits of participating in the research, and whether they consent to participate. The questions also pertain to how information about brain anomalies is interpreted and communicated to research subjects by radiologists and other health professionals. In addition, there are concerns about confidentiality of this information and whether it warrants clinical follow-up. Finally, Illes et al. propose creating a national database for incidental findings, including false positives.

In "Legal and Ethical Issues in Neuroimaging Research: Human Subjects Protection, Medical Privacy, and the Public Communication of Research Results" Jennifer Kulynych focuses on investigators' obligations to subjects participating in nontherapeutic neuroimaging research. This refers to research from which participants are not expected to benefit medically. Kulynych cites many legal cases involving similar ethical and legal issues to frame her discussion. She pays particular attention to how researchers may be legally liable for any harm to subjects by being negligent in disclosing and communicating information from brain scans. As in other areas of medical research, Kulynych points out that subjects in neuroimaging research must give voluntary and fully informed consent to participate in this research. Researchers also have an obligation to respect the privacy and confidentiality of research subjects. Specifically, Kulynych asserts that neuroimaging researchers should anticipate detecting signs of potentially harmful conditions of which a subject is unaware. They can protect subjects and avoid liability by indicating in consent forms whether subjects would be told about any diagnostic significance of brain images and whether clinicians would review such findings as part of a follow-up.

In "Incidental Findings on Research Functional MR Images: Should We Look?" Alex Mamourian reports that diagnostically significant brain anomalies occur in one to two percent of asymptomatic control subjects in imaging trials. He explains how these images are often ambiguous in terms of their clinical significance because it can be difficult to separate normal from abnormal findings. This difficulty is further complicated by the fact that, unlike in the doctor-patient relationship, it is not always clear what ethical and legal obligations investigators have to research subjects. Mamourian asks whether a diagnosis of early brain disease or malformation based on incidental findings would add quality or years of life to these subjects. His own experience of having an experimental MRI scan reveal a cerebral aneurysm shows that in some cases one can answer this question affirmatively.

Illes and Eric Racine elaborate on Illes's earlier views on the ethical, social, and legal implications of neuroimaging in "Imaging or Imagining? A Neuroethics Challenge Informed by Genetics." Neuroimaging data pose an ethical and epistemological challenge because the meaning of the data is shaped not only by a scientific framework but also by cultural and anthropological frameworks. Addressing concerns about privacy and identity, Illes and Racine describe the parallels and differences between genetics and neuroimaging. They maintain that genetics may be a useful starting point for discussion of the ethics of neuroimaging. Yet they assert that the relationship between the brain and personal identity or the self is more direct than the relationship between genes and identity. It is because the brain is so fundamental to our selves that we must be cautious in how we interpret findings from neuroimaging studies.

Lynette Reid and Francoise Baylis argue in "Brains, Genes, and the Making of the Self" that Illes and Racine fail to appreciate the full extent of the differences in the ethical implications of genetics and neuroimaging. These differences are most significant with respect to privacy and confidentiality, especially privacy of thought. Unlike genetics, images of the brain can reveal our thoughts and can tell us who we are by displaying the basic components of our identities. Genetics cannot do this and therefore may not be a useful point of departure for addressing the most fundamental ethical issues in neuroimaging.

Recall Donald Kennedy's concluding remarks in his article in Part I. He said that he did not worry that neurobiology would change our views on free will and responsibility. However, he expressed concern about what neurobiology might mean for the privacy of his inclinations and his thoughts. Consistent with the principle of respect for autonomy, individuals have the right of privacy of their thoughts and other sensitive information about their brains. Whether or not neuroimaging can read minds, there is potential for misuse or abuse of findings from brain scans. Policies and legislation need to be formulated, implemented, and enforced to prevent this from occurring.

*Chapter 9*

# Neuroethics in a New Era of Neuroimaging*

*Judy Illes*

Although investigations about brain, mind, and behavior date back to the ancient philosophers, a new discipline called *neuroethics* has emerged formally only during the past year to embody theoretical and practical issues in the neurologic sciences that have moral and social consequences in the laboratory, in health care, and in the public domain. The first specific references to neuroethics in the literature were made a little more than a decade ago. They described, for example, the role of the neurologist as a neuroethicist faced with patient care and end-of-life decisions[1] and philosophical perspectives on the brain and the self.[2] As a discipline, per se, neuroethics was launched in a conference sponsored by the Dana Foundation called "Neuroethics: Mapping the Field" held in San Francisco in May 2002.[3] Bringing together approximately 150 neuroscientists, scholars in biomedical ethics and the humanities, lawyers, public policy makers, and representatives of the media, the conference emphasized four major areas: "Brain Science and the Self" (or "Our View of Ourselves"), devoted to issues of human freedom and responsibility, the biologic basis of personality and social behavior, choice and decision making, and consciousness; "Brain Science and Social Policy," including issues of personal and criminal responsibility, true and false memory, education and theories of learning, social pathology, pri-

*From *American Journal of Neuroradiology* 24 (2003), pp. 1739–1740.

vacy, and the prediction of future brain pathology; "Ethics and the Practice of Brain Science," spanning topics of pharmacotherapy, surgery, stem cells, gene therapy, neuroprosthetics, and parameters for guiding research and treatment; and "Brain Science and Public Discourse," including the development of broad and informed public discourse, mentoring of young trainees, and encouragement of responsible understanding and reporting in the media.

The ethical challenges introduced by advanced capabilities in neuroimaging were recognized as a priority for the new discipline, taking into consideration significant concerns and potentially thorny issues that have surfaced both in research and in the clinical environment. The research imaging issues are the focus of the present editorial; clinical neuroethics issues will be the focus of a forthcoming *American Journal of Neuroradiology* editorial.

## Functional Neuroimaging: Behavior, Reasoning, Thought

In a recent report, Illes et al.[4] provided empirical validation of the expanding terrain of brain imaging studies by using measurements of regional blood flow from functional MR imaging. Through an analysis of the more than 3,400 peer-reviewed papers examining the application of functional MR imaging, alone or in combination with other neuroimaging modalities in the decade between 1991 (the genesis of functional MR imaging) and 2001, a steady growth in studies with evident ethical and social implications was shown. These included studies of social attitudes, human cooperation and competition, brain differences in violent people, religious experience, genetic influences, and variability in patterns of brain development.

Imagine, for example, a moral reasoning experiment in which you could choose to save the lives of five people on a runaway trolley car by pulling a switch to send it on an adjacent track where one person stands (and who would not survive).[5] Alternatively, you could choose to push one of the people off the trolley and onto the track, thereby blocking the movement of the trolley and saving the remainder of the group. Most people respond that the "switch" option is morally acceptable, while the "push" option is not.[6] Functional MR imaging studies of healthy adult participants engaged in resolving such dilemmas,[5] making decisions about statements that have moral content (e.g., "The judge condemned the innocent man" or "The elderly are useless") versus neutral content ("The painter used his hand as a paintbrush"),[7] or making decisions about race and stereotypes[8] have begun to probe such uniquely human processes and have pushed the envelope well beyond the lines of where neuroradiology and cognitive neuroscience have traditionally intersected.

Extending well beyond cortical maps of sensorimotor function, language, and attention, maps that include the medial frontal and orbitofrontal gyri, posterior cingulate gyrus, angular gyrus, amygdala, and fusiform area for moral reasoning, emotion, and judgment—arguably among the deepest forms of human thought—have now been described. No doubt, the diagnostic and predictive validity for real-world behaviors, especially those that are potentially value-laden or culturally determined, is still unsolved.[9] However, as functional MR imaging and other advanced neuroimaging technologies continue to mature, the issue of validity becomes steadily addressed.[10] Therefore, with a growing regard for the novelty and breadth of information that neuroimaging can deliver about the complexity of human behavior, ethical concerns regarding the potential data misuses or abuses have come to the foreground. These range from the creation of a personal sense of stigma to discrimination in health coverage or employment.

The prima facie question for advanced neuroimaging, in fact, is moral and social acceptability of research topics and study design. We must ask, for example, whether all studies of normative neurobehavioral phenomena are ethically acceptable. How might social or racial biases affect applications of the technology, the conditions under which imaging is performed, or the way interpretations are made? What does a statistically normal activation pattern of *moral behavior* really mean, and, by extension, what would the implication of an *abnormal* brain activation pattern be in a healthy person normally (i.e., within predicted behavioral or physiological norms) performing a task that involves moral judgment, deception, or even sexual responsiveness?[11] Dilemmas posed by incidental findings of structural anomalies in medical research have been raised in the past and have surfaced recently for research MR images specifically.[12,13] However, incidental findings of *functional* anomalies may give rise to an entirely new kind of challenge related to both the interpretation and the appropriate use of data. Ensuing questions relate to what protocols may need to be put in place for the discovery of such findings and how (or if) they should be communicated to a participant.[14,15] It is imperative to consider the clinical significance of a finding, what a participant would want to know, and the risks of inadvertent disclosure or exploitive use of such information. Although one may debate whether these risks are significant, in this century marked by technological innovation and a society quick to embrace high technology, it would be imprudent to think that they do not exist at all. Just as the regulations of the new 2003 Health Insurance Portability and Accountability Act extend the Belmont Report's principles and guidelines for the protection of human participants in research, what will protect the quantitation of human thought in 2010?

In 1932, Aldous Huxley wrote in *Brave New World*,[16] "The ethical issues raised by. . . feats of human engineering are qualitatively no different from those we shall have to face in the future. The difference will be quantitative: in scale and rate. Even so, the individual steps may still go on being so small that none of them singly will bring those issues forcibly to light: but the sum total is likely to be tremendous. That is why we have to look for those issues now."

We have, in fact, entered an era in which issues surrounding the *ethics of neuroimaging* and the *neuroimaging of ethics* (i.e., ethical reasoning and behavior) are now both at hand.[17] Neuroradiologists have a vital role to play in identifying the issues as the new discipline of neuroethics continues to evolve and in ensuring that the enthusiasm for and benefits of neuroimaging information outweigh associated risks in any of the areas in which neuroimaging may be used practically. Knowledge harnessed from lessons of the past in genomics and other areas of biomedical research, and from the multidisciplinary perspectives of all stakeholders, can provide essential information for delineating priorities for neuroimaging and ethics in research and education for the short term and for the allocation of sustainable resources and infrastructure over the long term.

## Acknowledgments

The author gratefully acknowledges Dr. Scott W. Atlas, chief of neuroradiology, Department of Radiology, and senior fellow, Hoover Institution, Stanford University, for thoughtful feedback on this review, and the Greenwall Foundation for its generous support of this work.

REFERENCES

1. R. E. Cranford, The neurologist as ethics consultant and as a member of the institutional ethics committee: The neuroethicist. *Neurologic Clinics* 7(1989):7:697–713.
2. P. S. Churchland, D. J. Roy, B. E. Wynne, R. W. Old, eds., *Our brains, Our selves: Reflections on neuroethical questions, bioscience-society* (New York: John Wiley; 1991), 77–96.

3. Marcus, S. J., ed., *Neuroethics: Mapping the field* (New York: Dana Foundation; 2002)

4. J. Illes, M. Kirschen, and J. D. Gabrieli, From neuroimaging to neuroethics. *Nature Neuroscience* 6 (2003): 205

5. J. D. Greene, R. B. Sommerville, L. E. Nystrom, J. M. Darley, and J. D. Cohen. An fMRI investigation of emotional engagement in moral judgment, *Science* 293 (2001): 2105–2108.

6. L. Helmuth, Moral reasoning relies on emotion, *Science* 293 (2001): 1971–1972.

7. J. Moll, R. de Oliveira-Souza, I. E. Bramati, and J. Grafman. Functional networks in emotional moral and nonmoral social judgments. *Neuroimage* 16 (2002): 696–703.

8. A. J. Golby, J. D. Gabrieli, J. Y. Chiao, J. L. Eberhardt, Differential responses in the fusiform region to same-race and other-race faces, *Nature Neuroscience* 4 (2001): 845–850.

9. A. Beaulieu, Images are not the (only) truth: Brain mapping visual knowledge, and iconoclasm, *Science, Technology, and Human Values* 27 (2002): 53–87.

10. J. E. Desmond, S. H. Annabel Chen, Ethical issues in the clinical application of fMRI: Factors affecting the validity and interpretation of activations. *Brain and Cognition* 50 (2002): 482–497.

11. B. A. Arnow, J. E. Desmond, L. L. Banner, et al., Brain activation and sexual arousal in healthy heterosexual males, *Brain* 125 (2002): 1014–1023.

12. J. Illes, J. Desmond, L. F. Huang, T. A. Raffin, and S. W. Atlas, Ethical and practical considerations in managing incidental findings in functional magnetic resonance imaging, *Brain and Cognition* 50 (2002): 358–365.

13. B. S. Kim, J. Illes, R. T. Kaplan, A. Reiss, and S. W. Atlas, Incidental findings on pediatric MR images of the brain, *American Journal of Neuroradiology* 23 (2002): 1674–1677.

14. D. Steinberg, What information should be disclosed to patients? *Medical Ethics* 9 (2002): 1–2

15. J. Kulynych, Legal and ethical issues in neuroimaging research: Human subjects protection, medical privacy, and the public communication of research results, *Brain and Cognition* 50 (2002): 345–357.

16. A. Huxley, *Brave new world* (1932), as cited in M.L.T. Stevens, *Bioethics in America: Origins and cultural politics* (2002).

17. A. Roskies, Neuroethics for the new millennium, *Neuron* 35 (2002): 21–23.

JUDY ILLES, PH.D. is senior research fellow at the Stanford Center for Biomedical Ethics and director of the program in neuroethics at Stanford University, Palo Alto, California.

# Chapter 10

# Ethical and Practical Considerations in Managing Incidental Findings in Functional Magnetic Resonance Imaging*

*Judy Illes, John E. Desmond, Lynn F. Huang,*
*Thomas A. Raffin, and Scott W. Atlas*

Over the past decade, functional magnetic resonance imaging (fMRI) has emerged as a powerful tool for mapping sensory, motor, and cognitive function and has reaffirmed and expanded upon clinical neurologic and neuropsychological studies, and neuroimaging studies based on other modalities such as electroencephalography (EEG), positron emission tomography (PET), single photon emission tomography (SPECT), and magnetoencephalography (MEG). Several recent fMRI studies have even begun to elucidate the mechanisms of complex and abstract phenomena such as decision making and moral behavior. Using event-related fMRI, for example, Van Veen, Cohen, Botvinick, Stenger, and Carter[1] have demonstrated a highly specific contribution of the anterior cingulate cortex to executive functions at response-related levels of processing, and in a series of fMRI studies using moral dilemmas as stimuli, Greene, Sommerville, Nystrom, Darley, and Cohen[2] have probed the influence of variations in emotional engagement

*From *Brain and Cognition* 50 (2002): 358–365.

on moral judgment. Thus, in this second decade of functional neuroimaging with MRI, it has become possible to detect an alignment of neural activation with some of the highest forms of human cognition; new perspectives and moral questions about rational thinking, intention, and individuality are certain to follow. Much of our thinking from neuroimaging studies, however, carries assumptions or presumptions about the subject population under study and, in particular, the neurologic status of subjects entered into studies as either healthy controls or belonging to a specific neurologic disease group. Despite reports of unexpected MRI findings in clinical and research brain imaging in both adult and pediatric cohorts,[3-5] many, if not most, research fMRI studies involving volunteers are performed by non-physicians; unanticipated findings on these imaging studies may go unrecognized and thereby leave subjects without appropriate referral.

The detection, significance, and management of incidental findings are keys to the welfare of the research subject as well as to the integrity of the studies. We will focus on both of these issues here and consider short-term and long-term strategies involving informed consent, expert physician participation, archiving of incidental findings, and the adoption of guidelines for handling unusual variation in neural activations or performance as a first step in a call for debate and consensus in the fMRI community to address these issues.

## Incidental Findings in Clinical Studies and Research

### Clinical Domain

Incidental neuroradiological findings are not uncommon in the clinical domain and have been reported variously in the literature. Imaging findings consistent with sinusitis have been reported, for example, in adult patients presenting with symptoms including chronic cough, seizures, head injuries, and various intracranial diseases in the adult population.[6-9] Other studies have reported substantial variation of incidental findings involving white matter lesions[10, 11] and sinus disease,[12] across the lifespan in symptomatic subjects.

### Research Domain

To our knowledge, a systematic analysis of incidental findings in research EEG, PET, SPECT, or MEG has not been conducted to date. Two studies, the first by Katzman et al.[4] and a second by our own group,[3] have been conducted for MRI. A third study of our own is ongoing. These studies are summarized here.

*NIH incidental MRI findings*[4]. A systematic analysis of incidental neuroradiological findings in the research domain and in MRI, in particular, was described first in the literature by Katzman et al.[4] The Katzman group addressed the prevalence of incidental findings specifically in a healthy asymptomatic population in a retrospective study of approximately 1,000 MRI research subjects with ages across the life span. Their cohort of subjects was 54.6% male and 45.4% female with an age range of 3–83 years (mean age 30.6 years). They reported an 18% incidence of findings, of which 15.1% required no referral, 1.8% required routine referral, 1.1% required urgent referral, and 0.2% required immediate referral. In subjects categorized for urgent referral, at least two were primary tumors.

*Stanford pediatric incidental findings study*[3,5]. In a study of incidental findings in research MRI scans by our own group,[3,5] we conducted a retrospective review of 225 research MR scans obtained for functional brain imaging from a cohort of children presumed to be neurologically healthy. The age range of the cohort was newborn to 17 years, 100 boys (44%) and 125 girls (56%). MR scan parameters varied, but T1-weighted images were available for more than 90% of the subjects, T2-weighted images were available for 64%, proton density images for 52%, and SPGR images for 46%. All scans were read and then classified using the method described by Bryan, Manolio, and Schertz[13] by a board-certified neuroradiologist.

We found incidental abnormalities in the brain images of 47 (21%) subjects, with the other 79% of the scans found to be normal. Of the 47 abnormalities detected, 17 (36%) were determined to have required routine referral for further evaluation; in a single case (2% of the total abnormalities; 0.5% of the cases studied) a cerebellar lesion was detected and categorized as an urgent referral.

The occurrence of the neuroradiologic abnormalities in the male cohort was twofold that of the females (v2 ¼ 5:12, p < :05). The percentage requiring either routine or urgent referral showed no significant gender difference (39% and 37%, respectively).

These findings in children broadly replicate those of the Katzman study in adults and other findings, for example, consistent with ventricular asymmetries in neonates[14] and a predominance of arachnoid cysts in males.[15,16] While it is unlikely that the findings in our study compromised the interpretability or validity of the research brain mapping studies in any way, even the limited presence of anomalies in a cohort of young participants is a matter of medical concern.

Notwithstanding the rather low frequency of clinically important abnormalities, we believe that the presence of any such findings, particularly in a pediatric age group, is significant and highlights the possible need for routine involvement by trained radiologists in these studies, for both detection and appropriate follow-up.

*Stanford adult incidental findings*[17]. In an ongoing study of adult fMRI research subjects in our institution, we have retrospectively reviewed an initial 129 cases to date. Scans were obtained from research fMRI studies of language, spatial processing, and higher cognitive function, with subjects ranging in age from 18 to 81 years. Sixty-three subjects were male (49%, age 23–70); 66 subjects were female (51%, age 18–81). T1-weighted images, T2-weighted images, and thin-section, T1-weighted, three-dimensional volume (SPGR) images were available for almost all subjects.

We found incidental abnormalities in 48 (37.2%) cases. Of the 48 abnormalities detected, three (6.3% of the 48; 2.3% of the total) were classified as requiring routine referral, two were categorized as requiring routine referral for midbrain infarcts, and one was categorized for routine referral for nonspecific white matter lesions without volume loss or signs of aging (Table 1). Table 1 also shows incidental findings not classified as requiring referral. Of this latter group, four findings (44.4%) were due to volume loss with accom-

*Table 1*  Classification of MRI abnormalities in a cohort of presumed healthy adult subjects participating in fMRI studies of language, spatial processing, and higher cognitive function

| No referral | Prominent basilar artery | 1 |
|---|---|---|
| | Left frontal venous angioma | 1 |
| | Mild ethmoid sinus disease | 1 |
| | Asymmetric ventricles | 1 |
| | Volume loss | 5 |
| Routine referral | White matter changes without signs of volume loss | 1 |
| | Midbrain infarct | 2 |

panying white matter changes and were detected in subjects 68–81 years of age with a mean age of 74.3 years. No findings categorized as requiring either urgent or immediate referral were detected.

In this relatively small sample, the overall percentage of findings was 39.6% in the male cohort (25/63) and 34.8% in the female cohort (23/66). Of the total number of abnormalities, one finding in the male cohort and two findings in the female cohort were classified as requiring referral.

Figs. 1 and 2 illustrate the case of a 42-year-old male subject, outside the adult cohort reported above, in whom a lesion was detected by one of the authors during MRI pre-scan. In the absence of guidelines for handling

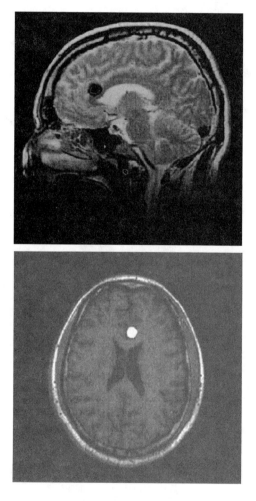

*Fig. 1.*    Sagittal MRI of a 42-year-old male with a vascular malformation initially entered as a healthy control into an fMRI protocol of picture naming. The lesion was detected and subject referred for immediate clinical follow-up. Fast spin echo 2D scan with TR = 2000ms, TE= 90 ms, flip angle = 90°, 5 mm slice thickness, and 24cm field of view.

*Fig. 2.*    Same subject as in Fig. 1. Fast spoiled GRASS (fspgr) axial 3D scan. TR = 14.3 ms, flip angle = 40°, TE = 5.4 ms, 1.5 mm thick slices, and 24 cm field of view.

such an unexpected finding—either frank, as in this case, or more subtle—the image was sent within 24 hours to a board-certified neuroradiologist known to the investigator but not part of the research team. The subject was diagnosed with a vascular malformation and referred for immediate clinical follow-up. Notwithstanding the appropriate medical management of the subject by the research team, this case highlighted dual concerns for us: (1) a primary need for well-established procedures for handling incidental findings and for communicating findings to subjects and (2) a secondary need for re-examining subject confidentiality, as this lesion was detected in the presence of the subject's companion, who was an observer during the scan.

*Discussion and Recommendations*

In research studies by other investigators and in our own work, incidental neuroradiologic findings have been detected in 20%–40% of cases, with a small percentage requiring referral. Although the majority of the cases are in fact normal, we believe that the presence of any significant clinical findings in an otherwise non-clinical research setting is still a matter of bioethical and medical concern. We propose a minimum standard for consenting research subjects for incidental findings in fMRI and call for, over a longer term, the adoption of universal guidelines for research neuroimaging that would serve the research community and our subjects.

*Informed Consent*

As a first measure toward articulating the clinical limitations of the research scans performed, the following text is being incorporated widely in fMRI consent forms used in our institution. The language was generated internally and IRB-approved:

> Incidental Findings: The investigators for this project are not trained to perform radiological diagnosis, and the scans performed in this study are not optimized to find abnormalities. The investigators and Stanford are not responsible for failure to find existing abnormalities in your MRI scans. However, on occasion the investigator may notice a finding on a MRI scan that seems abnormal. When this occurs, a neuroradiologist will be consulted as to whether the finding merits further investigation, in which case the investigator will contact you and your primary care physician and inform you of the finding. The decision as to whether to proceed with further examination or treatment lies solely with you and

your physician. The investigators, the consulting neuroradiologist, and Stanford are not responsible for any examination or treatment that you undertake based upon these findings. Because the images collected in this study do not comprise a proper clinical MRI series, these images will not be made available for diagnostic purposes.

The text represents a clear minimum standard for dissociating the research scan from a clinical scan. There are, however, several caveats to consider. Among them:

1. While it is assumed that informing prospective subjects about the risks and benefits of research protects their rights as autonomous decision makers, the literature suggests that subjects and patients defer substantially to the subject-investigator relationship to guide them through research participation decisions.[18–20]

2. The text "The decision as to whether to proceed with further examination or treatment lies solely with you and your physician" presumes that the subject has a primary care physician (PCP). However, an undue burden may be placed on individuals who are drawn from university or other subject pools and whose access to medical care may not specify a PCP per se.

3. When the subject is screened and entered into a protocol as a healthy control, "healthy" is a dominant attribute. However, even with extensive and methodical screening for medical and psychiatric history, there remains no verification of this in fMRI protocols in which a clinical reading of the image is not done.

Limitations of informed consent, evidence for incidence of neuroradiologic abnormalities in subject pools, and the nature of fMRI as advanced medical technology typically situated in a medical setting triangulate to make a compelling case for further deliberation of fMRI protocols with respect to clinical review of the image data.

### Working Toward a Universal Set of Principles and Consensus for fMRI Research

In light of the data presented here, we urge major professional societies representing the interests of fMRI researchers to actively support a platform for discussion and debate of the following questions:

1. Should all fMRI studies be conducted by a research team that includes a physician qualified to review images for clinical findings and provide fol-

low-up referral to a health care provider selected by or available to the subject?

2. What formal, standardized procedures are needed for facilitating the transfer of information, although not necessarily the transfer of the research-grade images, to either the PCP or a physician within a medical clinic?

3. If neuroradiologic readings of all fMRI studies are introduced, what are the new risks posed if subjects enroll in fMRI research studies motivated by possible clinical benefit? What revisions to the informed consent process will be needed to ensure protection of both the research team and the subject?

4. What are the implications for medical insurability if a benign lesion in detected in an asymptomatic subject not seeking medical attention who enrolled in a research study?

5. Even in the absence of any disagreement about the need for radiologic review of all fMRI scans, what are the practical considerations given the sheer volume of studies conducted?

6. How do we best harness the information about incidental findings across our laboratories? We can consider creating a national database for incidental findings, including false positives, for which the following data are recorded:

>   i. the circumstances in which a finding was detected (e.g., age of subject, gender, type of study);
>
>   ii. how the finding was handled;
>
>   iii. to whom the finding was reported;
>
>   iv. the manner in which the finding was reported in writing (as a clinical report, e-mail, memo), verbally;
>
>   v. how the finding was followed up;
>
>   vi. the outcome for the individual.

7. By extension to the incidental anatomical findings, what are the appropriate guidelines for handling and communicating variation in neural activations or performance variations that appear outside expected norms? This may be especially vital in the case of studies of sexual attitudes or behaviors in which the information may be poten-tially anxiety-provoking to subjects.[21]

8. Is there an imperative for handling research neuroimaging studies with the same confidentiality as clinical studies?

There is no doubt that standardized introduction of expert physician readers would impact the high cost and overall experimental burden of research MRI already inherent to these studies. However, with approximately 3,500 brain and behavior fMRI studies published between 1991 and 2001 alone (NLM PubMed search: keyword "FMRI"; at least 30,000 research participants assuming an average of 10 subjects per study), the number of studies and people we may impact with improved protocols is substantial. We can address added personnel costs with academic currency to some extent, by attracting more radiologists as active contributors to the intellectual challenges of fMRI research, for example, and through joint authorships on peer-reviewed publications and presentations. Alternatively, we can consider formally instituting center-wide expert physician consultation supported by operating budget, shared extramural funding, or both. Regardless of the strategy, any increase in costs will need to be recognized and supported by our sponsors in both the federal and the non-federal arenas and built into our research budgets.

## Conclusion

We have brought to the foreground ethical and practical dilemmas raised by findings of unexpected abnormalities in research fMRI. The issues are significant and complex. With both subject welfare and fundamental knowledge about the relationships between brain and behavior at stake, however, dialogue, debate, and consensus have become imperative.

## Acknowledgments

The generous support of the Greenwall Foundation is gratefully acknowledged.
    Thanks also to Dr. Allyson Rosen and Dr. H. F. Machiel Van der Loos.

REFERENCES

1. V. Van Veen, J. D. Cohen, M. M. Botvinick, V. A. Stenger, and C. S. Carter, Anterior cingulate cortex, conflict monitoring, and levels of processing, *Neuroimage* 14, no. 6 (2001): 1302–1308.

2. J. D. Greene, R. B. Sommerville, L. E. Nystrom, J. M. Darley, and J. D. Cohen, An fMRI investigation of emotional engagement in moral judgment, *Science* 293 (2001): 2105–2108.

3. J. Illes, B. S. Kim, R. T. Kaplan, A. Reiss, and S. W. Atlas, Neurologic findings in healthy children on pediatric fMRI: Incidence and significance, International Society for Magnetic Resonance Imaging, Honolulu, Hawaii, April 2002.

4. G. L. Katzman, A. P. Dagher, and N. J. Patronas, Incidental findings on brain magnetic imaging from 1000 asymptomatic volunteers, *Journal of the American Medical Association* 282, no. 1 (1999): 36–39.

5. B. S. Kim, J. Illes, R. T. Kaplan, A. Reiss, and S. W. Atlas, Neurologic findings in healthy children on pediatric MRI: Incidence and significance, *American Journal of Neuroradiology* 23 (2002): 1–4.

6. T. E. Havas, J. A. Motbey, and P. J. Gullane, Prevalence of incidental abnormalities on computed tomographic scans of the paranasal sinuses, *Archives of Otolaryngology–Head and Neck Surgery* 114, no. 8 (1998): 856–859.

7. Y. Iwabuch, Y. Hanamure, J. Hirota, and M. Ohyama, Clinical significance of asymptomatic sinus abnormalities on magnetic resonance imaging. *Archives of Otolaryngology–Head and Neck Surgery* 123 (1997): 602–604.

8. J. J. Jensen and F. T. Black, The prevalence and significance of incidental paranasal sinus abnormalities on MRI, *Rhinology* 38, no. 1 (2000): 33–38.

9. K. Patel, S. V. Chavada, N. Violaris, and A. L. Pahor, Incidental paranasal sinus inflammatory changes in a British population, *Journal of Laryngology and Otology* 110, no. 7 (1996): 649–651.

10. T. Autti, R. Raininko, S. L. Vanhanen, M. Kallio, and P. Santavuori, MRI of the normal brain from early childhood to middle age: Appearances on T2 and proton density weighted images and occurrence of incidental high signal foci, *Neuroradiology* 36, no. 8 (1994): 644–648.

11. F. E. De Leeuw, J. C. de Groot, E. Achten, M. Oudkerk, L. M. Ramos, R. Heijboer, A. Hofman, J. Jolles, J. van Gijn, and M. Breteler, Prevalence of cerebral white matter lesions in elderly people: A population-based magnetic resonance imaging study, The Rotterdam Scan Study, *Journal of Neurology, Neurosurgery, and Psychiatry* 70, no. 1 (1999): 9–14.

12. P. J. van der Veken, P. A. Clement, T. Buisseret, B. Desprechins, L. Kaufman, and M. P. Derde, CT-scan study of the incidence of sinus involvement and nasal anatomic variations in 196 children, *Rhinology* 28, no. 3 (1990): 177–184.

13. R. N. Bryan, T. A. Manolio, and L. D. Schertz, A method for using MR to evaluate the effects of cardiovascular disease on the brain: The Cardiovascular Health Study, *American Journal of Neuroradiology* 15 (1994): 1625–1633.

14. E. Y. Shen and F. Y. Huang, Sonographic finding of ventricular asymmetry in neonatal brain, *Archives of Disease in Childhood* 64, no. 5 (1989): 730–732.

15. R. W. Oberbauer, J. Haase, and R. Pucher, Arachnoid cysts in children: A European cooperative study, *Child Nervous System* 8, no. 5 (1992): 281–286.

16. K. Wester, Peculiarities of intracranial arachnoid cysts: Location, sidedness, and sex distribution in 126 consecutive patients, *Neurosurgery* 45, no. 4 (1999): 775–779.

17. J. Illes, L. Huang, A. Rosen, and S. W. Atlas, The burden of incidental findings in adult neuroimaging (in preparation).

18. A. M. Capron, R. R. Faden, and T. L. Beauchamp, (Almost) everything you ever wanted to knew about informed consent, *Medical Humanities Review* 1, no. 1 (1987): 78–82.

19. N. E. Kass, J. Sugarman, R. Faden, and M. Schoch-Spana, Trust, the fragile

foundation of contemporary biomedical research, *Hastings Center Report* 26, no. 5 (1996): 25–29.

20. J. Sugarman, N. E. Kass, S. N. Goodman, P. Perentesis, P. Fernandes, and R. R. Faden, What patients say about medical research. *IRB* 20, no. 4 (1998): 1–7.

21. Ethics and Humanities Subcommittee of the American Academy of Neurology, Ethical issues in clinical research in neurology: Advancing knowledge and protecting human research subjects, *Neurology* 50 (1998): 592–595.

JUDY ILLES, PH.D., is senior research fellow at the Stanford Center for Biomedical Ethics and director of the program in neuroethics at Stanford University, Palo Alto, California.

JOHN E. DESMOND, PH.D., is assistant professor of radiology at Stanford University Medical School.

LYNN F. HUANG, M.D., is in the Department of Radiology at Stanford University.

THOMAS A. RAFFIN, M.D., is Director Emeritus of the Biomedical Ethics Center and Professor Emeritus of pulmonary and critical care at Stanford University.

SCOTT W. ATLAS, M.D., is professor of radiology and chief of neuroradiology at Stanford University Medical School.

*Chapter 11*

# Legal and Ethical Issues in Neuroimaging Research: *Human Subjects Protection, Medical Privacy, and the Public Communication of Research Results** *

*Jennifer Kulynych*

Current efforts to delineate the field of "neuroethics" reflect an emerging view that ethical problems in the neurosciences merit a distinct domain within the broader arena of bioethics.[1] As evidenced by this special issue of the journal *Brain and Cognition*, neuroimaging techniques are a logical focus for this new "neuroethical" inquiry, given the capacity of imaging technology to animate mind-brain relationships in ways that compel us to reexamine concepts of identity, personal responsibility, and criminal culpability.

Although neuroimaging in medicine gives rise to important ethical concerns—many of which are addressed by other authors in this volume—the use of neuroimaging in clinical research merits the special attention of neuroethicists. At present, all research involving human subjects† is the focus of intense scrutiny by regulators and policy makers, following a spate of com-

*From *Brain and Cognition* 50 (2002): 345–357.

†Although the National Bioethics Advisory Commission and some patient organizations advocate use of the term "human participant," as opposed to "human subject," this paper will follow the convention of federal regulations in referring to "human subjects," in recognition of the imbalance of power between researcher and subject and the resulting ethical and legal obligations of the researcher toward the subject.

pliance enforcement actions at major research institutions and a wave of lawsuits by aggrieved research participants. Plaintiffs' attorneys, bioethicists, and politicians have proposed various reforms, but the credibility of clinical research must be sustained by those who conduct it. Unless busy investigators take the time to understand regulatory requirements, and, equally important, unless they participate in developing ethical standards specific to their fields of research, groundbreaking work will be impeded by the growing perception that research participation involves unmitigated conflict of interest and unreasonable risk.

This paper will consider neuroethics from the perspective of the neuroimaging investigator whose research involves human subjects. The focus will be on three issues of contemporary significance in clinical research: compliance with regulatory requirements for the protection of human subjects, safeguarding privacy and confidentiality, and ethical behavior in the public communication of research results.

## The Use of Human Subjects in Neuroimaging Research

Neuroimaging studies may involve multiple disciplines (e.g., physics, neuroanatomy, neurophysiology, neuropsychology, neuropsychiatry, and radiology). The field encompasses many different modalities (e.g., PET, SPECT, MRI, fMRI, and EEG). Neuroimaging researchers may be basic scientists who study brain anatomy and physiology or develop new imaging techniques, or they may be clinicians who seek to diagnose brain disorders and explore brain-behavior relationships. Yet, regardless of the nature of the research, whenever a researcher collects data from a human being or analyzes existing, identifiable data or records, with the intent to publish the findings or otherwise contribute to "generalizable knowledge," that researcher is likely to be engaged in human subjects research, with all of the associated duties and obligations (45 C.F.R. §§46. 101–102). For every member of a neuroimaging research team, understanding the complicated landscape of what has come to be termed "human research protections"[2,3] is a necessary first step toward the responsible conduct of research.

The investigator who uses human subjects in research bears a weighty responsibility: he or she must comply—conscientiously—with an ever-expanding body of laws and regulations intended to protect human subjects and to ensure that their participation in research is voluntary and fully informed. Many of these requirements apply not only when the investigator interacts with the subject but also when the investigator obtains medical information, biological samples, or test data, including brain images. At present the U.S. Congress, federal and state regulators, and attorneys for aggrieved research volunteers are seeking to increase legal protections for human subjects research, while enforcing existing requirements with new vigor.

## The Consequences of Non-Compliance

Failing to meet the obligations that human subjects research entails is fundamentally unethical because it may place research volunteers at unnecessary risk. In addition, when made public, compliance failures may seriously damage the reputation of the researcher and his or her institution. Increasingly, non-compliance invites litigation: compliance failures have been cited in a series of recent lawsuits brought by research subjects who suffered adverse events, allegedly as a consequence of ethical and regulatory lapses on the part of investigators and institutions.[4] Allegations of compliance failures in human subjects research are also likely to trigger government investigations that may result in severe regulatory sanctions.

*Recent research litigation.* Perhaps the most highly publicized suit to date followed the death in 1999 of Jesse Gelsinger, a participant in a gene therapy trial at the University of Pennsylvania. Gelsinger, who suffered from a relatively mild form of a rare liver disorder, died four days after receiving experimental gene therapy in an early-stage clinical trial. The Food and Drug Administration (FDA) concluded that the investigator in the trial violated the agency's regulations,[5] and Gelsinger's parents filed suit, alleging that the investigators and the institution misrepresented the risks of the gene therapy protocol and failed to disclose financial conflicts of interest.[6]

Gelsinger's case was resolved in a sealed settlement, but another suit was filed in connection with the same research protocol, and others are pending in connection with alleged compliance failures in clinical trials conducted at Fred Hutchinson Cancer Center, the University of Oklahoma, UCLA, and Ohio State University.[7] In 2001, Johns Hopkins avoided litigation with the parents of a normal volunteer who died in a pulmonary study by settling the matter for an undisclosed sum. In a subsequent investigation, government officials concluded that both the investigator and Johns Hopkins itself failed to meet regulatory requirements for human subjects research.[8]

*Expanding investigator liability.* Obtaining a subject's informed consent and safeguarding that subject's welfare are a researcher's most important obligations, but until recently, it appeared that only physician-investigators were likely to be sued by research subjects for failing to do so. It is a well-established tenet of tort law that the physician-patient relationship is a fiduciary or "special" relationship that gives rise to a heightened legal duty of care, the breach of which may be grounds for a lawsuit if the patient is harmed.

By contrast, until recently there has been little in the way of legal precedent to support a claim that a researcher who does not provide treatment to a research subject owes that subject a heightened duty of care and may be held liable for negligence. One reported case from 1986, *Whitlock v. Duke University*,[9] is a notable exception. In *Whitlock* a participant sued Duke University and faculty researchers for brain injuries sustained in research that simulated deep sea diving (637 F. Supp. 1463 (M.D.N.C., 1986), aff'd 829 F.2d 1340 (4th Cir. 1987)). The court's decision, upheld on appeal, recognized a duty of investigators in "non-therapeutic" research (in the court's view, research not intended to benefit the subject) to warn subjects of all "reasonably foreseeable" risks. The *Whitlock* decision failed to spark a trend toward similar litigation, perhaps because the plaintiff lost his case when the court ruled his evidence insufficient.

More highly publicized and perhaps more far-reaching in its impact was the recent (2001) case of *Grimes v. Kennedy-Krieger Institute*. The plaintiff in this case, a minor child, sued a highly regarded research institution affiliated with Johns Hopkins University. Researchers from this institution were studying the success of partial lead paint abatement strategies in reducing lead poisoning among low-income children living in highly lead-contaminated Baltimore housing stock (366 Md. 29, 782 A.2d 807 (2001)). The plaintiff and her mother resided in a house in which the landlord had partially abated the existing lead paint with the assistance of a government-funded program. The complaint alleged that the researchers, who promised in the consent form to test subjects' blood lead levels, were negligent in failing to promptly inform the plaintiff's mother of elevated lead in her child's blood. The plaintiff claimed that Kennedy-Krieger, by virtue of its association with the government-funded abatement program, created an experiment in which subjects would be purposefully exposed to known lead hazards. In a motion to dismiss the case, Kennedy-Krieger took the position that its research extended only to observing the impact of the government's lead abatement program, and thus its researchers owed no legal duty to the child subjects of the research.

The lower court granted Kennedy-Krieger's motion to dismiss the case, finding no legal duty on the part of the researchers that would give rise to tort liability. The Maryland high court reversed this decision, holding that researchers conducting "non-therapeutic" research do owe a duty to research subjects and may be held liable for negligence. The Maryland court viewed the scope of this duty as defined by federal research regulations and the Nuremberg Code (a set of international standards for human experimentation), as well as the promises made to subjects in the research consent form.

The *Grimes* opinion gained further notoriety when the court went beyond the duty issue to question whether the study design met applicable legal and ethical standards. In the body of its opinion the court stated that a parent of a minor child in Maryland may not legally consent to that child's participation in "non-therapeutic" research involving "any risk" to the child. This language has generated considerable anxiety and uncertainty within the state's pediatric research community, given that even research involving treatment often includes a "non-therapeutic" (e.g., control or placebo group) component, while no research may be characterized unequivocally as free of all risks.

It is unclear whether courts in other states will follow North Carolina and Maryland's lead in holding that researchers in a "non-therapeutic" study owe a special duty to research subjects and are therefore liable to those subjects for civil damages. Should this occur, it is important to remember that individual researchers employed by state institutions may have immunity from civil suit for state tort claims that arise out of acts or omissions committed within the scope of their employment. Many institutions will be obligated by either law or policy to indemnify their employees for acts committed within the scope of employment. Additionally, public and private research institutions carry insurance to cover the costs of defending their officers and employees in civil suits, although this coverage generally would not extend to acts of willful misconduct. Notwithstanding these protections, the *Grimes* case is likely to heighten the liability exposure of individual researchers and research institutions in Maryland and perhaps in other jurisdictions where research litigation has arisen.

Of particular note is the emphasis that the Maryland court placed upon researchers' alleged failure to promptly inform subjects and their parents of elevations in blood lead levels, even though these elevations arguably were not the direct result of the research procedures. Guided by this case, neuroimaging researchers should anticipate comparable disclosure issues that might arise in an imaging protocol, should researchers detect a potentially harmful medical condition of which the subject is unaware. For example, consent forms for neuroimaging studies should indicate whether brain images will be reviewed by clinicians for possible abnormalities. Consent forms should specify whether any diagnostic information will be conveyed to participants or their physicians, and if so, at what stage of the research. Researchers should also be aware that they may incur liability for negligence if the subject reasonably believes that a medical professional will review neuroimages obtained in the study and the subject is not informed promptly of foreseeable harms (e.g., diagnostically useful indications of risk or abnormality). Conversely, if a qualified medical professional (e.g., neurologist or radiologist) will not review the neuroimages, subjects should be informed of this fact in writing. Researchers who are not qualified medical professionals should not risk liability for the unauthorized practice of medicine by attempting to provide diagnostic information to research participants.

*Regulatory sanctions.* Researchers who do not learn the rules—or who simply disregard them—place themselves and their institutions at risk of severe regulatory sanctions. Institutions whose researchers fail to comply with federal research regulations may be disbarred from receiving federal grant funds (45 C.F.R. Part 76). Individual investigators may be disqualified from participating in (technically, from receiving investigational drugs or devices) in trials regulated by the FDA (21 C.F.R. §312.70). IRBs [institutional review boards] or institutions may be subject to administrative sanctions or be disqualified, meaning that the FDA will not approve conduct of the research at the institution or accept data generated from a trial overseen by the IRB (21 C.F.R. Part 56, Subpart E). Where an institution evidences a pattern of compliance failures, government sanctions may include the "death penalty"—a suspension of the institution's entire federally funded human subjects research program such as occurred in recent years at Duke, Johns Hopkins, the University of Pennsylvania, the University of Colorado, and elsewhere.[10] It is important to realize that the risk of regulatory sanctions is not confined to compliance failures in clinical trials: in 1999 complaints about survey research involving genetic information triggered a federal investigation that resulted in the temporary shutdown of all human subjects research at Virginia Commonwealth University.[11]

States are also creating regulatory regimes for human subjects research that impose obligations beyond those contained in federal regulations. For example, California's Human Subjects in Medical Experimentation Act requires that investigators provide all research subjects with an "Experimental Subject's Bill of Rights" containing specified disclosures (Cal. Health and Safety Code §§24170–24179.5). New York's Protection of Human Subjects Act imposes specific IRB review and informed consent requirements upon any research that is not already subject to federal research regulations (N.Y. Pub. Health Art. 24-A). In 2002 Maryland enacted a similar statute (Human Subject Research Act, HB 917, effective October 2002). Failure to meet any additional obligations imposed by the state in which research is conducted could result in both state and federal sanctions, because federal research regulations require researchers to comply with all applicable laws for the protection of human subjects (45 C.F.R. 46.101(e)).

Lastly, researchers whose work involves the use of patient information created in a clinical setting (e.g., test results and medical records) should be aware that, effective April 14, 2003, the new federal medical privacy regulation (45 C.F.R. Parts 160 and 164) will impose significant restrictions—enforced by regulatory sanctions, civil, and criminal penalties—upon the use of "protected health information" for research purposes.[12]

*Human Subjects Compliance Issues for Neuroimaging Researchers*

*Determining when a neuroimaging study is "Human Subjects Research".* To comply fully with applicable requirements, neuroimaging researchers must understand when their activities constitute "human subjects research" as defined by the federal government. Academic investigators should first consult the research policies of their home institution. These policies implement the promise—known as the "assurance"—that the institution has made to the federal Office for Human Research Protections (OHRP) that research will comply with the Public Health Service (PHS) regulation for the protection of human subjects (implementing a federal policy known as the "Common Rule") (45 C.F.R. §46.103). The institution must have a valid assurance on file with OHRP as a condition of receiving and spending NIH or other PHS grant funds. (The Common Rule applies in a slightly different fashion to research funded by other federal agencies such as the National Science Foundation or the Department of Education.) The OHRP is responsible for overseeing institutional compliance with federal regulations for the protection of human subjects and in cases of non-compliance, may suspend an institution's assurance, shutting down the institution's federally funded research.

The PHS research regulations apply to federally funded research involving "human subjects"—a broadly defined term that includes identifiable private information (e.g., medical records or clinical data) about an individual (45 C.F.R. 46.102(f)). Even if the data custodian removes identifiers such as name and Social Security number and replaces them with codes, research using the information remains "human subjects research" so long as the code exists and the identifiers could be reassociated with the data. The OHRP regards coded data as exempt (i.e., not "human subjects research") only when the investigator is prevented from ever accessing the code key, by written agreement with the data custodian or some other equally certain means.[13]

Because "research" is defined as an activity undertaken to contribute to generalized knowledge, data analyses conducted strictly for internal quality improvement or assessment purposes are exempt from the regulations. Surveys and interviews constitute human subjects research under the regulations if the subjects are identifiable (or the data contain coded identifiers) and the information collected is personal or sensitive (e.g., medical information). When the question arises as to whether a use of data is exempt from the PHS regulations, an institutional official, not the investigator, must make this determination.[14]

For research involving investigational medical products, the Food and Drug Administration (FDA) has adopted regulations that define a "human subject" as an individual who becomes a participant in research (21 C.F.R. §50.3(g)). These regulations apply to "clinical investigations," which are (with some exceptions) experimental uses of a drug or device regulated by the FDA for the purpose of demonstrating safety or efficacy of the product (21 C.F.R. §50.3(c)). In considering when neuroimaging research constitutes "human subjects research," it is important to appreciate that many academic institutions apply the PHS standards to all research conducted by faculty or employees, even when such research is sponsored by a private company. Thus, privately sponsored research that does not constitute a "clinical investigation" under FDA regulations would nonetheless be subject to the institution's requirements for IRB review and other human subjects protections when the investigators are faculty, staff, or students.

To illustrate, when an investigator asks a patient or a normal volunteer to participate in a prospective study evaluating a new imaging technique or testing a research hypothesis, that investigator is obviously engaging in human subjects research. Perhaps less obvious, but also within the scope of human subjects research, is the retrospective review of patients' clinical scans and medical records by a faculty radiologist, if done with the intent to publish any interesting findings in an abstract or manuscript. But what about a graduate student or fellow who is granted access to an archive of MRI scans maintained by another laboratory? If the student analyzes these scans for her dissertation, is she conducting "human subjects research"? The answer is yes, unless the scans and any accompanying demographic information and research data are stripped of all identifiers, including codes. Alternatively, if identifiers are encoded, the data recipient must have agreed unequivocally (preferably in writing) that she will never have access to the code key. Otherwise, the OHRP will consider the scans to be potentially "identifiable" and the research to be "human subjects research" requiring prior review and approval by an IRB.[13]

*Rules of the road.* Neuroimaging researchers should be aware of the host of regulations that may apply to their human subjects research activities. The (non-exhaustive) list below indicates some of the most important "rules of the road" for clinical neuroimaging studies:

1. Federal regulations and (in some cases) state laws requiring IRB approval and informed consent for all non-exempt human subjects research.

2. FDA regulations requiring an Investigational New Drug [IND] application or Investigational Device Exemption [IDE] for safety and efficacy studies involving unapproved uses of drugs or medical devices,* and FDA regulations imposing certain obligations upon principal investigators in clinical investigations (e.g., reporting requirements for adverse events).

3. Federal requirements, state laws, and institutional policies governing use of radioactive materials, including investigator training and mandatory pre-certification to handle radioactive materials by the institutional Radiation Safety Committee.

4. Federal, state, and institutional requirements for the disclosure of investigators' financial interests in human subjects research. These requirements may be embodied in state law, federal regulation, or institutional policy.

5. The new federal medical privacy rule, which in April 2003 will require researchers using "identifiable" health information obtained in clinical settings to seek a very specific form of written authorization from each patient, or alternatively, a waiver of authorization from an IRB or "privacy board."

---

*Pursuant to FDA regulations, a clinical investigation involving a marketed drug or biologic product will require submission of an *IND* unless specified exemption criteria are met. See 21 C.F.R. SS 312.2 (b)(1).

Certain of these requirements, such as the FDA's IND/IDE rules, apply only to clinical investigations involving regulated products (e.g., scanners, contrast agents, and nuclear medicine products), while others, such as the federal medical privacy rule, potentially apply to any use of patient information or medical records in research.

*Describing and disclosing risk.* Properly characterizing research risks is critical, from both a legal and an ethical standpoint. The investigator should consider carefully how the risks of a protocol are described to the IRB and disclosed to research subjects. A central issue in risk assessment concerns when the research may properly be characterized as "minimal risk." This issue often arises with respect to MRI protocols, because MRI scans performed in a clinical setting are commonly viewed as low risk. Proper characterization of risk is important because federal regulations require full and accurate disclosure of all risks and "discomforts" that are reasonably foreseeable, even if minimal (45 C.F.R. §46.116(a)(2)).

Under federal research regulations, protocols that present no more than minimal risk to the research subject are eligible for expedited IRB review (45 C.F.R. §46.110; 21 C.F.R. 56.110). The IRB chair (or chair's designee) may approve an expedited protocol without review by a convened meeting of the IRB. Expedited designation is desirable because the investigator may avoid significant delays associated with full IRB review. The OHRP has stated in guidance that "magnetic resonance imaging" procedures may generally receive expedited review;[15] however, the nature of the risk entailed may vary depending upon the nature of the research.

For research involving normal volunteers, risks of the research protocol must be assessed in relation to the risks that a normal person would encounter in daily life or during the performance of routine physical or psychological tests (45 C.F.R. 46.102(i)). It is important to query whether, when scans are performed solely for research purposes, the risks of MRI may exceed this standard. These risks include potentially acute discomfort and psychological distress from lying prone in the confined gantry of the MRI scanner, acoustic noise, which in some scanning procedures may cause hearing loss without the use of ear protection, and the possibility of serious physical injury if the subject and the nearby environment are not properly screened for metal implants or objects.[16] Such adverse consequences may be infrequent, but they can and do occur, as evidenced by the recent tragic death of a young patient struck by a metal canister during a routine MRI scan.[17]

Federal regulations permit a somewhat different approach to risk assessment when the research involves clinical procedures undergone in the course of treatment.[18,19] (Office for Human Research Protections, 2001a, 2001b). For example, if a subject would undergo an MRI scan even if not participating in research, the risk to be assessed by the IRB is simply the risk imposed by use of the scan for research. If little additional risk is posed by the secondary use of the images for research, the protocol may indeed qualify as minimal risk.

Of course, when an imaging protocol involves exposure to very high field strength magnets or novel pulse sequences, or the injection of contrast agents or radioactive materials, the risks may escalate beyond what might reasonably be considered "minimal." The researcher bears the ethical responsibility to fully explain these elevated risks to the IRB and to prospective subjects.

In addition, an often overlooked risk of imaging research is the real—albeit small—possibility that the imaging study might reveal a serious medical condition, such as a tumor, venous malformation, or presymptomatic evidence of a neurodegenerative disorder (e.g., Alzheimer's disease). When treatment options are limited, such knowledge may be psychologically distressing; if communicated to a provider and entered into the medical record the information may compromise the subject's insurability. Moreover, when the information results from an unvalidated analysis or technique, it may be unwise or unethical to use it for clinical purposes. If researchers do contemplate sharing images or results from an imaging study for clinical purposes, the subjects should be advised in advance of these considerations.

*Financial conflicts of interest.* With commercial sponsorship of academic research increasing rapidly, the issue of financial conflicts of interest in human subjects research has drawn the attention of legislators, policy makers, and plaintiff's attorneys. Once viewed primarily as a threat to the integrity of scientific data, financial interests in research are now regarded as a potential threat to the safety of human subjects.[20] There is widespread concern that existing financial disclosure regulations are insufficient, prompting calls for more expansive federal regulation of financial relationships between research sponsors, investigators, and institutions. Two major academic organizations, the Association of American Universities[21] and the Association of American Medical Colleges[22], have issued new guidance calling for academic institutions to voluntarily exercise greater oversight of financial interests in research.

Neuroimaging researchers may have financial conflicts of interest when they conduct research funded in whole or in part by a pharmaceutical company, medical technology company, or other commercial sponsor, while at the same time receiving speaking fees from the sponsor or compensation for consulting or service on the sponsor's scientific advisory board. The potential for conflict also arises when a researcher owns an equity interest in the sponsoring company, receives royalties from a licensee of the technology that is the subject of the research, or otherwise stands to benefit financially from the success of the research.

Individual investigators must comply fully with the financial disclosure policies of their institutions. Current federal regulations require institutions to collect certain information about an investigator's financial interests in federally funded research. Institutions must then determine which financial interests constitute conflicts of interest that must be managed or eliminated. Some institutions now review faculty financial interests in privately sponsored research, in accordance with the recommendations of the OHRP and the AAMC. In addition, some states have their own statutory conflict-of-interest disclosure requirements applicable to employees of state institutions.

Beyond financial disclosure to the institution, investigators must consider their legal and ethical obligation to disclose potentially conflicting financial interests to research participants. At present federal research regulations do not expressly mandate such disclosure unless it is required by the IRB. Because institutions need not share investigators' financial disclosure information with the IRB, practices are highly variable. In California, however, the state's highest court held in the case of *Moore v. Regents of the University of California* that physician-investigators must disclose conflicting financial and other interests to patients who are also research subjects (793 P.2nd 479 (1990)). Elsewhere, in the absence of any legal precedent, respect for the decisional autonomy of the research subject arguably requires disclosure of any information that might be material to that subject's decision to participate[23]—including the researcher's competing financial interests.

## Privacy and Confidentiality

Investigators have an ethical obligation to respect the privacy and safeguard the confidentiality of research subjects. Federal research regulations charge IRBs with the formal responsibility of assessing the risks to privacy and confidentiality in research (45 C.F.R. §46.111(a)(7)). When an investigator wishes to use identifiable records, test results, or other patient data without the consent of the data subjects, an IRB must review the proposed project and determine that it constitutes legitimate research, the potential benefits of which outweigh the invasion of privacy (45 C.F.R. §§46.111(a)(2), 46.116(d)). IRBs must also assess an investigator's ability to maintain the confidentiality of subjects' data. Standard methods of safeguarding the confidentiality of research records include replacing names and other identifiers with codes and storing paper and electronic research records securely.

*Privacy and Confidentiality Concerns with High-Resolution Anatomical Images*

With the advent of MRI techniques that permit high-resolution anatomical scanning, neuroimaging researchers must confront new threats to research subjects' privacy and confidentiality. Readily available computer software used for rendering surface views of the brain from a high-resolution volumetric MRI scan (Fig. 1A) may easily be employed to generate recognizable images of a subject's facial features (Fig. 1B). As a result, even when all other identifying information is removed from the scan header and any accompanying data files, the MRI scan remains potentially identifiable.

Because new federal regulations for the protection of medical privacy (described below) expand current notions of identifiability, researchers and IRBs must assess the extent to which a facial rendering from a volumetric MRI scan could be identified if compared against other publicly available information, such as photographs. The perceived "identifiability" of MRI data files is likely to increase with the development and widespread deployment of facial identification software and systems in response to new terrorism threats.[24] These systems operate by comparing anatomical dimensions of faces in two-dimensional images and might conceivably be used to match a volume-rendered facial image against a database of photographs or digital images of known persons.

The potential identifiability of otherwise anonymous image files is of great concern to those in the field who are anxious to encourage electronic data sharing. This issue is most pressing in the field of functional neuroimaging (fMRI) research. The neuroscience literature is awash in fMRI studies reporting newly discovered neural substrates for cognitive processes. The rapid growth of this literature reflects the popularity of fMRI as a technique for exploring brain-behavior relationships.[25]

Unfortunately, the lack of standard protocols for image acquisition, data

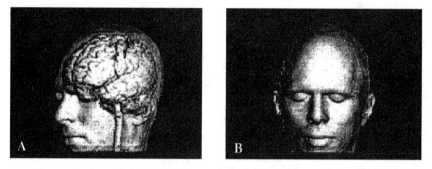

*Fig. 1.* Surface renderings of brain and facial features from volumetric MRI data set. Images courtesy of The National Institutes of Health Division of Intramural Research Programs.

formatting and analysis, and behavioral task parameters makes the comparison of published results and the replication of findings difficult. Additionally, neuroimaging research is expensive and requires access to specialized facilities, limiting the number of potential investigators. To address these concerns, several groups are attempting to construct open archives of raw image files that could be made available to the research community, in the model of the large databases created to share sequence data from the human genome.[26–29]

The possibility that high-resolution MRI data sets may be inherently identifiable is an obstacle to creation of these large anonymous neuroimaging databases. One possible solution is for investigators to strip from the MRI data set any image elements (pixels) that could be reconstructed into craniofacial features. Such stripping can be accomplished through laborious manual effort or via automated software programs, but there are no standard protocols for stripping image data that ensure that the resulting data set retains its scientific utility. Under the regime of the new federal medical privacy rule, the development of such protocols will be essential if data-sharing efforts are to succeed.

## The New Federal Medical Privacy Rule

Neuroimaging researchers face a difficult challenge in designing research protocols to accommodate the requirements of the aforementioned new federal medical privacy rule. This regulation (45 C.F.R. Parts 160 and 164), which implements a provision of the Health Insurance Portability and Accountability Act of 1996 (HIPAA), will necessitate a new and far more complicated approach to most research involving human subjects or identifiable medical information.

Effective April 14, 2003, the federal medical privacy rule will dictate how "covered entities"—hospitals, health plans, and health care providers—may use "individually identifiable" medical information (45 C.F.R. §164.104). Although the administration has proposed to modify certain aspects of the privacy rule, the modifications under consideration would not eliminate many of the requirements that affect research. As of this writing (June 2002) the outcome of the modification proposal is uncertain and covered entities must prepare to comply with the rule as published.

The rule's restrictions will apply whenever a provider (either hospital or physician) uses or discloses identifiable records and information for research. Researchers who wish to obtain identifiable records or clinical information without specific written consent from each patient must seek a new "waiver of authorization" from an IRB, in addition to the waiver of consent that is required under federal research regulations.

The privacy rule will impose new restrictions upon the collection of identifiable health information for concurrent use in research and treatment (what the rule terms "research involving treatment"). Hospitals, physicians, and other covered entities that fail to meet the rule's exacting requirements will face the prospect of regulatory audits and civil or criminal penalties. Researchers who are not covered health care providers but who wish to use health information in research will also need to become familiar with the rule's new requirements for authorization forms and IRB waiver of authorization.

It is critical to understand that "individually identifiable" in the lexicon of the privacy rule has a broad yet precise meaning, in contrast to the use of the term "identifiable" in federal research regulations. The latter deem information "identifiable" if the identity of the subject may be "readily associated" with the information but permit IRBs discretion to determine when this is in fact the case (45 C.F.R. §46.102(f)). Information that might be considered sufficiently anonymous to be exempt from federal research regulations (e.g., a patient record stripped of name, address, and phone number) is not likely to meet the standard of the privacy rule, which applies a strict legal test. The privacy rule deems information identifiable unless there is "no reasonable basis" to believe that the information can be used to identify an individual (45 C.F.R. §164.514(a)).

This exacting standard for identifiability requires the covered entity to remove far more than simply direct identifiers such as names, addresses, and Social Security numbers. In fact, to presume that information has been successfully "de-identified" under the privacy rule, the covered entity must remove a list of eighteen specific identifiers, including most zip codes, birth dates, admission dates, biometric identifiers, full-face photographic or "comparable" images, and any other unique identifying number, characteristic, or code, except for codes that are specially derived so as to be incapable of translation into identifying information. Alternatively, the covered entity must engage a statistical expert to study the potential identifiability of the data and to document, through accepted methods and analyses, that the risk of reidentification is "very small" (45 C.F.R. §164.514(b)).

As noted above, it is possible that anonymous high-resolution anatomical image data sets may not meet the privacy rule's standard for deidentification. In either case, the rule separately imposes a requirement that a covered entity limit uses and disclosures of patient information to the "minimum necessary" (45 C.F.R. §164.514(d)). Applying this standard, a covered entity might conceivably require that high-resolution image data sets be stripped of craniofacial pixels before permitting the MRI data to be used or disclosed for research purposes.

## Public Communications

When applied to brain imaging research, "neuroethics" involves far more than regulatory compliance. The field of neuroimaging must also concern itself with the interface between the scientist and society. As policy makers, the courts, and the public become aware of imaging techniques and intrigued by this window into the living brain, researchers must avoid inadvertently fueling misconceptions about the power and promise of neuroimaging. This task is complicated by media accounts that portray brain imaging technology as the functional equivalent of a polygraph, a Rorschach test, or a Ouiji board.[30,31]

In their public statements and writings, neuroimaging researchers must communicate more than just dramatic pictures that purport to represent the brain in the throes of emotion, perception, or cognition.[32] Researchers have an ethical duty—and a professional responsibility—to explain the limits of current science, particularly with respect to the use of imaging techniques to diagnose addictive and psychiatric disorders or to inform decisions about criminal culpability, propensity for violence, or cognitive capacity. Research in many of these areas has revealed small group differences in brain structure or function between affected individuals and normal volunteers. Often, however, the literature does not support the use of brain imaging to draw diagnostic or legal conclusions about cognitive capacity or moral culpability in the individual case.[33]

The duty to communicate responsibly about science may be greatest when a researcher is advising the judiciary, serving as an expert witness in a legal proceeding, or offering advice to legislators or policy makers who will make decisions about reimbursement or health care resource allocation. Professional societies can facilitate responsible communication about neuroimaging research in these venues by adopting policy statements that indicate which techniques are regarded as sufficiently valid for particular diagnostic uses. One model for such guidance is the technology assessment of PET scanning published by the American Academy of Neurology.[34]

## Conclusion

This paper has explored ethical issues of significance to neuroimaging researchers, including protections for human subjects, issues of privacy and confidentiality, and a researcher's obligation to communicate science responsibly. Evolutions in the concept of what constitutes a "human subject" and the researcher's legal duties to human subjects are topics of current interest in bioethics, with clear implications for neuroimaging researchers. As a result of recent legal developments, particular attention is warranted to a researcher's duty to inform subjects of foreseeable risks arising in what courts refer to as "non-therapeutic" research, where the intent of the research is not to provide care or to benefit the subject directly. Neuroimaging researchers, whose studies may often be regarded as "non-therapeutic," must assess all risks, even those not specifically associated with the imaging procedure, and determine whether all necessary information has been or will be promptly communicated to research subjects.

Ethical conduct of research involves more than rote compliance with legal and regulatory requirements. Researchers serve a public role as identified experts and as representatives of the scientific community. Bioethicist Ruth Faden, director of the Johns Hopkins Bioethics Institute, has observed that investigators at the forefront of their disciplines "should also be at the cutting edge of the social and ethical debate about their work."[35] This admonition is especially pertinent to neuroimaging, a field in which the technological capacity to generate brain images far exceeds scientists' current ability to interpret what imaging data reveal about the mind and the brain. The new field of neuroethics may further Faden's laudatory aim, both by promoting education about the responsible conduct of brain imaging research and by bringing the unique perspective of neuroimaging researchers to bear in policy debates about medical privacy, human subjects protection, and issues of equal importance to the research community.

## REFERENCES

1. Stanford Center for Bioethics, *Neuroethics: Mapping the field* (New York: Dana Foundation; 2002), information available: www.scbe.stanford.edu/neuroethics_conference.html.
2. Association for the Accreditation of Human Research Protection Programs, *Principles and standards* (2002), available: http://www.aahrpp.org/principles.htm.
3. Bureau of National Affairs, Protection of human subjects key in new research accreditation program, *Health Care Daily* 7 (2002), ISSN 1091–4021.
4. G. Blumenstyk, Crusader for the rights of research volunteers: Alan C. Milstein mixes novel legal theories with a strategy of suing everyone in sight, *Chronicle of Higher Education,* January 11, 2002, A4.
5. Food and Drug Administration, Notice of initiation of disqualification proceeding to James M. Wilson, M.D., Ph.D. (2000), available: www.fda.gov/foi/nidpoe/nl21.pdf.
6. *Gelsinger v. Trustees of the University of Pennsylvania et al.* (2000), complaint available: http://www.sskrplaw.com/links/healthcare2.html.
7. S. Sherman, R. Kohl, and P.A. Podolsky, Web site listing of current clinical trials litigation and source documents (2002), available: http://www.sskrplaw.com/gene/.
8. R. Steinbrook, Protecting research subjects: The crisis at Johns Hopkins, *New England Journal of Medicine* 346 (2002): 716–720.
9. *Whitlock v. Duke University,* 637 F. Supp. 1463 (M.D.N.C. 1986), aff'd 829 F.2d 1430 (4th Cir. 1987).
10. J. Brainard, Spate of suspensions of academic research spurs questions about federal strategy, *Chronicle of Higher Education* 46 (2000$_{[JM8]}$): A29.
11. D. Amber, Case at VCU brings ethics to forefront, *Scientist* 14 (2001): 1.
12. J. Kulynych and D. Korn, The effect of the new federal medical privacy rule on research, *New England Journal of Medicine* 346 (2002): 201–204.
13. Office for Human Research Protections, *Engagement of institutions in research* (1999), available: http://ohrp.osophs.dhhs.gov/humansubjects/assurance/engage.htm.
14. Office for Human Research Protections, *OPRR Reports: Exempt research and research that may undergo expedited review* (1995), available: http://ohrp.osophs.dhhs.gov/humansubjects/guidancec/hsdc95-02.htm.
15. Office for Human Research Protections, *Categories of research that may be reviewed through an expedited review procedure* (1998), available: http://ohrp.osophs.dhhs.gov/humansubjects/guidance/expedited98.htm.
16. F. G. Shellock, *Reference manual for magnetic resonance safety* (Salt Lake City: Amirsys, 2002).

17. C. Landrigan, Preventable deaths and injuries during magnetic resonance imaging, *New England Journal of Medicine* 345 (2001): 1000–1001.

18. Office for Human Research Protections, *IRB Guidebook*, Chap. III-A, Risk/Benefit Analysis (2001), available at http://ohrp.osophs.dhhs.gov/irb/irb_chapter3.htm#e1.

19. Office for Human Research Protections, Draft interim guidance: Financial relationships in clinical research (2001), available at http://ohrp.osophs.dhhs.gov/humansubjects/finreltn/finguid.htm.

20. Ibid.

21. Association of American Universities Task Force on Research Accountability, *Report on individual and institutional financial conflicts of interest* (2001), available: http://www.aau.edu/research/COI.01.pdf.

22. Association of American Medical Colleges Task Force on Financial Conflicts of Interest in Clinical Research, *Protecting subjects, preserving trust, promoting progress: policy and guidelines for the oversight of individual financial interests in human subjects research* (2001), available: www.aamc.org/coitf.

23. National Commission for the Protection of Human Subject of Biomedical and Behavioral Research, *Belmont report: Ethical principles and guidelines for the protection of human subjects* (1979).

24. R. O'Harrow, Racial recognition system considered for U.S. airports, *Washington Post*, September 24, 2001, A14.

25. E. Russo, Debating the meaning of fMRI, *Scientist* 14 (2000), 20.

26. Dartmouth Brain Imaging Center (2001), Web URL http://dbic.dartmouth.edu.

27. FMRI Data Center (2002), Web site URL www.fmridc.org.

28. M. Marshall, A ruckus over releasing images of the human brain, *Science* 289 (2000): 1458–1459.

29. Organization for Human Brain Mapping, Neuroimaging databases, *Science*, 292 (2001): 1673–1676.

30. Brain imaging technology can reveal what a person is thinking about, *ScienceDaily* (2000), available: www.sciencedaily.com/releases/2000/11/001110073236.htm.

31. S. Vedantam, The polygraph test meets its match: Researchers find brain scans can be powerful tool in detecting lies, *Washington Post*, November 12, 2001, A02.

32. R. Monatersky, Land mines in the world of mental maps, *Chronicle of Higher Education* 48 (2001): A20.

33. J. Kulynych, Psychiatric neuroimaging evidence: A high-tech crystal ball? *Stanford Law Review* 49 (1997): 1249–1270.

34. American Academy of Neurology, Assessment: Positron emission tomography, *Neurology* 41 (1991): 163–167.

35. D. Keiger and S. De Pasquale, Trials and tribulation, *Johns Hopkins Magazine* (2002), available: http://www.jhu.edu/~jjhumag/0202web.

JENNIFER KULYNYCH, J.D., is an attorney with the health care group Ropse & Gray LLP in Washington, D.C.

*Chapter 12*

# Incidental Findings on Research Functional MR Images: *Should We Look?*

*Alex Mamourian*

In the Marx brothers movie *A Day at the Races*, Margaret Dumont, in the role of a wealthy matron, praises Groucho as one of the finest doctors she has known, saying, "Why, I didn't know there was a thing the matter with me till I met him." Although radiologists frequently find themselves in a similar role, "incidental findings" present difficult medical and ethical questions when they appear in research imaging studies.

This is, of course, not a new problem. Even a cursory search with the keywords "incidental findings" will direct you to numerous articles. They will cover a range of topics such as sinus MR findings, cervical spine abnormalities shown on dental radiographs, and nephrolithiasis disclosed on emergency CT scans.[1] In daily practice, physicians who review screening studies for cardiac calcification or lung cancer must make a decision whether they will even look at the soft tissues of the mediastinum or upper abdomen.

What brings this subject to the forefront is the expanding role of functional MR imaging (fMR imaging) for neuroscience research. This technology has proved to be a powerful tool for investigators who study brain function, and it has captured the interest of the general public. These studies, however, present some potential pitfalls, because they make high-resolution MR images of apparently healthy subjects the responsibility of investigators who may

*From American Journal of Neuroradiology* 25 (2004): 520–522.

not have formal training in image interpretation. Although obtaining these images is effortless, management of the images presents some difficult choices, with significant implications for the subjects and the researchers.

Because these functional studies require advanced MR systems with powerful gradients, there is always the option to create high-quality structural images. Some limited T1-weighted images are always acquired for coregistration of the functional data, but these studies usually do not include fluid-attenuated inversion recovery (FLAIR) or T2-weighted images. The question that each research center must deal with at some level is whether a radiologist should review some, all, or none of the images. And if the investigator is to have the studies reviewed, who would be responsible for contacting the patient or his/her physician when there are abnormal but often equivocal findings in terms of clinical significance? From a practical viewpoint, how will the center even get a radiologist to review these studies at a time when radiologists are in short supply?

Do not expect to find the answers to these questions here. I can only offer the evidence, defend our rationale, and describe the approach that has been used at our institution. First, the data. Although there are fewer than a dozen articles on this topic, large and small studies of young and old all report medically significant findings in 1%–2% of their subjects. What justifies medical intervention is not simple to define, but nearly all would agree that findings of a CNS tumor or aneurysm deserve further attention. Katzman et al.[2] reported two confirmed brain tumors and one unconfirmed in a study group of 1,000 largely young but all asymptomatic volunteers imaged with MR. Two other articles—one by Yue et al.[3] and another by Mirza et al.[4]—have more diverse subject groups, yet the incidence of significant findings again fall in this range of 1%–2%. An article by Lubman et al.[5] included 98 healthy controls and 242 subjects with psychotic mental illness. Although there were more findings in those with psychoses, particularly those with symptoms for two years, the incidence of significant findings in all groups was again 1.1%. A review of our own experience with 198 fMR imaging examinations with 97 controls and 101 subjects revealed three subsequently confirmed cerebral aneurysms, for an incidence of significant findings of 1.5%. One important feature common to all these studies is that a neuroradiologist reviewed the images.

Accepting that range of 1%–2% as the expected incidence across the board, what should be done with these abnormal imaging findings? Kim et al.[6] recommend that all research studies involving pediatric subjects should involve a radiologist who would be expected not only to identify the findings but also to ensure that there is appropriate follow-up. Although such an approach is probably not feasible at all centers, there should be a defined pathway for dealing with these abnormal imaging findings because some of these unexpected results are not appropriate to discuss casually with the subject, particularly when his or her medical history is unknown. There are also records and images that need to be filed in a fashion that protects privacy but allows for retrieval.

In the spirit of disclosure that we all hope pervades medical publications, I must admit at this point that I am not a dispassionate observer in this arena. Not long after finishing the review of our own experience with incidental findings, I volunteered to serve as a subject for a study we were finishing up for this past ASNR meeting. With no small degree of irony, one of our very skilled MR technologists found a cerebral aneurysm on my images. Am I grateful for this discovery of an asymptomatic aneurysm? Absolutely. Did I have surgery? Wouldn't you? Consider how you might feel, assuming that you could, after an aneurysm ruptured a year from now and it was evident but not recognized on a retrospective review of a research imaging study? I bring this to light largely to help put these questions into some perspective and emphasize that this is not some abstract concept that does not require your attention just now.

It is not entirely clear what the legal obligation of the investigator to the subject might be, because the traditional patient-physician relationship does not exist with its associated benefits and obligations. The article by Illes et al.[7] very nicely reviews the scope of this problem and suggests that informed consent might be the appropriate vehicle to limit the expectations of the subject. Illes et al. use a consent in which it is made clear that images are not reviewed by a radiologist and the study they undergo is not equivalent to a clinical imaging session. In one section of the informed consent quoted in the article, however, they state, "The investigators for this project are not trained to perform radiological diagnosis. . . . However, on occasion the investigator may notice a finding on a MRI scan that seems abnormal." This is the standard practice at many centers, but I wonder whether the subjects find this ambiguous? It seems reasonable that they would infer that, if there is no follow-up, their imaging findings must have been normal. What about those that are abnormal for which findings are too subtle to be noticed in a casual review of the images? The literature presents compelling evidence that there will be many significant imaging findings that will not be detected by an investigator who is not an experienced imager. For example, small meningiomas, which are among the most common CNS tumors, are often very difficult to discern on unenhanced T1-weighted images.

Because there are no obvious legal guidelines, it would seem appropriate to respond to the expectations of the subjects. Although some assume that the subjects in fMR imaging studies are motivated solely by their interest in forwarding medical research, in some cases the possibility of getting a "free" MR image of the brain is a powerful motivation. For example, as part of a study at our institution on aging and memory, subjects were asked about their motivation for volunteering for the study during a structured interview. Among 23 healthy controls determined to be cognitively normal on neuropsychological testing only half endorsed being "interested in helping research" as a key factor. In this sample, many said they wished to enroll because they were either concerned about their own memory or in response to a family member's concern about their memory. Although it is also tempting to assume that these incidental findings are less of an issue in controls as compared with subjects, in our small experience two of the three aneurysms (or three total, including me) were in the asymptomatic control group. In the study by Katzman et al., all the patients were young and asymptomatic, yet they found two brain tumors. Perhaps controls are more likely to volunteer for fMR imaging if they have some ill-defined concern that there is something wrong? For whatever reason, significant findings occur in all groups, and there is often a tacit expectation that someone with training will look at the subject's image.

It is worth considering in passing the question that underlies the whole topic: are we really adding quality or years of life to these subjects by making the early diagnosis of brain disease? How does this differ in principle from the "screening CT" scans, which have had a generally cool reception by the medical community? For example, it would seem that, at this time, even the neurosurgical community is not quite sure of the optimal approach to small, incidentally identified cerebral aneurysms. Does this uncertainty absolve us of responsibility? There is, however, at least one fundamental way that this situation differs from screening studies. It is the researcher who is asking the subject to get in the MR system. In this circumstance, we should meet the subjects' expectations or advise them in advance that there is no diagnostic value attached to their participation in the study.

In fact, many research centers are currently dealing with these questions on a case-by-case basis, depending on what is seen on the image. Is it reasonable to use an approach in which the image is sent to a radiologist only after being screened and considered abnormal by the investigator? This approach presumes that the task of separating normal from abnormal findings is simple, yet from practice I would argue that this step is often the most difficult task for an imager.

The approach used at our medical center was established because many of our subject groups have underlying diseases such as multiple sclerosis, dementia, traumatic brain injury, cancer, or mental illness, and we assumed at the outset that the incidence of abnormalities might be higher in these groups. What proved to be of interest and surprising to me over time was the equally high incidence of significant findings in the "normal" control groups. Once the decision was made to have all images reviewed by a neuroradiologist, we also decided to obtain either a FLAIR or T2-weighted image, in addition to their T1-weighted study needed for the coregistration of the fMR image. All of the aneurysms in our experience were evident on the T2-weighted or FLAIR images. These images are read weekly, usually in a conference setting with the investigators. Any previous images are available, and the reader has access to the participant's age and, after an initial blinded review, to some clinical information regarding the status of the subject. The reading is then transcribed and after approval by the neuroradiologist it is included in participants' research file and in a password-protected database of image readings. Subjects are asked at the outset whether they wish to have a copy of their MR imaging placed in the hospital archives for future medical reference. As part of the informed consent process, participants are asked to name their primary care provider and give consent for the investigators to notify their provider if any findings are evident that merit follow-up care. Participants also are asked to give consent to having a note regarding the findings placed in the hospital record. At these weekly meetings plans for follow-up are decided, and they may include the notification of the participant and/or the designated care provider or a request for further imaging such as contrast-enhanced images or MR angiography.

What incentives can be used to engage the radiologist? Apart from the perceived obligations to the subject group, there should be financial, and academic incentives. This topic is again well dealt with in the article by Illes et al., and we use a combination of academic involvement in these research projects and funding that reflects the time needed to make this a workable arrangement.

There are many approaches currently used by the research community to address, or not address, incidental findings. Although this variation is likely to continue, centers that use structural imaging only as a localizing framework should advise the subjects of this in the informed consent process; however, because the evidence makes it clear that significant findings will be encountered in 1%–2% of their cases, research centers should develop a consistent approach that involves trained imagers so that these subjects will have the opportunity to receive appropriate follow-up imaging or care.

## Acknowledgments

I would like to thank Andrew Saykin, Ph.D., William Black, M.D., and Jennifer Ramirez for their assistance and advice.

REFERENCES

1. W. A. Messersmith, D. F. M. Brown, and M. J. Barry, The prevalence and implications of incidental findings on ED abdominal CT scans, *American Journal of Emergency Medicine* 19 (2001): 479–481.

2. G. L. Katzman, P. A. Dagher, and N. J. Patronas, Incidental findings on brain magnetic resonance imaging from 1000 asymptomatic volunteers, *Journal of the American Medical Association* 282 (1999): 36–39.

3. N. C. Yue, W. T. Longstreth, A. D. Elster, et al., Clinically serious abnormalities found incidentally at MR imaging of the brain: Data from the cardiovascular health study, *Radiology* 202 (1997): 41–46.

4. S. Mirza, T. H. Malik, A. Ahmed, et al., Incidental findings on magnetic resonance imaging screening for cerebellopontine angle tumours, *Journal of Laryngology and Otology* 114 (2000): 750–754.

5. D. I. Lubman, D. Velakoulis, P. D. McGorry, et al., Incidental radiological findings on brain magnetic resonance imaging in first episode psychosis and chronic schizophrenia. *Acta Psychiatrica Scandinavica* 106 (2002): 331–336.

6. B. S. Kim, J. Illes, R. T. Kaplan, et al., Incidental findings on pediatric MR images of the brain, *American Journal of Neuroradiology* 23 (2002): 1674–1677.

7. J. Illes, J. E. Desmon, L. F. Huang, et al., Ethical and practical considerations in managing incidental findings in functional magnetic resonance imaging, *Brain and Cognition* 50 (2002): 358–365.

ALEX MAMOURIAN, M.D., is associate professor of radiology and director of the Magnetic Resonance Imaging Program at the Dartmouth-Hitchcock Medical Center.

*Chapter 13*

# Imaging or Imagining?
## A Neuroethics Challenge Informed by Genetics[*]

### Judy Illes and Eric Racine

*What a sensation stethoscopy caused! Soon we will have reached the point where every barber uses it; when he is shaving you, he will ask: 'Would you care to be stethoscoped, sir?' Then someone else will invent an instrument for listening to the pulses of the brain. That will make a tremendous stir, until in fifty years' time every barber can do it. Then, when one has had a haircut and shave and been stethoscoped (for by now it will be quite common), the barber will ask, 'Perhaps, sir, you would like me to listen to your brain-pulses?'* — Kierkegaard, 1846[1]

From a 21st-century partnership between bioethics and neuroscience, modern neuroethics has emerged. While neuroethical discussion and debate about psychological states and physiological processes date back to the ancient philosophers, advanced capabilities for understanding and monitoring human thought and behavior enabled by modern neurotechnologies have brought new ethical, social, and legal issues to the forefront. They draw on and extend anatomo-clinical approaches to cerebral localization and functional specialization that began in the 16th and 17th centuries, after a hiatus of more than two millennia from the days of Aristotle and Hippocrates (300 and 400 B.C.E.).[2] Some issues, akin to those surrounding modern ge-

[*]From *American Journal of Bioethics* 5, no.2 (2005): 5–18.

netics, raise critical questions regarding prediction of disease, privacy, and identity. However, with new and still-evolving insights to our neurobiology and previously unquantifiable features of profoundly personal behaviors such as social attitude, value, and moral agency, the difficulty of carefully and properly interpreting the relationship between brain findings and our own self-concept is unprecedented. Ways of tackling practical questions in neuroimaging will depend on how we deal with the fundamental one of interpretation—the principal reason that traditional bioethics analysis, as laid out in the ethics of genetics, will not suffice as a guide.

Consider, for example, the following sampling of article titles appearing in the scientific literature over the past two to three years: "The Good, the Bad, and the Anterior Cingulate,"[3] "Morals and the Human Brain: A Working Model,"[4] "Strategizing in the Brain,"[5] "The Medial Frontal Cortex and the Rapid Processing of Monetary Gains and Losses,"[6] or "The Neural Basis of Economic Decision-Making in the Ultimatum Game,"[7] as well as those appearing in popular print media, such as "How the Mind Reads Other Minds,"[8] "Tapping the Mind,"[9] "Why We're So Nice: We're Wired to Cooperate,"[10] and "There's a Sucker Born in Every Medial Prefrontal Cortex."[11] From these, we observe that quantitative profiles of brain function—"thought maps"—once restricted to the domain of medical research and clinical neuropsychiatry, may now have a natural relevance in our approach to daily life. This trend conceivably introduces possibilities—or at least desires—for using brain maps to assess the truthfulness of statements and memory in law, profiling prospective employees for professional and interpersonal skills, evaluating students for learning potential in the classroom, selecting investment managers to handle our financial portfolios, and even choosing lifetime partners based on compatible brain profiles for personality, interests, and desires. Further, these trends bring to the foreground what would appear to be a strict epistemological challenge at the core of neuroethics—proper interpretation of neuroimaging data. The challenge will prove to be twofold. First, at the scientific level, the sheer complexity of neuroscience research poses challenges for integration of knowledge and meaningful interpretation of data. Second, at the social and cultural level, we find that social interpretations of imaging studies are bound by cultural and anthropological frameworks. In particular, the introduction of concepts of self and personhood in neuroimaging illustrates the interaction of interpretation levels. Addressing the challenge will involve creative human imagination and conscious awareness of scientific and cultural presuppositions.

This paper, therefore, explores the evolution of functional brain imaging capabilities that have led to bold new findings and claims about behavior in health and disease. We draw on recent neuroimaging research and the proposed applications. Taking a closer look at how genetics has been analyzed from an ethical standpoint, we compare issues raised in genetics with issues in functional neuroimaging, using functional magnetic resonance imaging (fMRI) as our model. Finally, we discuss interpretation of neuroimaging data as a key epistemological and ethical challenge, inescapable for neuroethics and intertwined with the history of neuroscience.

## Functional Neuroimaging

From generations of work by neurotechnologically curious and skilled scientists and engineers, powerful functional neuroimaging tools have been introduced to the modern era. The most prominent tools to date—electroencephalography (EEG), magnetoencephalography (MEG), position emission tomography (PET), single photon emission computed tomography (SPECT), and functional magnetic resonance imaging (fMRI)—have provided a continuing stream of information about human behavior.

One of the oldest approaches dates back to 1929, when neuropsychiatrist Hans Berger announced the invention of the electroencephalogram and showed that the relative signal strength and position of electrical activity generated at the level of the cerebral cortex could be measured using placement of electrodes at the scalp.[12] With its exquisite temporal resolution, the stimulus-evoked EEG response "event related potential" (ERP) was the first tool to unveil fundamental knowledge about the working of the human brain in near real time. Over time, other imaging modalities have come to achieve this goal by capitalizing on brain signals such as extracranial electromagnetic activity (MEG), metabolic activity and blood flow (PET and SPECT), and regional blood oxygenation (fMRI) that yield different and complex measurements of functional activity (for a very readable anthropologic perspective on PET specifically, see Dumit).[13] By and large, all utilize comparison or subtraction methods between two controlled conditions, heavy statistical processing, and computer-intensive data reconstructions to produce the colorful maps with which we have become familiar. All have roots in the diagnosis and intervention of the wide range of psychiatric and neurological diseases known to us, including head trauma, dementia, mood disorders, stroke, cancer, seizures, and the impact of drug abuse, to mention but a few.

**Table 1** Characteristics and trade-offs of major functional neuroimaging technologies. Combined modalities systems such as EEG and fMRI are becoming increasingly common. PET and SPECT are forerunners to frontier technology in molecular imaging.

| | Measurement | Technology | Strengths | Limitations | Notes |
|---|---|---|---|---|---|
| EEG | Electrical activity measured at scalp | Electro-encephalogram; 8 to > 200 scalp electrodes | Noninvasive, well-tolerated, low cost, sub-second temporal resolution | Limited spatial resolution compared to other techniques | Tens of thousands of EEG/ERP findings reported in the literature |
| MEG | Magnetic fields measured at scalp, computed from source-localized electric current data | Superconducting quantum inter-ference device (SQUID); ~80–150 sensors sur-srounding the head | Noninvasive, well-tolerated, good temporal resolution | Cost, extremely limited market and availability | |
| PET | Regional absorption of radioactive contrast agents yielding measures of metabolic activity and blood flow | Ringed-shaped PET scanner; several hundred radiation detect-ors surrounding the head | Highly evolved for staging of cancers, meas-uring cognition function and evolving to a clinical imaging tool for predicting disease involving neurocognition, such as Alzheimer's Disease | Requires injec-tion or inhalation of contrast agent such as glucose or oxygen; lag time of up to 30 minutes between stimulation and data acquisition, limited availability (fewer than 100 PET scanners exist in US today) given short half-life of isotopes and few locations with cyclotrons to produce them; cost | |
| SPECT | Like PET, a nuclear med-icine technique that relies on regional absorp-tion of radio-active contrast to yield measures of metabolic activity and blood flow | Multidetector or rotating gamma camera systems. Data can be re-constructed at any angle, in-cluding the axial, coronal and sagittal planes, or at the same angle of imaging obtained with CT or MRI to facilitate image comparisons | Documented uses mapping psychiatric and neurological disease including head trauma, dementia, atypi-cal or unresonsive mood disorders, strokes, seizures, the impact of drug abuse on brain function, and atyp-ical or unrespons-ive aggressive behavior | Requires injection of contrast through intravenous line; cost | Currently available in two states (CA and CO) for purchase without physician referral; emphasis is on ADHD and Alzheimer's Disease (approx. out-of-pocket cost: $3000 per study) |
| fMRI | Surplus of oxygenated blood recruited to regionally activated brain | MRI scanner at 1 Tesla and higher filed strengths; 1.5T most common | Noninvasive, study repeatabil-ity, no known risks | Cost of equip-ment and physics expertise to run and maintain systems | Rapid proliferation because of its wide clinical availability of research studies using fMRI alone or in combination with other modal-ities, growing from 15 in 1991 (13 journals) to 2224 papers in 2003 (335 journals), represent-ting an average in-crease of 56% per year |

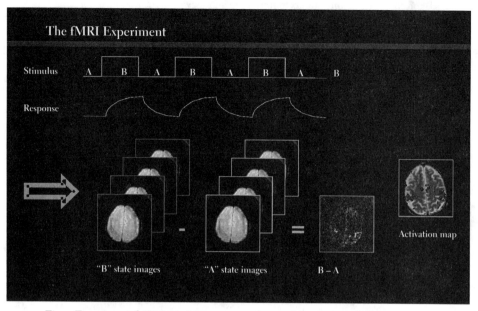

*Fig.* 1 Experimental ("B" state) images are subtracted from control ("A" state) images to achieve regional activation maps with fMRI. Courtesy of Gary H. Glover, Lucas MRS/MRI Center, Stanford University Medical Center.

The different techniques each have relative advantages and disadvantages; these are summarized briefly in Table 1 to provide reference for this discussion. Given their technical trade-offs, fMRI stands out as likely to have the greatest enduring impact on our society outside the realm of academia and medicine. It is the widespread availability of MR scanners today and the noninvasiveness of the imaging approach enabled by MR that have set fMRI apart from other neuroimaging tools and made it a model for neuroethical discussions. The activation maps produced by fMRI reflect indirect effects of neural activity on local blood flow under constrained experimental conditions. Like PET and SPECT paradigms, a typical fMRI experiment utilizes a stimulus designed for acquiring the relative difference in brain activity between an experimental and a control (baseline) task, as illustrated in Figure 1. The surplus of oxygenated blood recruited to relatively active brain regions produces the effects measured by MR.

*Applications of Functional Neuroimaging in Clinical Medicine,*
*Cognitive Science, and Law*

Beyond the use of fMRI in mapping of salient cortical areas prior to surgical intervention for epilepsy, tumors, or arteriovenous malformations, other active efforts to make the technology relevant in the clinical setting have focused on Alzheimer's disease (AD), mental illness in adults, and pediatric pathology such as attention deficit hyperactivity disorder (ADHD).[14] Applications of fetal MRI have also shown great promise in providing better diagnosis of structural central nervous system anomalies, and functional studies of fetal brain blood flow are not lagging far behind. At the opposite end of the life spectrum, first approaches using fMRI to determine levels of consciousness in patients in minimally conscious or vegetative states have also been attempted (e.g., Schiff et al.[15]).

Over the past ten years of fMRI development and expanding boundaries of cognitive neuroscience, the innovation has been applied to gain new non-health-related knowledge about human motivation, reasoning, and social attitudes. In a comprehensive literature review, we demonstrated a steady expansion of fMRI studies, alone or in combination with other imaging modalities, with evident social and policy implications, including studies of lying and deception, human cooperation and competition, brain differences in violent people, genetic influences, and variability in patterns of brain development.[14] In one intriguing but unpublished study, Beauregard et al. (reviewed by Curran)[16] used a combination of EEG, fMRI, and PET to probe neural underpinnings of religious experience. They are focusing on the phenomenon known as *unio mystica*, a joyous sense of union with God reportedly experienced by a group of cloistered Carmelite nuns in Montreal, Canada. Discussion of the potential meaning and practical uses of such deeply personal neuroprofiles is ripe for bioethical consideration.

Outside the arenas of medicine and cognitive science, the legal arena offers an obvious venue for attempting to translate neuroimaging into meaningful, real-world use. As Hank Greely[17] wrote,[11] "neuroscience may provide answers to some of the oldest philosophical questions, shedding light, for example, on existence limits, and meaning of free will. It may also provide new ways to distinguish truth from lies or real memories from false ones. This ability to predict behavior with the help of neuroscience could have important consequences for the judicial system as well as for society as a whole." Greely provides an extensive review of the legal issues in his chapter "Prediction, Litigation, Privacy, and Property: Some Possible Legal and Social Implications of Advances in Neuroscience."[18] A few cases suitable to elucidating where ethical models for brain imaging may intersect with or diverge from genetics are explored here.

Looking back to 1985, when a Supreme Court holding in *Ake v. Oklahoma*[19] imposed a constitutional requirement for states to provide psychiatric assistance in a criminal defense when the question of sanity is raised, criminal defendants began to argue that "psychiatric assistance" should include a complete neurological evaluation, including scans like PET or MRI. PET studies have shown that committed murderers, for example, as a group, have poor functioning in the prefrontal cortex, a locus of impulse control.[20] In some cases, PET images have been used to argue that a defendant was biologically predisposed to committing a crime and, therefore, should be spared a conviction or death sentence. In at least one court case *(People v. Jones)*[21] a homicide conviction was reversed because the state failed to provide brain scans.*

In a relatively new application of EEG, EEG-derived "brain fingerprinting" has been promoted as a tool for determining whether an individual is in possession of certain knowledge of a crime.[22] It is the possession—or lack of possession—of the relevant facts about a crime that brain fingerprinting attempts to quantify through measures of brain-wave responses to relevant words or pictures presented at rapid rates on a computer screen. When the brain recognizes significant information—such as crime scene details—it responds with a "memory and encoding related multifaceted electroencephalographic response." Unlike polygraph testing, which measures an individual's fear of getting caught in a lie by tracking relevant physiological markers, brain fingerprinting ostensibly measures brain waves emitted when information stored in the brain is recognized.**

New applications of fMRI that bridge cognitive science and law also have

---

*In a 1992 New York murder case, *People v. Weinstein*,[24] Weinstein was accused of strangling his wife to death and throwing her body from a twelfth-floor apartment. Weinstein's functional PET scans and structural MRI images revealed an arachnoid cyst, and the evidence was admitted in court for the purpose of establishing an insanity defense. The ruling was made despite evidence that such pathology has no known link to criminal behavior. The PET scan depicted the juxtaposition of a black lesion (the cyst) on the red and green colored areas of "normal" brain activity, and was considered so profound as to prove to the court that Weinstein's brain was not functioning within normal parameters. The prosecution in this case accepted a manslaughter plea.

**Brain fingerprinting played a significant role in the case of Terry Harrington, for example, whose murder conviction was reversed and a new trial ordered after he spent 22 years in prison *(State of Iowa v. Terry Harrington)*.[25] The case dates back to 1977, when Harrington, who was 17 years old at the time, was convicted of murdering a retired police the potential to change approaches to truth verification and lie detection. Langleben et al.,[23] for example, used fMRI to study neural patterns associated with deception. In their landmark experiment, volunteers were instructed to either truthfully or falsely confirm or deny having a playing card in their possession. When subjects gave truthful answers, the fMRI showed increased activity in visual and motor cortex. When they were deliberately deceptive, additional activations were measured in areas including the anterior cingulate cortex, to which monitoring of errors and attention has been attributed. Langleben et al. concluded that "essentially, it took more mental energy to lie than to tell the truth."[26] These results are consistent with those of Moll et al.[4] and Heilman[27] that implicate the temporo-polar cortex, insula, precuneus, and their connections in an extended neural circuit that attributes conscious emotions and feelings, especially those with a social context, to perceptions and ideations.[28–30] In the future, therefore, we may be able to discern not only whether an individual is being deceptive but also whether the deception was premeditated or not.*

## Pathways from Genetics to Neuroimaging Science

Ellen Wright Clayton[31] provides a comprehensive review of the impact that advancements in genetics and molecular biology have had on society, and she argues for genomics as a complex phenomenon that presents specific challenges for clinicians and patients alike. Neuroscience is no less ethically complex, and neuroscientists, like geneticists and even nuclear physicists before them, are increasingly gaining awareness about the potential implications of their research at the bench, in medicine, and in the *public domain.*[32]* Drawing on some major ethical, legal, and social (ELSI) variables, this section examines the extent to which the ethics of genomics can serve as a model for ethical analysis of neuroimaging.

In another fMRI experiment related to deception and more broadly to ly-

---

officer. When, in 2000, Harrington underwent fingerprinting, his brain did not emit the expected electroencephalographic patterns in response to critical details of the murder. The results were interpreted to suggest that he was not present at the murder site, a conclusion corroborated by the fact that his brain did emit the requisite patterns in response to details of the event used as his alibi in the case. When confronted with the brain fingerprinting evidence, the original prosecution witness recanted his testimony and admitted that he had lied during the original trial, falsely accusing Harrington to avoid being prosecuted himself.

*Even in these early stages, some ire about who is conducting what kind of research and with what motivation has already surfaced, especially with respect to research that would seem only to yield financial rewards and not deliver either scholarly or medical benefit.

ing, Schacter et al.[33] demonstrated the potential to discern false from truthful memory. In an even more recent study, Anderson et al.[34] showed areas of neural activation in the dorsolateral prefrontal cortex associated with the active suppression of memory.

### Discrimination, Stigma

Given the growing recognition that health information is not entirely private, Clayton[31] and others (e.g., Rothenberg and Terry[35]) have suggested that the most common fear about genetic information is its potential use in justifying denial of access to health insurance, employment, education, and even financial loans to people with particular genetic characteristics or diagnoses. While these issues may not be the immediate ones for neuroimaging, as we have seen, little stands in the way for similar concerns about neuroprofiling with functional imaging to arise even as neuroimaging techniques continue to mature. While neuroscientists tease out artifacts masquerading as neural effects and develop analytic methods that provide more intuitive ways of interpreting the data than possible today, there already exists a healthy regard for the novelty and breadth of information that neuroimaging can deliver about human health, behavior, and cognitive fitness. How will such technology be used advantageously to benefit people and society? Could it be used harmfully for ill-intentioned purposes? Will Canli's paradigms for imaging personality become adopted for triaging team players or weak decision makers in the workplace[36] or, in this post-Columbine era, at the door of our high schools to triage out students with a predisposition to unruly or violent behavior? Perhaps screening for good humor would be more acceptable.[37, 38]

It will be the moral obligation of bioethicists and neuroscientists alike to think proactively about the impact that such effects might have on people, from the point of view both of benefits such as self-knowledge[39] and personal choice, as well as risks, especially for children and adolescents at critical stages in their personal and educational development.[40] By what means will anyone resist coercive uses of such technology if employment or educational opportunity is at stake? If the paradigms of Golby et al.,[41] Phelps et al.,[42] or Richeson et al.[43] for studying race and social attitudes could be adopted for determining eligibility to become a police officer, a school principal, or even a national leader, would this be a legitimate allocation of public funds? Much mischief beyond discrimination and stigma may be created by over-interpretation of any such results (see also "Scanning the Social Brain").[44]

*Privacy of Human Thought*

Functional neuroimaging poses pivotal challenges to thought privacy. Should thought information have similar privacy status as genetic information? Probably not less, but perhaps more. No doubt increased information about the neurobiology of how we think, and potentially why we think what we think, is likely to cause significant ethical dilemmas for clinicians and researchers, especially as measured thought patterns may vary as much with the hemodynamic properties of the relevant vasculature[45] as with gender and day-to-day variations in mood and attention.[46] Moreover, they are highly subject to variability in the culture and values of the people interpreting them.[47, 13] Watson (as cited in Mauron[48]) has stated that the human genome is, at least in part, "what makes us." The "brainome,"[49] then, touches more upon who we fundamentally "are"— gnarly territory, at best.

With a small leap of faith for real-world validity, the way in which these studies are edging toward biologic measurements of personhood is illustrated, for example, in a now-landmark trolley car study by Greene et al.[50, 51] In this moral reasoning experiment, subjects were scanned while they made decisions about scenarios in which they could, for example, choose to save the lives of five people on a runaway trolley car by pulling a switch to send it onto an adjacent track where one person stands (and who would not survive) or to push one of the people off the trolley and onto the track, thereby blocking the movement of the trolley and saving the remainder of the group. Other studies have required research participants to resolve statements of moral content (e.g., "The judge condemned the innocent man" or "The elderly are useless") versus neutral content ("The painter used his hand as a paintbrush").[52] All such studies touch upon human thought processes that push the envelope of cognitive neuroscience into a domain of significant social concern in which privacy is a vital ingredient.

*Genetic Versus Neuro Determinism*

Despite its probabilistic nature, genetic information tends to be viewed as a definitive form of health data. Individuals feel a sense of inevitability with regard to their genes, Clayton[31] argues, as well as a sense of genetic determinism. Such determinism, or genomic essentialism,[48] has become popular in our culture, especially in the way that results of behavioral genetics studies are communicated to the public. We have read reports about genes for violence, homosexuality, alcoholism, and even language. The essentialist stance is strengthened by the fact that we have a tendency to believe that we are our brains. However, long-standing studies of developmental brain plasticity and new activation studies of reorganization after injury have amply demonstrated that any such reductionist view of complex phenotypes is incomplete without consideration of intervening external and cultural factors.[53] Some have argued that the biological sciences, which deal with open systems, are improperly fit for universal and deterministic laws, as are physics and chemistry.[54] Arguments that make neuroscience deterministic could well be flawed conceptually and empirically.[55] However, given the tendency to oversimplify complex genetic and brain data, discussion of meaning and practical use is a clear imperative.

### Prediction of Disease, Public Health

Countless medical examples exist of symptomatic individuals who, with an inherited genetic defect for a given disorder, must carefully monitor their daily activities to ensure their own health and safety and that of others. But what of the asymptomatic individual who learns from a functional neuro-image of a predisposition to a disease of the central nervous system that ultimately affects cognitive performance and lifelong independence? What are the implications for third parties, as in the case of neurogenetic disorders, for which functional patterns may surface as sensitive predictors of disease? What is the new role of physicians in the entrepreneurial world of self-referred imaging services?

Work on imaging-based diagnosis holds enormous promise for providing new, quantitative evidence for otherwise qualitative diagnoses based on clinical findings, however it also raises compelling questions about what cautions are needed as patients yearn for earlier and earlier diagnosis about diseases for which cures or even treatments do not yet exist. Will such data provide welcome new information or impose new burdens on families, physicians and allied health care professionals? Who will have access to this technology? How will physicians and patients incorporate these new types of data into their reasoning about treatment, compliance and life planning? Access to advanced technologies by the privileged only, whether for diagnosis, medical intervention or for a competitive neurocognitive edge, will only further upset an already delicate and hardly acceptable status quo.

*Confidentiality and Responsibility*

We are in an era of neuroinformatics in which sharing of genetic, brain, and other data is encouraged. In some cases of large federally sponsored research, brain data sharing is even required.[56] With only partial brain information now needed to identify research participants, and new imaging genomics studies[57] coupling genetic information with brain mapping ("genotyped cognition"[58]), major new uncertainties exist about the safety and confidentiality of data stored in cyberspace. Other issues concern the protection of human subjects, including confidentiality and responsibility related to incidental findings. What if pathology is discovered unexpectedly in a shared data set? With whom does the burden of disclosure and care lie—the primary or secondary laboratory? Beyond the laboratory, how shall commercial use of freely shared brain imaging data in the for-profit sector be defined? No doubt countless other examples exist in research beyond these few.

As we have seen, our concept of legal responsibility may also be changed by neuroimaging. In the United Kingdom, cautions about the use of neuroimages such as PET in the courtroom have already been expressed. In 2002, for example, at a debate titled "Neuroscience and the Law," hosted by the Royal Institution of Great Britain, forensic psychiatrist and criminal barrister Nigel Eastman argued that neuroimaging science is still too imprecise to make "an unequivocal connection between brain structure and behavior." "Even if . . . psychopaths have physically different brains from other people, does it mean anything? Does an abnormal brain automatically mean abnormal behavior? Does it mean a loss of control sufficient to impact on legal responsibility? Do you abolish free will on the basis of an odd brain scan?" Even while neuroimaging cannot establish moral culpability[59] of where, when, or how a crime occurred or individual guilt,[60] the constant stream of innovative scientific approaches is aimed at deriving biologic correlates for behaviors committed in the past is unrelenting.[61] As we seek to understand responsibility of others through their biology, it is incumbent upon us to contemplate, yet again, our own responsibilities in interpreting such information, and in protecting access and appropriate use.

*Crossroads*

In Table 2 we summarize areas where ELSI in genetics converge with and diverge from neuroimaging. These are the crossroads at which we can begin to transition to our thinking about the ethics of making brain maps from our experience with genetics. As the discussion above and this table both show, similarities are striking and span the domains of both research and clinical ethics. Among the profound practical benefits are new knowledge about the human condition and knowledge that informs self-determination and life planning. They include negatives as well, such as the potential for personal and legal discrimination, inequities of access, risks to confidentiality, inaccuracies inherent to predictive testing of any nature and associated anxiety,[62] and commercial use.[63] Pressing issues unique to genetics but not to neuroimaging are not apparent, but the reverse is noteworthy inasmuch as the wide range of technical and subjective factors, including paradigmatic, physiologic, and investigator biases in research play into the interpretation of results and the global potential for biologizing human experience that reaches far beyond any previous window on individual traits.

Table 2  Comparison of ethical, legal and social issues in genetic testing and neuroimaging.

| ELSI Variables | Genetic Testing | Functional Neuroimaging |
| --- | --- | --- |
| Risk of discrimination, stigma, coercion | Yes | Not at present, but growing concern exists for the evolution of the technology and expanding use |
| Privacy | Yes | Yes |
| Distributive justice | Yes | Yes, once the technology moves into mainstream clinical medicine |
| Diagnostic uses | Yes | Emerging |
| Prediction | Yes | Emerging |
| Commercial use | Yes | Emerging; some limited availability already exists in the direct-to-consumer marketplace |
| Results subject to variability in test used | Potentially but not considered a significant risk | Highly significant |
| Results subject to physiologic and day-to-day variations | No | Highly significant |
| Results subject to variability in disposition of the interpreter | No | Highly significant |
| Biologization of personal thought | Possibly in mental illness and neurodegenerative disease | Highly significant |
| Secondary uses of data | Yes | Yes |

## Interpretation as a Key Neuroethical Challenge

The idea that the genome is the "secular equivalent of the soul" has been legitimately criticized.[64] New to neuroethics will be the need to tackle responsibly—with the inevitable and omnipresent working hypothesis (or the "astonishing hypothesis," to quote Crick[65])—that the mind is the brain. Responsible and careful interpretation of data will therefore become a crucial issue as we wrestle to untangle what we image from what we imagine. Here, genetics as a model is limited, and bioethicists will have the greatest role in bringing critical thinking to the field. Fundamentally, the challenge is twofold, as proper ethical interpretation is a crucial concern at both the scientific and the social levels.

While fMRI today may surpass other neuroimaging techniques in its use for understanding human behaviors that may have practical relevance, we are witnessing a dynamic stream of new applications and new technical possibilities. Like genetic testing, models for minimizing harm that may result from false positives and inappropriate attributions of cause and effect to otherwise correlative results are critical. Apart from genetic testing, brain maps can be readily portrayed as iconic proof of pathology to people at any level of literacy. Yet, as we have seen, the brain image represents unparalleled complexity—from the specialized medical equipment needed to acquire a scan, to the array of parameters used to elicit activations and the statistical thresholds set to draw out meaningful patterns, to the expertise required for the objective interpretation of the maps themselves. Moreover, an absence of standards of practice in the laboratory (in fact, innovation and creativity still define the state of the art in neuroimaging today) and in the medicolegal setting creates another layer of complexity for drawing conclusions about behavior, responsibility, and cognitive well-being[59, 66] that will need to be penetrated with appropriately responsive ethical approaches. With dynamic images in hand, we may forget the epistemological limits of how the images were produced, including variability in research designs, statistical treatment of the data, and resolution. It is worth recalling that, in the past, various models of the brain have been proposed by great minds only to be seen later as mere imagination of the brain's real functioning. Descartes used pneumatics as a paradigm to explain how the "animal spirits" were produced by the flow of blood from the heart to the brain.[67] Later on, the eminent anatomist Franz Joseph Gall proposed phrenology to the courts for establishing facts and choosing appropriate sentences for convicted criminals.[68] In the twentieth century, Moniz's psychosurgery procedures certainly left behind an "unhappy legacy."[69]

Today, some scientists and philosophers urge that we adopt the computer metaphor, neural networks, or other models to understand brain function. However different and in some sense far apart, these examples highlight cautions needed in the interpretation of brain findings and their intended applications. In an issue of the journal *Brain and Cognition* that represented a pioneering venture into ethical issues in neuroimaging,[70] the community of authors who contributed to it already then cited cautions of interpretation as a common concern. These cautions have been reiterated by others,[71] and alone justify the increasing attention to frontier neurotechnologies, their capabilities and limitations, and new ethical approaches for thinking about them.[72]

When links are made between neuroimaging findings and our self-concepts in particular, it is even clearer that the ethics of genetics can only partially help settle ethical issues. Genetics and genomics have provided fertile ground for many ethical reflections on human nature, but the relationship between the brain and the self is far more direct than the link between genes and personal identity.[48] The locus for integrating behavior resides in the brain, even if discrete features are determined by our genes. Whether neurotechnology measures that behavior through imaging, or manipulates it through implants of neural tissue or devices, it will fundamentally alter the dynamic between personal identity, responsibility, and free will in ways that genetics never has. Indeed, neurotechnologies as a whole are challenging our sense of personhood and providing new tools to society for judging it.[73]

Interpretation of neuroimaging studies is bound by not only scientific frameworks but also cultural and anthropological ones. Consider concepts such as "moral emotions" that are based on assumptions that some emotions are moral and others not. They illustrate the cultural aspect of the interpretation challenge, which is based on the fact that the self is defined in diverse ways. For example, central to Buddhism is the Doctrine of No-Soul, whereas in Hinduism, the self is a religious and metaphysical concept.[74] Even within Western traditions, which may appear to be monolithic, various beliefs have served as "sources of the self."[75] As Winslade and Rockwell[76] wrote, "Humans are forever prone to make premature and presumptuous claims of new knowledge. . . . One may think that brain imagery will reveal mysteries of the human mind. But it may only help us gradually comprehend the organic, chemical and physiological features of the brain rather than provide the keys to unlock the secrets of human behavior and motivation." Whatever the outcomes of imaging turn out to be, they will depend on scientific as well as cultural scrutiny of neuroimaging research.

In a study for which neuroscientists teamed up with Buddhist monks to understand the mind and test for insights gained by meditation,[77] culturally laden concepts such as "person," and "emotions" are being questioned by imaging. Some argue that consciousness and spirituality could be changed by such findings on the brain. Therefore not only does culture penetrate neuroimaging; neuroimaging is increasingly penetrating non-scientific culture. This is why neuroethics needs to consider not only ethics of neuroscience but also a neuroscience of ethics[78] and, we may add, reflection on their scientific and cultural implications.

Regardless of the functional neuroimaging technology du jour, we are left with lingering interpretation and other questions that any new ethical approach for brain imaging will have to address. Time, scholarly dedication, and collaboration across the vested disciplines will help resolve them. Some of these questions, which will inevitably raise the bar and challenges of interpretation, are:

- Revisiting a classic dichotomy in the conduct of research, is there neuroimaging research that we can do (or will be able to do) but ought not to? If we were to accept that the biology of social processes studied within the constraints of the laboratory translates seamlessly to real-world validity, should the contents of our minds even be studied this way?[79] Who should decide and according to what scientific and cultural parameters?
- What are the trade-offs between cautious research with public oversight versus the potential for overregulation in a reaction to adverse or reckless events? Many lessons may be learned from genetics, but "reckless" will surely take on new meaning in discussions about the neurobiology of moral reasoning and social behavior.
- How can the reductionist approach of neuroimaging to human behavior be made compatible and complementary to approaches represented by philosophy, sociology, and anthropology? How will applications based upon this approach interact with wider cultural perspectives on the self?
- Will the large investment in neuroimaging be justified by new knowledge? Are there some forms of funding that should be eschewed because they may lead to methods for thought control or personal financial gain?

- What new ethical challenges will neurotechnologies bring us in the future? What will the portability of near infrared optical imaging offer? Shall transcranial magnetic stimulation be transferred from the medical arena for treating depression to the open market for boosting or fine-tuning cognition like caffeine or other over-the-counter stimulants? With advances in reporter probes,[80] what ethical approaches will be needed for managing new information and therapies brought forward from the coupling of molecular imaging, targeted either in the central nervous system or elsewhere, and gene therapy trials?

The answers to these questions will surely not be binary. As in the past, they will depend fundamentally on the individuals involved and the context in which they are confronted. Fresh thinking, especially about the relationship between the self and the brain, will have to be elaborated for these new types of brain data, as the layers of complexity of interpretation and the overall stakes are arguably far greater than ever before. Commenting on Aldous Huxley's *Brave New World* (1932) written approximately 100 years after Søren Kierkegaard's foresight on brain pulses, Pontecorvo wrote:

> The ethical issues raised by . . . feats of human engineering are qualitatively no different from those we shall have to face in the future. The difference will be quantitative: in scale and rate. Even so, the individual steps may still go on being so small that none of them singly will bring those issues forcibly to light: but the sum total is likely to be tremendous."[81]

Pontecorvo was partly right: There is no doubt that the sum total is tremendous. He could never have predicted, however, the extent to which changing qualities would parallel changing quantities.

## Addressing the Challenge

This paper explored the evolution of functional brain imaging capabilities that have led to bold new findings and claims about behavior in health and disease. Recent neuroimaging findings and their proposed applications show that a great number of new ethical issues will be raised. ELSI variables have given us an invaluable starting point, but neuroethics will need to address issues of data interpretation in great depth at both the scientific level and the cultural level. Neuroimaging illustrates this double challenge remarkably since imaging technologies and methodologies are grounded in scientific assumptions. Meanwhile, imaging is an area where investigation of social behavior and selfhood is rapidly increasing and becoming a legitimate endeavor. Indeed, at the heart of imaging is an effort to make sense of an image in need of interpretation. Right alongside the new concept of "imaging neuroethics," others like "neuromarketing," "neuroeconomics," "neuroenablement,"[82] "neurotheology," and even "neurocorrection"[83] have been spawned. All raise concerns about scientifically warranted and culturally sensitive interpretation and application.

With the existence of many views about mind and brain, neuroethics will have to foster discussions among neuroscientists whose methods may vary and interpretation of results differ. These discussions will have to extend to include meaningful dialogue with scholars in the humanities about concepts like morality, moral judgments, and moral emotions— concepts in need of critical appraisal before we can seriously investigate their neural correlates. Open dialogue with the public is no less necessary, given that different cultural and religious perspectives subject findings to different interpretations and ethical boundaries. Responsible dissemination of information through the media and public education are also essential in closing the gap between scientists and concerned citizens, especially as the complexity and abstractness of results increase.

Interpretation necessitates creative human imagination and conscious awareness of scientific and cultural presuppositions. Hence, the new generation of neuroethicists must be committed to openly examining the epistemological limits of imagery,[84] interdisciplinary appraisal, and public perspectives on these issues. Bioethicists will continue to bring ethical knowledge to the discussion and identify and clarify moral quandaries; bioethicists, however, will also have to work as facilitators of a broader dialogue where different perspectives can meet and contribute to a deeper understanding of the issues. Therefore, while in the past technology and ethics may have leapfrogged each other, in this new era, bioethicists and neuroscientists will be well served by working gracefully together to understand the power of a visual image and the impact it can have on people and collectively on society.

## Disclosures

Supported by the Greenwall Foundation, NIH/NINDS RO1 #NS045831, and the Social Sciences and Humanities Research Council of Canada, #756-2004-0434.

## Acknowledgments

We are indebted to Dr. David Magnus, Dr. H. F. M. Van der Loos, Ms. Kim Karetsky, and Connie Stockham for their invaluable input to this paper.

REFERENCES

1. S. Kierkegaard, *Papers and journals: A selection*, trans. A. Hannay (1846; New York: Penguin Books, 1996).
2. J. C. Marshall and G. R. Fink, Cerebral localization then and now, *Neuroimage* 20 (2003): S2–S7.
3. G. Miller, The good, the bad, and the anterior cingulate, *Science* 295 (2002): 2193–2194.
4. J. Moll, R. de Oliviera-Souza, and P. J. Eslinger, Morals and the human brain: A working model, *Neuroreport* 14, no. 3 (2003): 299–305.
5. C. F. Camerer, Strategizing in the brain. *Science* 300 (2003): 1673–1675.
6. W. J. Gehring and A. R. Willoughby, The medial frontal cortex and the rapid processing of monetary gains and losses, *Science* 295 (2002): 2279–2282.
7. A. G. Sanfey, J. K. Rilling, J. A. Aronson, et al., The neural basis of economic decision-making in the ultimatum game, *Science* 300 (5626) (2003): 1755–1758.
8. C. Zimmer, How the mind reads other minds, *Science* 300 (2003): 1079–1080.
9. I. Wickelgren, Tapping the mind, *Science* 299 (2003): 496–499.
10. N. Angier, Why we're so nice: We're wired to cooperate, *New York Times*, 2002, 1.
11. C. Thompson, There's a sucker born in every medial prefrontal cortex, *New York Times Magazine*, October 26, 2003, 54.
12. K. Karbowski, Sixty years of clinical electroencephalography, *European Neurology* 30, no. 3 (1990): 170–175.
13. J. Dumit, *Picturing personhood: Brain scans and biomedical identity* (Princeton, NJ, Princeton University Press, 2004).
14. J. Illes, M. P. Kirschen, and J. D. E. Gabrieli, From neuroimaging to neuroethics, *Nature Neuroscience* 6, no. 3 (2003): 250.
15. N. D. Schiff, D. Rodriguez-Moreno, A. Kamal, K. H. S. Kim, J. T. Giacina, F. Plum, and J. Hirsch, FMRI reveals large-scale network activation in minimally conscious patients, *Neurology* 64 (February 2005): 514–523.
16. P. Curran, Soul search : Emotions, spirituality, and transcendence: Scientist gains notoriety for work with nuns, *Montreal Gazette*, October 19, 2003, A14.
17. H. T. Greely, Neuroethics? *Health Law News* (Health Law and Policy Institute, University of Houston Law Center, July 2002), 5.
18. H. T. Greely, Prediction, litigation, privacy, and property: Some possible legal and social implications of advances in neuroscience, in *Neuroscience and the law: Brain, mind, and the scales of justice*, ed. Brett Garland (New York: Dana Press, 2004).
19. *Ake v. Oklahoma*, 470 U.S. 68, 74 (1985).
20. A. Raine, M. S. Buchsbaum, J. Stanley, S. Lottenberg, L. Abel, and J. Stoddard, Selective reductions in pre-frontal glucose metabolism in murderers, *Biological Psychiatry* 36 (1994):365–373.
21. *People v Jones* (620 N.Y.S.2d 656); 1994 N.Y. App. Div.
22. L. A. Farwell and S. S. Smith, Using brain MERMER testing to detect concealed knowledge despite efforts to conceal, *Journal of Forensic Sciences* 46, no. 1 (2001): 1–9.

23. D. D. Langleben, L. Schroeder, J. A. Maldjian, R. C. Gur, S. McDonald, J. D. Ragland, C. P. O'Brien, and A. R. Childress, Brain activitiy during simulated deception: An event-related functional magnetic resonance study, *Neuroimage* 15, no. 3 (2002): 727–732.

24. *People v. Weinstein* (591 N.Y.S. 2d 715); (Sup. Ct. 1992).

25. *State of Iowa v. Terry Harrington* (284 N.W.2d 244; 1979 Iowa Sup.).

26. J. W. Evans, Functional magnetic resonance images and lie detection, http://www.law.uh.edu/healthlawperspectives/HealthPolicy/021231Functional.html, access date: January 31, 2005.

27. K. H. Heilman, The neurobiology of emotional experience, in *The neuropsychiatry of limbic and subcortical disorders*, ed. S. Salloway, P. Malloy, and J. L. Cummings, 133–142 (Washington, DC: American Psychiatric Association, 1997).

28. R. Adolphs, D. Tranel, H. Damasio, and A. Damasio, Fear and the human amygdala, *Journal of Neuroscience* 5 (1995): 5879–5892.

29. A. R. Damasio, *Descartes' error.* (Netcong, NJ: Penguin Putman Publications, 1994).

30. J. LeDoux, The emotional brain, fear, and the amygdala, *Cellular and Molecular Neurobiology* 23, nos. 4–5 (2003): 727–738.

31. E. W. Clayton, Ethical, legal, and social implications of genomic medicine, *New England Journal of Medicine* 349, no. 6 (2003): 562–569.

32. S. M. Mariani, Neuroethics: How to leave the cave without going astray, *Medscape General Medicine* 5, no. 4 (2003): 33.

33. D. L. Schacter, R. L. Buckner, and W. Koutstaal, Memory, consciousness, and neuroimaging, *Philosophical Transactions of the Royal Society of London B* 353 (1998):1861–1978.

34. M. C. Anderson, K. N. Ochsner, B. Kuhls, J. Cooper, E. Robertson, S. W. Gabrieli, G. H. Glover, and J. D. E. Gabrieli, Neural systems underlying the suppression of unwanted memories, *Science* 303 (2004): 232–235.

35. K. H. Rothenberg and S. F. Terry, Human genetics: Before it's too late—addressing fear of genetic information, *Science* 297 (2002): 196–197.

36. T. Canli and Z. Amin, Neuroimaging of emotion and personality: Scientific evidence and ethical considerations, *Brain and Cognition* 50, no. 3 (2002): 431–444.

37. T. Canli, H. Siver, S. L. Whitfield, I. H. Gotlib, and J. D. E. Gabrieli, Amygdala response to happy faces as a function of extraversion, *Science* 296 (2002): 2191.

38. D. Mobbs, M. D. Greicius, E. Abdel-Azim, V. Menon, and A. L. Reiss, Humor modulates the mesolimbic reward centers, *Neuron* 40, no. 5 (2003): 1041–1048.

39. R. F. Weir, S. C. Lawrence, and E. Fales, *Genes and human self-knowledge: Historical and philosophical reflections on modern genetics* (Iowa City: University of Iowa Press, 1994).

40. J. Savulescu, Predictive genetic testing in children, *Medical Journal of Australia* 175, no. 7 (2001): 379–381.

41. A. J. Golby, J. D. E. Gabrieli, J. Y. Chiao, and J. L. Eberhardt, Differential responses in the fusiform region to same-race and other-race faces, *Nature Neuroscience* 4 (2001): 845–850.

42. E. A. Phelps, C. J. Cannistraci, and W. A. Cunningham, 2003. Intact performance on an indirect measure of race bias following amygdala damage, *Neuropsychologia* 41, no. 2 (2003): 203–208.

43. J. A. Richeson, A. A. Baird, H. L. Gordon, et al., 2003. An fMRI investigation of the impact of interracial contact on executive function, *Nature Neuroscience* 6, no. 12 (2003): 1323–1327.

44. Scanning the social brain (editorial), *Nature Neuroscience* 6, no. 12 (2003): 1239.

45. M. D'Esposito, L. Y. Deouell, and A. Gazzaley, Alterations in the BOLD fMRI

signal with aging and disease: A challenge for neuroimaging, *Nature Reviews Neuroscience* 4 (2003): 863–872.

46. R. C. Gur, L. H. Mozley, P. D. Mozley, et al., Sex differences in regional cerebral glucose metabolism during a resting state, *Science* 267 (1995): 528–531.

47. A. Beaulieu, Images are not the (only) truth: Brain mapping, visual knowledge and iconoclasm, *Science, Technology and Human Values* 27 (2002): 53–87.

48. A. Mauron, Renovating the house of being: Genomes, souls, and selves, *Annals of the New York Academy of Sciences* 1001 (2003): 240–252.

49. D. Kennedy, Neuroethics: An uncertain future, *Proceedings of the Society for Neuroscience Annual Meeting, New Orleans, Louisiana*, 2003.

50. J. D. Greene, R. B. Sommerville, L. E. Nystrom, et al., An fMRI investigation of emotional engagement in moral judgment, *Science* 293 (2001): 2105–2108.

51. J. D. Greene, From neural "is" to moral "ought": What are the moral implications of neuroscientific moral psychology? *Nature Reviews Neuroscience* 4 (2003): 847–850.

52. J. Moll, R. de Oliviera-Souza, I. Bramati, and J. Grafman, Functional networks in emotional and nonmoral social judgments, *Neuroimage* 26 (2002): 696–703.

53. N. S. Ward and R. S. Frackowiak, Towards a new mapping of brain cortex function, *Cerebrovascular Disease* 17, Suppl 3 (2004): 35–38.

54. E. Mayr, *Toward a new philosophy of biology* (Cambridge, MA: Harvard University Press, 1988).

55. E. Racine, *Pourquoi et comment tenir compte des neuroscience en 'ethique? Esquisse d'une approche neurophilosophique 'emergentiste et interdisciplinaire* (Why and how to take into account neuroscience in ethics? Toward an emergentist and interdisciplinary neurophilosophical approach) (Laval Theologique and Philosophique, in press).

56. S. H. Koslow and S. E. Hyman, Human brain project: A program for the new millennium, *Journal of Biology and Medicine* 17 (2000): 7–15.

57. A. R. Hariri and D. R. Weinberger, Functional neuroimaging of genetic variation in serotonergic neurotransmission, *Genes, Brain, Behavior* 2, no. 6 (2003): 341–349.

58. S. Hammann and T. Canli, Individual differences in emotion processing, *Current Opinion in Neurobiology* 14, no. 2 (2004): 233–238.

59. J. Kulynych, Psychiatric neuroimaging evidence: A high tech crystal ball? *Stanford Law Review* 49 (1997): 1249–1270.

60. Committee to Review the Scientific Evidence on the Polygraph, *The polygraph and lie detection* (Washington, DC: National Academy Press, 2003).

61. J. Illes, A fish story: Brain maps, lie detection and personhood, *Cerebrum, Special Issue on Neuroethics* (New York: Dana Press, 2005).

62. S. Michie, J. Weinman, J. Miller, et al., Predictive genetic testing: High risk expectations in the face of low risk information, *Journal of Behavoral Medicine* 25, no. 1 (2002): 33–50.

63. J. F. Merz, D. Magnus, M. K. Cho, et al., Protecting subjects' interests in genetics research, *American Journal of Human Genetics* 70, no. 4: 965–971.

64. A. Mauron, Is the genome the secular equivalent of the soul? *Science* 291 (2001): 831–832.

65. F. Crick, *The astonishing hypothesis: The scientific search for the soul.* (London: Simon and Schuster, 1995).

66. D. Nelkin and L. Tancredi, *Dangerous diagnostics: The social power of biological information* (New York: Basic Books, 1989).

67. J.-P. Changeux, *Neuronal man*, trans. Laurence Garey (Princeton, NJ: Princeton University Press, 1997).

68. G. Lanteri-Laura, Examen historique et critique del' éthique en neuropsychiatrie, dans le domaine de la recherché sur le cerveau et les thérapies (Historical and critical examination of neuropsychiatry in brain research and therapy), in *Cerveau et psychisme humains: Quelle e'thique?* (The human brain and psyche: Ethical considerations) ed. Gérard Huber (Paris: John Libbey, 1996).

69. L. O. Gostin, Ethical considerations of psychosurgery: The unhappy legacy of the pre-frontal lobotomy, *Journal of Medical Ethics* 6, no. 1 (1980): 149–156.

70. J. Illes, ed., Ethical challenges in advanced neuroimaging, *Brain and Cognition* 50, no. 3 (New York: Academic Press, 2002).

71. J. C. Gore, R. W. Prost, and W. R. Hendee, Functional MRI is fundamentally limited by an inadequate understanding of the origin of fMRI signals in tissue, Point/Counterpoint, *Medical Physics* 30, no. 11 (2003): 2859–2861.

72. R. H. Blank, *Brain policy: How the new neurosciences will change our lives and our politics* (Washington, DC: Georgetown University Press, 1999).

73. R. Wolpe, The neuroscience revolution, *Hastings Center Report* (July–August 2002).

74. B. Morris, 1994. *Anthropology of the self. The individual in cultural perspective* (Pluto Press, CO: Neuroscience and the Law, 2002); Proceedings from "Neuroscience and the Law," Royal Institution, London, http://www.abc.net.au/rn/science/mind/s598783.htm.

75. C. Taylor, *Sources of the self: The making of modern identity.* (Cambridge, MA: Harvard University Press, 1989).

76. W. J. Winslade and J. W. Rockwell, Bioethics, *Health Law News* (Health Law and Policy Institute, University of Houston Law Center, 2002), 1.

77. Global News Wire—Asia Africa Intelligence Wire, The resonance of the mind, November 20, 2003.

78. A. Roskies, Neuroethics for a new millennium, *Neuron* 35, no. 1 (2002): 21–23.

79. K. R. Foster, P. R. Wolpe, and A. L. Caplan, Bioethics and the brain, *IEEE Spectrum* (2003): 34–39.

80. E. E. Kim, Targeted molecular imaging, *Korean Journal of Radiology* 4, no. 4 (2003): 201–210.

81. M. L. T. Stevens, *Bioethics in America: Origins and cultural politics* (Baltimore, MD: Johns Hopkins University Press, 2000). Quoting G Pontecorvo, Prospects for genetic analysis of man, in *The control of human heredity and evolution*, ed. T. M. Sonneborn, 81–82 (New York: Macmillan, 1965).

82. Z. Lynch, Neurotechnology and society, *Annals of the New York Academy of Sciences* 1013 (2004): 229–233.

83. M. Farah, J. Illes, R. Cook-Deegan, H. Gardner, E. Kandel, P. King, E. Parens, B. Sahakian, and P. R. Wolpe, Neurocognitive enhancement: What can we do? What should we do? *Nature Reviews Neuroscience* 5 (2004): 421–425.

84. E. Racine, J. Illes, *Is neuroethics the heir of the ethics of genomics?* (Calgary, Alberta: Canadian Bioethics Society, 2004).


OTHER REFERENCES FOR THIS ARTICLE WERE:

A. W. J. Gehring, A. Karpinski, and J. L. Hilton. 2003. Thinking about interracial interactions, *Nature Neuroscience* 6, no. 12 (2003): 1241–1239.

B. H. Gardner, There's a sucker in every prefrontal cortex (opinion editorial), *New York Times*, November 30, 2003, 26.

C. T. Klingberg, M. Hedehus, E. Temple, T. Salz, J. D. Gabrieli, M. E. Moseley, and

R. A. Poldrack, Microstructure of temporo-parietal white matter as a basis for reading ability: Evidence from diffusion tensor magnetic resonance imaging, *Neuron* 25, no. 2 (2000): 257–259.

JUDY ILLES, PH.D., is senior research fellow at the Stanford Center for Biomedical Ethics and director of the program in neuroethics at Stanford University, Palo Alto, California.

ERIC RACINE, PH.D., is director of the Neuroethics Research Unit at the Institut de Recherches Cliniques de Montreal (IRCM).

*Chapter 14*

# Brains, Genes, and the Making of the Self[*]

## *Lynette Reid* and *Francoise Baylis*

In an effort to shed light into the "black box" of the human brain, the U.S. Congress proclaimed the 1990s the "Decade of the Brain." The funding that flowed from this declaration encouraged a convergence of disciplines and knowledge in basic and applied neuroscience, as well as collaboration between neuroscientists and philosophers of mind. At the end of this decade, following dramatic advances in our technical ability to study the brain and newfound enthusiasm for various techniques of surgical and pharmaceutical manipulation of the brain, William Safire called for a new discipline—neuroethics—to examine critically some of the ethical issues with brain interventions.[1,2]

The growing neuroethics literature includes discussions of many ethical issues that bear a family resemblance to traditional bioethical problems and the problems of the ethics of genetics; other ethical issues, however, are much less familiar. In "Imaging or Imagining: A Neuroethics Challenge Informed by Genetics," Illes and Racine[3] survey the vast emerging landscape of neuroethics and mark it with indications of similarity to, and difference from, the realm of ethics in genetics. What they gain in breadth, however, they lose in depth, one place where this is evident is in their treatment of privacy and confidentiality.

[*]From *American Journal of Bioethics* 5, no. 2 (2005): 21–23.

Privacy can be distinguished from confidentiality insofar as privacy is (claimed to be) a matter of discretion in the process of collecting information about or from a person, while confidentiality concerns a promise to respect the wishes of that same person in regards to storage and possible disclosure of such information. For example, in the context of genetics, I may respect your privacy by gathering information (and samples) with your permission (respecting your wish not to have certain tests performed), communicating test results with you alone, and doing so in a location adequately protected from other ears. I would then respect your confidentiality by not sharing the results of these tests with others without your permission. The potentially harmful consequences of unwanted disclosure of personal genetic information (such as a diagnosis of genetic illness, an increased susceptibility to genetic disease, or a finding that challenges an assumption of kinship) are familiar and include the risks of discrimination in health insurance, life insurance, employment, and education, as well as the risk of stigmatization. Such risks extend to family and community when there is disclosure without consent of all parties involved.

By analogy, in neuroimaging there are issues of privacy and confidentiality related to managing information about diagnosis of disease and unanticipated clinical findings. Discussions of such issues would not constitute particularly novel or interesting work in bioethics, however, if they did no more than rehearse familiar arguments and assert familiar ethical obligations but for a new audience (such as radiologists and neurologists). What is novel and particularly interesting about privacy and confidentiality with neuroimaging (and for which there is no adequate precedent in the ethics and genetics literature) is the predicted—and unprecedented—access to human thought.

The assumption of privacy in relation to genetic information is that others should not collect or access "our information" without our knowledge or permission. In relation to thought, the starting assumption is that, intentions aside, others cannot "read our thoughts" and so invade our privacy. Indeed, it is not uncommon for general discussions of privacy to refer to "privacy of thought" as a fundamental paradigm of privacy. It follows that the potential violation of this realm of privacy—and here we do not mean its contingent violation in a particular instance but the technological and cultural violation of this old impossibility—is the fundamental issue of privacy in neuroethics. This critical point is neither identified nor explored by Illes and Racine. Indeed, the text that follows under the heading "privacy of human thought" bears little relation to the issue of privacy. While there is a broad gesture to the "gnarly" character of privacy, the discussion focuses on imaging studies of moral reasoning and emphasizes problems with interpreting socially and culturally laden data.

The "gnarly" issue of privacy has no meaningful precedent in prior "ethical reflection on genetics" because "genes" and "thoughts" are fundamentally different constructs. Our genes are part of our biology and one aspect of our lineage. Our thoughts—that is, our reasoning, our motivations, our attitudes, beliefs, and values—are our selves and our personal identity. Our "brains are us" in ways that our genes never could be. In parallel, our thoughts are "private" in ways that are distinct from our shared genetic heritage. To be clear, we are neither hereby advocating a metaphysical identity,[4] nor reasserting a discredited form of metaphysical privacy.[5] Rather, we are advocating attention to important distinctions between genetic data and thought processes, all the while recognizing that both are (in different ways) relevant to identity formation. In our view, genes certainly have an influence on who we are, but our thoughts are significantly more central to the self.

Our genes, inherited from our biological parents, are in an important sense not "ours" to begin with: they are the endowment of our parents—though, by the same token, they are as little "our parents'" as they are "ours." Indeed, they are information about "who we are" but only in the sense of who "our people" are: who our parents are, our grandparents, our siblings, our descendents. To be sure, genes may be of considerable personal significance, but there is a very real sense in which they are not exclusively personal information—which is why we sometimes speak of genomes and genetic data as familial or communal information.[6] The gene that is read in testing or sequencing is what it is mostly because of my heritage and ancestry. It becomes part of my personal identity through the stories I tell, and others accept, about my genes. In sharp contrast to the "family endowment" of genes, our thoughts, while they may be subject to parental influence, are crucially a function of the influences of many others, including (centrally) ourselves. Our thoughts are the basis of all our stories, which in turn are the basis of our socially constructed identities. Our lives are collections of stories "acquired, refined, revised, displaced and replaced"[7] over time through introspection and continued lived experience. Our stories (products of our thoughts and our experiences) are constructed and maintained through intimate personal relations as well as more-distant social relations of mutual recognition. These stories, crucially expressed in our thoughts, tell us who we are, where we are from, where we have been, and where we are going in ways that are entirely different from what our genes alone can tell us.[8]

Thoughts (including sometimes, for some people, thoughts about our genes) are both the raw materials of our stories and the language in which we constitute and tell ourselves our stories, as well as the ways in which we prepare these stories for communication to others. The relevance of the brain to identity is not, as some philosophers argue,[4] that we are the contents of our consciousness. Rather, in constructing our identity, our personal life narrative, we arrange and order our experiences and thoughts, giving greater weight to some thoughts and discarding others. We rework (for instance) a gut response into a step in a narrative—a narrative that could go in many different directions, partly subject to choice, partly subject to circumstance, partly subject to desire and need, and in no small measure subject to the responses of others. A crucial aspect of the development of these personal narratives is the play of their communication: the process whereby thoughts, more or less worked over internally, are tested against the responses of others, who may accept, reject, confirm, or remain neutral to our stories. We register these responses and respond to them in turn. The private phases of this process no doubt are fundamental to identity formation, as Sissela Bok has argued.[9] The prospect that one or more neurotechnologies (used alone or in combination) might bypass this play of communication and, one day, potentially expose private moments in the process of identity formation portends a fundamental shift in personal identity. Illes and Racine suggest that "while the ethics of genetics provides a legitimate starting point . . . for tackling ethical issues in neuroimaging, they do not suffice" (this issue). We agree that a careful examination of ethical issues in neuroimaging will call on our thinking to be alive to dimensions untouched by the ethics of genetics. In our view, however, it is not at all clear that attention to ethical issues in genetics is a useful, let alone legitimate, starting point for tackling ethical issues in neuroimaging. Privacy is a case in point. Whether or not we give permission (informed consent, in traditional bioethics) for our thoughts to be read "directly"—in a way that bypasses the identity-forming play of communication—this very possibility will constitute a fundamental reconfiguration of personal identity. There simply is no equivalent threat to personal identity with the rapidly expanding array of genetic technologies, even as these technologies increase the risk that intimate information about our genome will be in the public domain.

REFERENCES

1. W. Safire, Stem cell hard sell, *New York Times*, 2001, A19.

2. W. Safire, Never retire. *New York Times*, 2005, A19.

3. J. Illes and E. Racine, Imaging or imagining? A neuroethics challenge informed by genetics, *American Journal of Bioethics* 5, no. 2 (2005): 5–18.

4. D. Parfit, *Reasons and persons* (New York: Oxford University Press, 1984).

5. L. Wittgenstein, *Philosophical investigations* (Oxford, UK: Blackwell, 1953). Copyright © Taylor & Francis, Inc. DOI: 10.1080/15265160590960401

6. F. Baylis and J. S. Robert, Radical rupture: Exploring biological sequelae of germ-line genetic intervention, in *A dividing line? Towards an ethics of germline gene therapy*, ed. J. Rasko, G. O'Sullivan, and R. Ankeny (Cambridge, UK: Cambridge University Press, 2005.)

7. M. Walker, *Moral understandings: A feminist study in ethics* (New York: Routledge, 1998).

8. F. Baylis, Black as me: Narrative identity, *Developing World Bioethics* 3, no. 2 (2003): 142–150.

9. S. Bok, *Secrets: On the ethics of concealment and revelation* (New York: Pantheon, 1983).

FRANCOISE BAYLIS, PH.D., is professor and Canada Research Chair in Bioethics at Dalhousie University, Halifax, Nova Scotia, Canada.

LYNETTE REID, PH.D., is assistant professor in the Department of Bioethics at Dalhousie University, Halifax, Nova Scotia, Canada.

*Part IV*

# Free Will, Moral Reasoning, and Responsibility

ADVANCES IN NEUROSCIENCE have shed considerable light on the neural correlates of our thought and behavior. These advances raise questions about free will, moral reasoning, and moral and legal responsibility. Most philosophers debating free will have focused on factors external to persons. Some have worried that natural laws and past events jointly entail a unique causal chain into the future, which presumably would mean that our actions could not be free. Other philosophers insist that free will and responsibility require that we be the ultimate source of our actions, which cannot be true if our actions are caused by prior events. Still others maintain that our actions are free when they do not result from coercion, compulsion, or constraint. Neuroscience shifts questions about free will and responsibility from such external references to an internal focus. The fundamental question is not what natural laws or past events allow us to do, but what we can do given the relation between the brain and the mind. What role does the brain play in our capacity for moral reasoning? Is brain science compatible with our conviction that we are free and responsible agents? Or does brain science suggest that this conviction is an illusion?

Free will, moral reasoning, and moral and legal responsibility rest on mental capacities that enable us to control our behavior by guiding attention, thought, emotion, and action in accord with intentions and goals. The authors in this section explore how the brain grounds the relevant mental capacities, the extent to which persons can control their behavior, and how knowledge about the brain may influence moral reasoning and moral and legal judgments about our actions.

In "The Neural Basis of Social Behavior: Ethical Implications," Antonio Damasio claims that there is no moral center in the brain. Rather, ethics depends on complex interactions among different systems in the brain. What Damasio calls the "bioregulation" of ethics is not just a function of the brain structures alone and thus is not a reductionist concept. The neuronal systems that ground our capacity to act in accord with rules and norms emerge from the combined influence of evolution, gene expression, and the needs of individuals within a social environment. Damasio cites cases of individuals with damage to certain regions of the brain making them unable to experience emotions, such as shame, guilt, and sympathy, that are necessary for moral sensibility. This was the fate of Phineas Gage, who sustained a brain injury in 1848 from an explosion while working on a railroad in Vermont. The injury transformed Gage from a rational and morally upright individual to one who displayed impulsive, irrational, and socially inappropriate behavior. Damasio distinguishes between patients with brain damage later in life who lost the ability to conform to social rules and control their behavior and patients whose brain damage earlier in life made them unable to learn these rules in the first place. He concludes by raising the hypothetical question of whether ethics would have evolved if evolution had somehow made us incapable of having and expressing social emotions.

In "Neuroscience: Reflections on the Neural Basis of Morality," Patricia Smith Churchland proposes that questions about rational free choice be framed in terms of being in control or out of control of our behavior. She defines "control" in terms of a "parameter space," which refers to a threshold of mental capacity depending on certain structural and functional features of the brain. We may choose to set the threshold separating the controllable from the uncontrollable at the level of neurotransmitters such as serotonin or dopamine. Or we could set the threshold at the level of interaction between the ventromedial frontal cortex and the amygdala in the limbic system. Churchland notes that further progress is necessary in understanding the dynamic properties of neural networks underlying moral reasoning and decision making for us to have a clearer sense of when a person is in or out of control of his or her behavior. She asserts that at the neuronal level the brain is a causal machine and brain events cause our behavior. But she adds that this by itself does not imply that we cannot be responsible for what we do. The determination of criminal responsibility, for example, depends on social and institutional factors in addition to the brain.

In "My Brain Made Me Do It," Michael Gazzaniga critically examines and rejects the idea that the brain determines our behavior and makes our experience of free will and moral responsibility an illusion. Distinguishing brains from persons, Gazzaniga claims that brains are determined devices that cannot threaten or explain away human freedom and responsibility. Persons are free by virtue of their interaction in the social world, and responsibility is a social construct that is also the product of this interaction. Gazzaniga points out the uncertainty of correlations between certain brain states and certain types of behavior. A brain disorder involving neural systems associated with reasoning and decision making by itself is not enough to excuse a person from responsibility for his or her actions. Gazzaniga claims that neuroscience can offer little to our understanding of free will and responsibility because the social aspects of these notions do not exist in the neural structures of the brain.

Expressing some of the same skepticism as Gazzaniga, Stephen Morse considers the possible influence of neuroscience in law in "New Neuroscience, Old Problems: Legal Implications of Brain Science." Morse asserts that the legal concept of the person as an agent capable of acting intentionally and for reasons cannot be reduced to biological causes in the brain. Just because some abnormal biological condition played a causal role in a person's behavior does not mean that the condition compelled the person to act. Causation is not equivalent to compulsion or lack of capacity for rationality and intentionality, and thus it cannot negate legal responsibility. Morse does admit, however, that neuroscience will discover conditions that can compromise rationality, and in this regard it may help to adjudicate claims for excuse and mitigation. At the same time, he emphasizes that unless neuroscience shows that no one is capable of minimal rationality—which he thinks is highly implausible—criteria of legal responsibility will be intact. Morse predicts that the influence of neuroscience in the legal domain will not be as widespread or profound as some might claim.

In "Moral Cognition and Its Neural Constituents," exploring the neurobiological basis of moral reasoning and decision making, William Casebeer offers a more positive view of the relation between brain and mind. He claims that the neurobiological, psychological, and normative aspects of morality have co-evolved due to widely distributed networks in cortical and subcortical regions of the brain. The neural mechanisms underlying moral cognition are also sensitive to social context. Casebeer points out that moral norms may not arise from purely natural processes and may not be a natural kind—a biologically based concept—in the same way that other cognitive phenomena are. He notes that of all the ethical theories in the history of philosophy, Aristotle's pragmatic virtue theory is the most compatible with neuroscience. According to virtue theory, particular actions are evaluated as reflections of a person's character, or a general disposition to act in a certain way. Casebeer claims that further experimental work in neuroscience will yield a better understanding of how ethical reasoning is embodied in the brain.

Joshua Greene continues Casebeer's discussion of the possibility of morality as a natural kind in "From Neural 'Is' to Moral 'Ought': What Are the Moral Implications of Neuroscientific Moral Psychology?" Although he is skeptical of attempts to equate moral properties with natural properties, Greene believes that knowledge of neurobiology has the potential to profoundly influence our moral thinking. He claims that our moral judgments are based largely on intuitions and considers how these intuitions have evolved in terms of their psychological and neural bases. To explore the neural mechanisms underlying our moral intuitions and judgments, Greene and colleagues conducted a brain imaging study that measured brain activity in participants as they responded to different moral dilemmas. The results of the study help to explain why we act or refrain from acting in situations where we are confronted with difficult choices. In particular, brain imaging may shed light on what is known as the "Trolley Problem." In this hypothetical scenario, one must choose between allowing a runaway trolley to kill five people on one track or redirect the trolley to a different track and thereby kill one person. Images showing increased activity in brain regions mediating this choice may explain our moral reasoning about permitting greater or lesser harms. They also help to explain why we have evolved altruistic instincts to help others in need, and why these instincts are stronger and why we are more likely to help others the more familiar we are with them.

The articles in this section display different opinions on what neuroscience can or cannot tell us about our capacity for free choice and moral reasoning. Significantly, none of the authors suggests that the brain determines our behavior or that our belief in free will, moral decision making, and responsibility is an illusion. Even with increasing knowledge of the brain, it is not neuroscientists but society that will decide how information about the brain should be used in judging whether a person acted freely, was capable of moral reasoning, and could be responsible for his or her actions.

*Chapter 15*

# The Neural Basis of Social
# Behavior: *Ethical Implications**

*Antonio Damasio*

Ethical behaviors are a subset of *social* behaviors; it's not possible to conceive of ethics outside the concept of society. And because there are nonhuman societies, the essence of ethical behavior does not begin with humans. There is evidence from primates and other species—from vampire bats to wolves—of conduct that appears, to our cultivated eyes, as moral conduct. Altruism, censure, recompense for certain actions, and compassion are evident examples in nonhuman and even nonprimate species.

Moreover, because the expression of ethical behavior is associated to a great extent with "social" emotions—such as sympathy, shame, embarrassment, guilt, and the form of social anger we call moral indignation—there is also evidence for such emotions not only in humans but in other species. This realization—it is a fact, not a hypothesis—may come as a shock for those who believe that ethical behavior is a distinctive human trait. As if it were not enough to have Copernicus tell us we are not at the center of the universe, Darwin tell us that we have humble origins, and Freud tell us that we are not masters of our own house, now we are being told that even in the realm of ethics there is forerunner behavior.

For those who feel humiliated by this revelation, however, please note that human ethics has a degree of complexity that makes it distinctly hu-

---

*From *Neuroethics: Mapping the Field* (Dana Press, 2002), pp. 14–20.

man. The refinement is human; the codification we have enacted on ethical behaviors is human; the narratives we have built around the situation are obviously human. It is only the basic behaviors behind the situations that are not uniquely human.

This fact—the relation of seemingly moral conduct with the emotions—has inspired in my work the following hypothesis: *The construction we call ethics began with the edifice of bioregulation.* By "bioregulation" I mean the set of automated mechanisms that allows us to balance our metabolism, maintain life, and achieve well-being, and that also produces drives and motivations, emotions of diverse kinds, and feelings.

Note that I am not reducing ethics to a simple matter of evolution, or of gene transmission or expression, or of brain structures alone. As conscious, intelligent, and creative creatures inhabiting a cultural environment, we humans have been able to shape the rules of ethics, shape their codification into law, and shape the application of the law into what we call justice. And we continue to do so. In fact, one purpose of conferences like this is to discuss ways in which we may shape the rules of ethics in keeping with the new problems posed by advances in science and technology.

So ethics is not just about evolution, even if I am suggesting that it starts with evolution. And it is not just about the brain. Culture does the rest, and the rest may be most of it.

Similarly, elucidating the biological mechanisms underlying ethics does not mean that those mechanisms, or their dysfunction, ensures certain behaviors. There certainly are determinants of behavior that come from our evolutionary biology—from the way our brains get set, and from the ways they get set both by genes and by the culture in which we develop—but there is still a degree of freedom that allows an individual to intervene. As far as I can see, there is free will—though not for all behaviors, and not for all conditions, and sometimes not to the full extent in any condition.

Unsurprisingly, I believe that what we call ethics today depends on the workings of certain brain systems. But now come a few additional disclaimers.

First of all, I am talking about *systems*, not centers. On a number of occasions I've pleaded with science writers not to talk about a brain "center" of anything—not of language, or memory, or morality—but the plea is often disregarded by the headline editor. Talking of "centers" gives the false impression that there's some kind of clearinghouse in the brain, in charge of a certain set of behaviors. Nothing could be further from the truth. We are in fact dealing with systems made up of several components that maintain complex interactions among themselves. It is only when those systems operate, in a given context, that certain kinds of behaviors emerge, along with certain kinds of cognition related to those behaviors.

The second disclaimer is just as important: Although certain systems in the brain are clearly related to moral behavior, they are not set by genes to operate for the purposes of morality and ethics. These systems are indeed dedicated, for example, to memory of particular kinds, or to decision making, or to creativity. But they respond to certain needs of an individual living in a social collective—to help the person harmonize with the conditions in which he or she is living—and these needs and conditions arise independently from what evolution has equipped us with.

The upshot is that ethics is a wonderful by-product. We could not have it if we did not have a capacity to learn, if we did not have a capacity to recall, if we did not have a capacity to imagine, reason, and create. But I doubt that there is a dedicated moral system in the brain, and certainly no moral center.

This is not to say that damage to the brain will not result in moral impairment. We now have a large collection of data on patients who are Gagelike in some way. They learn, they recall, they preserve their language, they manage logic quite nicely. And yet, even if they recall social conventions and rules of ethical behavior, they are no longer able to apply them effectively. Though they know what is "right" and "wrong" and "good" and "bad," they are impaired in a whole class of social emotions such as embarrassment, shame, guilt, and sympathy. This, in turn, impairs the decision-making mechanism that is needed for appropriate social management, and subsequently impairs any new learning of this sort of social knowledge.

The cases I am referring to are of adults who have been upstanding members of society up to the point when their brains sustained damage. What happens if the brain is damaged much earlier? Recently we reported the cases of two patients who had sustained lesions of the same sector of the brain but at very young ages—one of them in the second year of life, the other even before the end of the first year.

We were able to study these patients in their twenties. In many respects they were entirely comparable in their behaviors to those who had sustained their lesions in adulthood, but there was an important distinction: they had not been able to learn the social conventions and ethical rules the adults had learned. It was not just a matter of not acting on the rules—they had not learned the rules to start with. And this, predictably, had led to a much worse quality of behavior.

So I would like to close by posing a question. Just imagine that by a quirk of fate, evolution had gone in a different way and humanity had come into being with the kind of losses that the Gage-type patients have sustained, and that we would not have the possibility of expressing the social emotions. What would the world have been like? Would ethics have ever developed? Would we be here to tell?

ANTONIO DAMASIO, M.D., PH.D., holds the David Dornsife Chair in Neuroscience and is professor of psychology and director of the Brain and Creativity Institute of the USC College of Arts and Letters at the University of Southern California.

*Chapter 16*

# Neuroscience: *Reflections on the* *Neural Basis of Morality**

## Patricia Smith Churchland

When we think about "the self " from a neurobiological perspective, it appears that we really should be talking not about one particular thing—some single entity that is the self—but rather about a multidimensional affair. "The self " is a set of capacities that involve not only representation of the body itself but also representation of internal aspects of the brain—the brain's mental life. This set encompasses such disparate things as our autobiography, what we currently feel about our body configuration, where we are in space and time, where we rate in the social order, and the status of our relations to other humans and nonhumans.

It has been argued, particularly by Hanna and Antonio Damasio, that the platform for an animal's most basic self-representational capacities is in the brain stem. This circuitry handles the fundamental problem of coordinating one's needs with one's internal milieu so that the body can move appropriately—to feed, flee, fight, or reproduce. Movement decisions must be elaborated so that you aren't feeding when you should be fleeing and to ensure that you do not try to do incompatible things.

Also within that basic brain-stem platform—in mammals at least, but probably in birds and other vertebrates as well—is a capacity to do motor planning. Organisms need to do some of the figuring out of how to solve a

*From *Neuroethics: Mapping the Field* (Dana Press, 2002) pp. 20–26.

particular motor problem offline—to conduct much of the trial-and-error business in a safe environment—namely, within the brain itself. This sort of motor planning appears to involve the development of an inner model. The work of David Wolpert and also of Rick Grush suggests that increasingly fancy inner modeling gives us the basis for imagining what can happen not only in a complex motor situation but in a social one as well.

Here's a brief sketch of what an inner model might look like: A goal state will be specified. It might be something as simple as "Can I reach that plum?" or something slightly more complicated, like "How do I hide from that predator?" The inner model basically proposes a quick and dirty suggestion about how best to achieve the goal. It then sends a signal to the "emulator," which essentially says, "If you do that, these will be the consequences." This information is then cycled back to the inner model, which can upgrade the initial solution: "Well, then, let's make a modification." The new plan will go to the emulator, which may then suggest consequences that are more self-serving. Ultimately—after this kind of back-and-forth iteration converges on a satisfactory solution—a signal is sent to the body, and there is a behavioral outcome as the plan is executed.

In brief, the wiring yields self-simulation with respect to the things in the world. But some of those things—at least for those of us who are social creatures—will entail the simulation of other *selves*. What will that organism do if I display anger? What will it do if I chase it? If I try to eat it? The simulation within this initially rather simple emulator structure can get very elaborate, as wiring permits.

Emulators may also, of course, involve self-control, so that the organism can make a decision that best serves its interests. Evaluation of self-interest will take into account, of course, not only immediate needs but also the long-term consequences of each considered action. It is perhaps not too surprising that one can conceive of the development of *conscience* within this very general structure of the emulator or inner model. In order for an animal to come to a fast decision about whether to do one thing or to reject that option and try to formulate another, relevant perception and relevant memory have to be fed into the emulator, and relevant computation must ensue. And all that seems to have a lot to do with, and to be greatly guided and optimized by, the presence of feelings generated in response to the inner modeling of an option. To a first approximation, what we call conscience is the negative feeling evoked by emulation of a social action.

How exactly any of this is done remains puzzling. In particular, we have little idea at this stage of the exact nature of the relevant computation. For example, instead of running through all *possible* options, which the organism clearly does not do, the brain manages to confront and deliberate on only the *pertinent* options. How "relevance computation" works is not well understood.

My sense is that the details of decision making, of choice, of acquisition of character and temperament, and of development of such things as moral character are going to elude us until we have made more progress on certain fundamentals of neuroscience—namely, the dynamic properties of neural networks. At present, there is an enormous gap between what we know about how *neurons* work, and what we know about *networks* of neurons.

Still, let me say just a little bit more about neuroconscience, though necessarily at a very general level. Whatever else it is, if the neuroconscience is connected to the emulator, it has to somehow also be connected in a very profound way to the reward-and-punishment system. It must involve simulation of injunctions and warnings, in the way that Socrates said that he heard a little voice telling him not to do immoral things. It must also involve what's sometimes called the theory of mind—the recognition of others as having beliefs, feelings, and desires. In other words, it involves the manipulation and use of those social emotions that Antonio Damasio talked about. Let me turn now to the related issue of making rational or self-interested choices.

Ultimately, we'd like to have some general understanding of the neural difference between someone who is operating with what we might loosely call free choice and someone who is not. Another way of putting this is that we want to understand the neural difference between someone who, roughly speaking, is *in control* and someone who, also roughly speaking, is *not in control*. We are beginning to understand some of the relevant parameters: levels of serotonin, levels of dopamine, hormones, the wiring between the amygdala and ventromedial frontal structures, leptin concentrations in the blood. For example, low levels of serotonin are associated with reckless behavior in monkeys. Leptin-receptor deficits correlate with obesity. Ventromedial frontal damage correlates with failure to evaluate consequences.

When we come to better understand these parameters and their role in rational choice, even if it's only at a general level, we can begin to think about the in-control versus out-of-control distinction in terms of a "parameter space." That is, each of the parameters (whatever they turn out to be) constitutes an axis in that space. This means we can start characterizing the volume within that parameter space wherein live the in-control brains. The boundary is probably not well defined.

There are undoubtedly many different ways of being in control; different combinations of parameter values will work equally well. Some people may manage to be in control when their serotonin levels are here and their dopamine levels are there, even while they have a rather tenuous connection between the amygdala and ventromedial frontal structures. Others might have different profiles but still be within the in-control volume of the parameter space. The relationships between the parameters are also a target for research.

In the long run, I suspect that we will be able to find general and, ultimately, highly detailed ways of distinguishing between the in-control brain and the out-of-control brain. Notice that in all instances the behavior is *caused* by brain events. At the level of the neuron and the neural network, the brain is a causal machine. Nevertheless, the fact of causality in the brain does *not* imply that there is no responsibility. The determination of responsibility within the criminal justice system depends on many factors, including efficacy of punishment, public safety, and the social importance of retribution.

PATRICIA SMITH CHURCHLAND, B.PHIL, is the UC President's Professor of Philosophy at the University of California, San Diego.

*Chapter 17*

# My Brain Made Me Do It[*]

*Michael Gazzaniga*

Imagine being a juror on a horrific murder case. As a juror you know, or should know, some things about America's judicial system. First, 95% of criminal cases never come to trial; most cases are either dismissed or plea-bargained. The latter resolution is in part driven by the fact that courts tell defendants that should they be found guilty at trial, the punishment will be more severe. Second, there is a huge probability that the defendant is guilty.

When you take your seat in the jury box, you also know you'll have to decide the case with eleven of your peers, people who may not be up on the latest scientific understanding about human behavior. You know most jurors don't buy excuses, facts presented about a defendant in an effort to claim that he or she is exculpable for the crime at hand. Jurors are tough, practical people. That is the profile of the American jury system. Nothing fancy, just twelve people trying to make sense out of a horrible event. Most have never heard the word "neuroscience" or given a moment's thought to the concept of "free will." They are there to find out whether the defendant committed the crime, and if they determine he did, they will probably throw the book at him. Very few juries are asked to consider whether a defendant is exculpable for reasons of insanity, and when they do hear such a defense, they usually don't buy it.

Against this real backdrop of what life is like in the American courthouse,

*From *The Ethical Brain* by Michael Gazzaniga (Dana Press, 2005) pp. 87–102.

a new wrinkle is appearing in the form of the perennial question, Do we as a species have "free will"? Did the defendant carry out the horrible crime freely and by choice, or was it inevitable because of the nature of his brain and his past experiences? As with so many issues where modern scientific thinking confronts everyday realities, the people in the jury box are not rushing to embrace this one. Yet it is my contention that even those tough jurors will have no choice, because someday the issue will dominate the entire legal system.

Brain mechanisms are being explored that help us understand the role of genes in building our brains, the role of neuronal systems in allowing us to sense our environment, and the role of experience in guiding our future actions. We now understand that changes in our brain are both necessary and sufficient for changes in our mind. Indeed, an entire subfield of neuroscience, called cognitive neuroscience, has arisen in recent years to study the mechanisms of this occurrence.

With this reality of 21st-century brain science, many people find themselves worrying about those old chestnuts—free will and personal responsibility. The logic goes like this: The brain determines the mind, and the brain is a physical entity, subject to all the rules of the physical world. The physical world is determined, so our brains must also be determined. If our brains are determined, and the brain is the necessary and sufficient organ that enables the mind, we are then left with these questions: Are the thoughts that arise from our mind also determined? Is the free will we seem to experience just an illusion? And if free will is an illusion, must we revise our concepts of what it means to be personally responsible for our actions?

These dilemmas have been haunting philosophers for decades. But with the advent of brain imaging, neuroscientists are exploring these questions, and, increasingly, the legal world is demanding answers. Defense lawyers are looking for that one pixel in their client's brain scan that shows an abnormality, a predisposition to crime or a malfunction in normal inhibitory networks, thereby allowing for the argument "Harry didn't do it. His brain did it. Harry is not responsible for his actions."

There is evidence that certain brains are more aggressive than others. Whether through neurochemical imbalances or lesions, brain function can become distorted, perhaps explaining certain violent or criminal behavior. Neuroscience also tells us that by the time any of us consciously experiences something, the brain has already done its work. When we become consciously aware of making a decision, the brain has already made it happen.

This raises the question, Are we out of the loop? It is one thing to worry about diminished responsibility due to insanity or brain disease, but now the normal person appears to be on the deterministic hook as well. Should we abandon the concept of personal responsibility? I don't think so. We need to distinguish among brains, minds, and personhood. People are free and therefore responsible for their actions; brains are not responsible.

Neuroscience will offer us some new ways to understand behavior, but ultimately we must realize that even if the cause of an act (criminal or otherwise) is explainable in terms of brain function, this does not mean that the person who carries out the act is exculpable. Based on the modern understanding of neuroscience and on the assumptions of legal concepts, I believe the following axiom: Brains are automatic, rule-governed, determined devices, while people are personally responsible agents, free to make their own decisions. Just as traffic is what happens when physically determined cars interact, responsibility is what happens when people interact. Personal responsibility is a public concept. It exists in a group, not in an individual. If you were the only person on earth, there would be no concept of personal responsibility. Responsibility is a concept you have about other people's actions and they about yours. Brains are determined; people (more than one human being) follow rules when they live together, and out of that interaction arises the concept of freedom of action.

The evidence that suggests that our brains drive our actions involves the way percepts result in movement, activities, actions in life. It also involves the way emotional states in the brain might bias all of our neural networks to make a certain kind of decision, such as those made under stress or sexual excitement. What it does *not* suggest is that brain mechanisms underlie the relationships that exist in a social structure, the rules that enable us to cohabitate, or a rule or value like personal responsibility. Those aspects of our personhood are—oddly—not in our brains. They exist *only* in the relationships that exist when our automatic brains interact with other automatic brains. They are in the ether.

## The Philosophical Stance on Free Will

Philosophers have long debated the nature and existence of free will, a seemingly essential concept if we are to hold and value the idea of personal responsibility. Without getting into the academic details of these views, we can identify two primary and opposing views: that we have free will and that we don't. Those who believe in free will (indeterminists) believe that some *x* factor—whether it's the "ghost in the machine," the soul, the mind, or the spirit—allows us to make choices and determine our actions and even our destiny by acting upon and changing the physical world and our path in it. Those who don't accept free will (determinists) believe we live in a predetermined world—whether it's caused by fate, preordination, or genetic hardwiring—where every action, human and otherwise, is inevitable.

In the rational world of science, the question arises, If determinism is true, what does the determining? Traditionally, genes have been implicated as the predictors of our destiny. Stephen Jay Gould, by no means an advocate for the idea of genetic determinism, explained the theory by stating that "if we are programmed to be what we are [by our genes] then [our] traits are ineluctable. We may, at best, channel them, but we cannot change them either by will, education or culture."[2] Some processes are largely determined by our genes (for example, if someone has the gene for Huntington's disease, he or she will almost certainly contract the disease—"good living, good medicine, healthy food, loving families or great riches can do nothing about it"[3]), but many of our traits are not entirely encoded in our genes. Our environment and chance also play a role in determining our traits and behaviors.

While genes build our brains, it is our brains, actively making millions of decisions per second, that ultimately enable our cognition and behavior. So it would seem that if we are to look at the issue of free will today, the brain is the place to look. Is the brain a deterministic organ, genetically hardwired to carry out actions over which we have no control? Or is the brain—the home to the mind, the ghost in the machine—something capable of free will?

## General Arguments for Free Will

If the brain carries out its work before one becomes consciously aware of a thought, it would appear that the brain enables the mind. This is the underlying idea of the neuroscience of determinism. The argument came to everyone's attention with the work of Benjamin Libet, in the 1980s.[4]

Libet measured brain activity during voluntary hand movements. He found that before we actually move our hand (between 500 and 1,000 milliseconds before), there is a wave of brain activity (the readiness potential). Libet set out to determine "the notorious 'time *t*,'"[5] somewhere in those 500 to 1,000 ms, when we make the *conscious decision* to move our hand.

Libet measured the activity of his subjects' brains using a technique known as event-related potentials (ERPs) while the subjects made a conscious and voluntary hand movement. A subject would stare at a clock, and at the very moment he made the conscious decision to flick his wrist, he would note the position of a black dot (for example, at the 5 or 53 mark) and would then report it to the experimenter. Libet would then correlate that moment with the time at which a readiness potential from the subject's brain waves was recorded.

Libet found that even before "time *t*," when the subject first became consciously aware of his decision to make a hand movement, the subject's brain was active—the readiness potential was present. The time between the onset of the readiness potential and the moment of conscious decision making was about 300 milliseconds. If the readiness potential of the brain begins before we are aware of making the decision to move our hand, it would appear that our brains know our decisions before we become conscious of them.

The evidence therefore seemed to favor the illusion and not the actuality of free will. Libet argued that because the time from the onset of the readiness potential to the actual hand movement is about 500 ms, and it takes 50 to 100 ms for the neural signal to travel from the brain to the hand to actually make the hand move, 100 ms are left for the conscious self to either go with the unconscious decision or veto it. That, he said, is where free will comes in—in the veto power.[6] Vilayanur Ramachandran, in an argument similar to John Locke's theory of free will,[7] suggests that "our conscious minds may not have free will but rather 'free won't'"![8]

Michael Platt and Paul Glimcher at New York University study the monkey brain and examine the activity of neurons in the "inferior parietal lobule." Their experiments fortify the notion that the brain acts on its own before we become consciously aware of its actions. Each neuron in the inferior parietal lobule has a receptive field that prefers one area of the visual world over all others. For example, when a monkey fixates on a central location and the experimenter moves a bar of light around on the wall, there will be one single place—say, five inches over and five inches up from the fixated spot—where the neuron sends, or "fires," signals at a higher rate than anywhere else. Anytime the bar of light is in that zone, the cell fires signals like a machine gun, but it stops firing when the bar of light is outside the zone, outside the neuron's receptive field.

Through a long series of experiments, these neuroscientists showed that the neurons know a great deal about their receptive fields. They found that the neurons in this area of the brain change their firing patterns depending on the size of the reward likely to be given, which shows that this part of the brain plays a critical role in deciding whether or not to act. The neurons are not passively attending to a specific part of the visual world; rather, they appear to be tied to the purpose of the movement and may actively aid in the decision process. All of this happens long before there is even a hint that the animal is deciding what to do. The automatic brain is at work once again.

Many experiments underline how the brain gets things done before we know about it. Another example is derived from my own research.[9] Our brains are wired in such a fashion that if you fix your gaze on a point in space, everything to the right of the point is projected to the visual areas in the left side of the brain, and everything to the left of the fixation point is projected to the visual areas in the right side of the brain. The two sides of the brain, however, are connected through the large fiber tract called the corpus callosum.

When we present the word "he" to the left of your fixation point and the word "art" to its right, you perceive the word "heart." This integration is achieved without your conscious awareness. Electrophysiological recordings carried out in my lab in collaboration with Ron Mangun and Steven Hillyard helped us decipher how this is accomplished in the brain and shed light on another example of how the brain acts and makes decisions well before we are aware of our deciding, or integrating "he" with "art."

Electrical potentials in the brain evoked by stimuli presented to a subject's senses can be measured by using event-related potentials. This procedure enables one to track, over time, the activation pattern of neurons in the cortex of one side of the brain and their cross-hemispheric connections through the corpus callosum to the opposite side of the brain. We found that after a stimulus (for example, "he") is presented to the left visual field, the right visual cortex quickly activates (the opposite is true for stimuli presented to the right visual field). About 40 milliseconds later the activity begins to spread to the left hemisphere, and after another 40 or so milliseconds the information arrives in consciousness and "heart" appears. The integration of "he" and "art" occurs long before we become consciously aware of the output "heart."

## Free Will and Violence

If the brain makes many decisions before we are aware of them, what about a real-life problem of free will: violent criminal behavior? In many ways neurobiology is used to argue reduced culpability for certain behaviors. Crimes of passion, temporary insanity, even self-defense are basically defenses that claim a person is not responsible for his or her actions because of a temporary (or permanent) abnormal brain state. The supposition is that any normal person would never commit a violent crime; therefore, an abnormality must cause the violent behavior. Those interested in the neurological mechanisms of violence could — and some do — use current neuroscientific knowledge to argue that if we can show that neurochemical imbalances or brain lesions cause violence, people suffering from them should not be held accountable for their actions. In dozens to hundreds of cases throughout the history of criminal behavior and the law, this argument has been made and defendants have been exonerated from charges. But the fact remains that not every person who has lesions, and not every person with, say, schizophrenia, is violent. Where do personal responsibility and free will have a role? They don't.

Consider some of the research on the neuroscience of violent behavior. Criminals who commit repeated acts of violence often have antisocial personality disorder (APD), a condition characterized by deceitfulness, impulsivity, aggressiveness, and lack of remorse.[10] People with APD have abnormal social behavior and lack the inhibitory mechanisms often associated with normal frontal lobe functioning. Ever since the famous case of Phineas Gage in 1848, psychologists have known of the critical role the frontal lobe plays in normal social behavior. Without a normally functioning frontal lobe, it seems the capacity to use "free won't" is impaired.

Phineas Gage is one of the best-known neuropsychological patients of all time. While working on the construction of a railroad, he survived an explosion that drove a tamping iron through his head, damaging areas of the frontal regions of his brain.[11] After he recovered, he appeared normal, but those who had known him before the accident noticed some changes. His friends said that Mr. Gage was "no longer Gage."[12] Indeed, his personality had changed drastically. He was impulsive, lacked normal inhibitions, and demonstrated inappropriate social behaviors (he would swear and discuss things of a sexual nature in situations where it was not appropriate). Unfortunately, no autopsy was done to determine the precise area of the frontal part of his brain that was damaged, but contemporary reconstruction of his brain based on the skull damage indicates that the lesion was localized to medial and orbital regions of the prefrontal cortex.[13]

Evidence from other patients with lesions to the prefrontal regions of the brain confirms that the prefrontal cortex plays a critical role in normal so-cial behavior.[14] The question then arises, Do criminals with APD who demonstrate abnormal social behaviors like those of patients with prefrontal lobe damage also have abnormalities in the prefrontal areas of their brain? To find out, Adrian Raine and his colleagues at the University of Southern California imaged the brains of twenty-one people with APD and compared them with the brains of two control groups, healthy subjects and subjects with substance dependence. (Because substance dependence often occurs along with APD, the experimenters wanted to ensure that they could discern brain differences associated with APD alone, and not with the substance abuse.[15]) Raine found that people with APD had a reduced volume of gray matter and a reduced amount of autonomic activity in the prefrontal areas of their brain, as compared with both control groups. The findings indicate a structural difference (in the amount of gray matter in the prefrontal lobe relative to the rest of the brain) between the brains of criminals with APD and the brains of the normal population. This also suggests that the volume difference in gray matter may lead to the difference in social behavior between the two groups.

Further support for this theory comes from the case study of a boy who displayed characteristics of APD from a young age. When playing Russian roulette, the boy shot himself in the head and damaged his medial prefrontal cortex.[16] Amazingly, he survived, and those who knew him well before the injury reported little or no change in his personality following the brain damage. His pre-injury behavior suggests that the boy's medial frontal cortex was not working properly (possibly due to a reduced volume of gray matter), and the continuation of the behavior problems post-injury indicates that damage to the already malfunctioning medial prefrontal cortex had little or no effect on the child's behavior.

It is possible that these people do not inhibit their impulses even though they *could* (and therefore should be held responsible for their actions). Further research will be needed to determine how much prefrontal damage or gray matter loss is necessary for the cessation of inhibitory function in the brain (and thus perhaps for the mitigation of responsibility). Neuroscientists must realize, however, that when considering the specific example of violence in the brain, the argument might be made that for any given brain state the correlation of nonviolent behavior could be just as high as the correlation of violent behavior. Specifically, most patients who suffer from Gage-type lesions involving the inferior orbital frontal lobe do not exhibit antisocial behavior of a type that would be noticed by the law. Even though a patient's spouse may be able to sense changes in the patient's behavior, the patient is still constrained by all the other forces in society, and the frequency of antisocial behavior is no different from that in the normal population. The same fact is true for people with schizophrenia. The rate of aggressive criminal behavior is no greater among them than among the healthy population. If people with Gage-type lesions or with schizophrenia are no more likely to commit violent crimes than anyone else, it seems that merely having one of these brain disorders is not enough to remove responsibility.

These facts, however, leave pure determinists unimpressed. Their position is based in the theory of causation, and because they believe that all actions have definable inputs, they also believe that our current shortcomings in understanding causation are just that—shortcomings. There is other evidence, however, for the existence of free will in a deterministic world. The prevalence of mechanistic descriptions of how the physical brain carries out behavior adds fuel to the general idea of determinism, but philosophers and scientists alike have argued that the concept of free will can exist, even if one assumes the brain is as mechanical as clockwork. These views challenge the idea that using the explanation of mechanism leads to exculpation.

In 1954, A. J. Ayer put forth the theory of "soft determinism." He argued, as had many philosophers, such as David Hume, that even if determinism exists, a person can still act freely. Ayer posited that free actions result from desires, intentions, and decisions without external compulsion or constraint. He made the distinction between free actions and constrained actions (not between uncaused and caused actions). Free actions are those that originate in oneself, by one's own will (unless one is suffering from a disorder), whereas constrained actions are those caused by external sources (for example, by someone or something forcing you physically or mentally to perform an action under hypnosis, or by disorders like kleptomania). When someone performs a free action A, he or she could have done B. When someone is constrained to do A, he or she could have done *only* A.[17] Ayer thus argued that actions are free as long as they are not constrained. Free actions depend not on the existence of a cause but rather on the source of the cause. Though Ayer did not explicitly discuss the brain's role, one could put it in terms of the brain: The brain is determined, but the person is free.

## Perspectives

Scholars have thought of many clever ways to address the issue of free will, though many also remain controversial. In my view, ghosts in the machine, emergent properties of complex systems, logical indeterminacies, and other characterizations miss the fundamental point: Brains are automatic, but people are free. Our freedom is found in the interaction of the social world.

Let's reconsider Harry and his crime of murder. Under our legal system, a crime has two defining elements: the *actus reus,* or proscribed act, and the *mens rea,* or guilty mind. For Harry to go to jail, the prosecution has to prove *both* beyond a reasonable doubt. In general terms, the courts and the legal system work hard to determine the agency of the crime. Where they want help from neuroscience is to determine whether Harry should be held "personally responsible." Did Harry do it, or did his brain? This is where the slippery slope begins, because, in truth, neuroscience can offer very little to the understanding of responsibility. Responsibility is a human construct that exists only in the social world, where there is more than one person. It is a socially constructed rule that exists only in the context of human interaction. No pixel in a brain scan will ever be able to show culpability or nonculpability.

In practice, legal authorities have had great difficulty in crafting standards to divide the responsible from the nonresponsible. The various rules for a finding of legal insanity, from the 1843 M'Naghten decision to the 20th-century Durham and ALI Model Penal Code tests, have all been found lacking.[18] Experts for the defense and prosecution argue different points from the same data. Again, the idea here is to have neuroscience come to the rescue of the current scheme.

At the crux of the problem is the legal system's view of human behavior. It assumes that Harry is a "practical reasoner," a person who acts because he has freely chosen to act. This simple but powerful assumption in the law drives the entire legal system. Even though all of us might conceive of reasons to break the law, we can decide not to act on such thoughts because we have free will. If a defense lawyer can provide evidence that a defendant had a "defect in reasoning" that led to his inability to stop from committing the crime, then the defendant—in this case Harry—can be exculpable. The defense wants a brain image or a neurotransmitter assay to show beyond a reasonable doubt that Harry was not thinking clearly, indeed *could* not think clearly and stop his behavior.

The view of human behavior offered by neuroscience is simply at odds with this idea. In many ways it is a tougher view, and in many other ways, a more lenient one. Fundamentally, however, it is different. Putting aside the caveats mentioned above, neuroscience is in the business of determining the mechanistic actions of the nervous system. The brain is an evolved system, a decision-making device that interacts with its environment in a way that allows it to learn rules to govern how it responds. It is a rule-based device that works, fortunately, automatically. And as the howls go up in reaction to this kind of formulation, let me quote from an earlier work:

> "But," some might say, "aren't you saying that people are basically robots? That the brain is a clock, and you can't hold people responsible for criminal behavior any more than you can blame a clock for not working?" In a word, no. The comparison is inappropriate; the issue (indeed, the very notion) of responsibility has not emerged. The neuroscientists cannot talk about the brain's culpability any more than the watchmaker can blame the clock. Responsibility has not been denied; it is simply absent from the neuroscientific description of human behavior. Its absence is a direct result of treating the brain as an automatic machine. We do not call clocks responsible precisely because they are, to us, automatic machines. But we do have other ways of treating people that admit judgments of responsibility—we can call them practical reasoners. Just because responsibility cannot be assigned to clocks does not mean it cannot be ascribed to people. In this sense human beings are special and different from clocks and robots.[19]

This is the fundamental point. Neuroscience will never find the brain correlate of responsibility, because that is something we ascribe to humans—to people—not to brains. It is a moral value we demand of our fellow, rule-following human beings. Just as optometrists can tell us how much vision a person has (20/20 or 20/40 or 20/200) but cannot tell us when someone is legally blind or has too little vision to drive a school bus, so psychiatrists and brain scientists might be able to tell us what someone's mental state or brain condition is but cannot tell us (without being arbitrary) when someone has too little control to be held responsible. The issue of responsibility (like the issue of who can drive school buses) is a social choice. In neuroscientific terms, no person is more or less responsible than any other for actions. We are all part of a deterministic system that someday, in theory, we will completely understand. Yet the idea of responsibility, a social construct that exists in the rules of a society, does not exist in the neuronal structures of the brain.

REFERENCES

1. Adapted from M.S. Gazzaniga and M. S. Steven, Free will in the 21st century: A discussion of neuroscience and the law, *Neuroscience and the Law*, B. Garland, ed., (New York: Dana Press, 2004).

2. S.J. Gould, *Ever Since Darwin* (New York: W. W. Norton, 1997).

3. D. C. Dennett, *Freedom Evolves* (New York: Viking Press, 2003, p. 157).

4. For a review, see B. Libet, Conscious vs. neural time, *Nature* 352 (1991): 6330: 27–28.

5. D.C. Dennett, *Freedom Evolves*, p. 228.

6. B. Libet, Do we have free will? *Journal of Consciousness Studies* 6 (1999):8–9, 45.

7. See J. Locke, *An Essay Concerning Human Understanding*, Book II, (1690): Chapter XXI, paragraph 47.

8. V. Ramachandran, Quoted in "The Zombie Within" *New Scientist*, September, 1998.

9. M.S. Gazzaniga and J. E. LeDoux, *The Integrated Mind* (New York: Plenum, 1978).

10. For diagnostic criteria of antisocial personality disorder, see *Diagnostic and Statistical Manual of Mental Disorders*, 4th ed. (Washington, D.C.: American Psychiatric Association, 1994).

11. J. Nolte, *The Human Brain: An Introduction to Its Functional Anatomy*, 5th ed. (St. Louis: Mosby, 2002), pp. 548–549.

12. Harlow, H. M. (1868). "Recovery from the Passage of an Iron Bar Through the Head," *Massachusetts Medical Society Publication* 2: 327.

13. For the details of the reconstruction, see H.T. Damasio, R. Grabowski, R. Frank, A. M. Galaburda, and A. R. Damasio, The return of Phineas Gage: clues about the brain from the skull of a famous patient," *Science* 264 (1994): 1102–1105.

14. A.R. Damasio, A neural basis for sociopathy, *Archives of General Psychiatry* 57 (2000):128.

15. It should be noted that some critics of the findings of this study question whether Raine et al. (A. Raine, T. Lencz, S. Bihrle, L. LaCasse, and P. Colletti) Reduced prefrontal gray matter volume and reduced autonomic activity in antisocial personality disorder, *Archives of General Psychiatry* 57 (2000):119–127) were really able to weed out the effects of substance abuse by means of their control group. These critics suggest that a more cautious conclusion would state that "APD combined with [substance use] is associated with pre-frontal cortical volume reductions." For more on this critique, see E. Seifritz, K.M. Dursteler-MacFarland and R. Stohler, Is prefrontal cortex thinning specific for antisocial personality disorder? *Archives of General Psychiatry* 58 (2001): 402 (comment on A. Raine, et al. [2000] [see above]). For a response, see E.D. Bigler, A. Raine, L. LaCasse, and P. Colletti, Frontal lobe pathology and antisocial personality disorder, *Archives of General Psychiatry* 58 (2001): 609–611.

16. E.D. Bigler, et al., Frontal lobe pathology and antisocial personality disorder, 2001.

17. A.J. Ayer, Freedom and necessity, *Philosophical Essays*, A. J. Ayer, ed. (London: Macmillan, 1954).

18. J.R. Waldbauer and M. S. Gazzaniga, The divergence of neuroscience and law, *Jurimetrics* 4 (2001): 357 (Symposium Issue).

19. Ibid.

MICHAEL GAZZANIGA, PH.D., is director of the Sage Center for the Study of the Mind at the University of California, Santa Barbara. He is also a member of the President's Council on Bioethics.

*Chapter 18*

# New Neuroscience, Old Problems:

*Legal Implications of Brain Science*\*

*Stephen J. Morse*

During the 1982 trial of John W. Hinckley Jr. for the attempted assassination of President Reagan and others, Hinckley introduced evidence of a brain abnormality to buttress his claim that he was legally insane. An expert witness used a computerized axial tomography (CAT) scan to demonstrate that Hinckley's sulci (the grooves in the folds of the brain's top layer) were wider than normal, a finding the expert then believed was linked to schizophrenia. None of the experts or commentators was sure of the legal significance of this finding, however, and we do not know whether that evidence influenced the jury's verdict that Hinckley was legally insane.

Today, as the Supreme Court considers whether it is unconstitutional to execute killers who were 16 or 17 years old when they committed homicide—a practice approved by a substantial minority of the states—opponents of executing juvenile killers are using more sophisticated brain imaging techniques to demonstrate that the brains of late adolescents are not fully developed. The opponents claim that such findings indicate that the death penalty would be unfair for teenagers who commit capital crimes.

Although the brain science of today is vastly advanced compared with its 1982 precursors, the legal implications are still not clear. No necessary connection exists between the findings of neuroscience and legal or moral poli-

\*From *Cerebrum* 6 (2004): 81–90.

cies or decisions. The law's concept of the person and the nature of law itself are both so fundamental to our understanding of ourselves and society that the new neuroscience may have fewer implications for law and society than popular imagination and even many scientists believe.

## The Law's Concept of the Person

The legal concept of the person is that of an agent who is capable of acting intentionally and for reasons. We are social creatures whose interactions are not governed primarily by innate repertoires; we are able to guide our behavior in light of reasons we may have for acting and do not solely and blindly follow instinct.

Law is a practical system of rules and institutions that evaluate, guide, and govern human action. It gives people good reason to behave one way or another, by making the consequences of noncompliance clear or through people's understanding of the reasons that support a particular rule. Law shares many characteristics with other sources of guidance, such as morality and custom, but is different because its rules and institutions are created and enforced by the state.

Physical causes explain the structure and mechanisms of the brain and nervous system (and all the other moving parts of the physical universe), but only human action—intentional bodily movements and other intentionally produced states—can also be explained by reasons. Law views human action as reason-governed and treats people as intentional agents, not simply as part of the biophysical flotsam and jetsam of the causal universe. It could not be otherwise. It makes no sense to ask a bull that gores a matador, "Why did you do that?" But this question makes sense and is vitally important when it is addressed to a person who sticks a knife into the chest of another human being. It makes a great difference to us if the knife wielder is a surgeon who is cutting with the patient's consent or a person who is enraged at the victim and intends to kill him.

Only human beings are fully intentional creatures. To ask why a person acted a certain way is to ask for reasons, not for reductionist biophysical, psychological, or sociological explanations. I am not positing the existence of non-natural properties, such as a soul; I assume that a perfectly naturalistic set of causes can explain intentionality and consciousness. But only persons can deliberate about what action to perform and can determine their conduct by practical reason.

Today we have no idea how the brain enables the mind (and scant information about precisely how it disables it), but when we solve this problem—if we ever do—the solution will revolutionize our understanding of biological processes. Our view of ourselves and all our moral and political arrangements are likely to be as profoundly altered as our understanding of biological processes. For now, however, despite the impressive gains in neuroscience and related disciplines, we still do not know mechanistically how action happens even if we are convinced, as I am, that a physicalist account of some sort must be correct.

For the law, then, a person is a practical reasoner. It assumes simply that people are capable of acting for reasons and are generally capable of minimal rationality according to mostly conventional, socially constructed standards of rationality.

## The Touchstones of Responsibility and Excuse

Rationality is the touchstone of responsibility. Only agents capable of rationality can use legal and moral rules as potential *reasons for action*. Only by its influence on practical reason can law directly and indirectly affect the world we inhabit. In order to maximize liberty and autonomy, the law presumes that adults are capable of minimal rationality and responsibility and that the same rules may be applied to all, but this presumption can be rebut- *CPL* ted in appropriate cases.

No uncontroversial definition of rationality exists in the disciplines that study it, such as philosophy, economics, and psychology. In various legal contexts, how much and what type of rationality is required for responsibility is a social, moral, and political issue that divides people. For example, the U.S. Supreme Court was asked to decide if the criteria for competence to stand trial should be different from the criteria for competence to plead guilty. In a closely split decision, the Court ruled that the same criteria should apply. Science could not answer this question because it is not a scientific issue; the debate is about human action. But the rationality criterion for responsibility is perfectly consistent with the facts—most adults are capable of minimal rationality virtually all the time—and with moral theories concerning fairness and justice that we have good reason to accept.

Conversely, lack of the capacity for rationality is the touchstone of excuse. Unless people were reasonably capable of understanding and using legal rules as premises in deliberation, law would be powerless to affect hu-

*lack of capacity for rationality*

man behavior and it would be unfair to hold them responsible. What rationality and consequent responsibility demands will differ across contexts. For example, a person is incompetent to contract if he or she is incapable of understanding the nature of the bargain; a person is criminally nonresponsible if the agent did not know the nature of his or her conduct or the applicable law.

The law contains coercion or compulsion criteria for nonresponsibility, but these criteria are demanding and only infrequently provide an excusing condition. Properly understood, coercion occurs when the person is placed through no fault of her own in a threatening "hard choice" situation from which she cannot readily escape and in which she yields to the threat. The classic example in criminal law is the excuse of duress. This requires that the person must be threatened with death or serious bodily harm unless she commits the crime and that a person of "reasonable firmness" would have yielded to the threat. The agent thus has acted intentionally and rationally to avoid death or grievous bodily harm. The crime is excused, because requiring human beings not to yield to some threats is simply too much to ask of creatures such as ourselves. How hard the choice must be can vary across contexts; a compulsion excuse for crime might require a greater threat than a compulsion excuse for a contract.

A persistent, vexing question is how to assess the responsibility of people who seem to be acting in response to some inner compulsion, or, in more ordinary language, seem to have trouble controlling themselves. What does it mean to say that an agent who is acting cannot control himself? I have explored this puzzle in my professional writing and have arrived at the conclusion that defects in rationality best explain these cases and that the law does not need an independent compulsion excuse in these "one party cases."[1] People who act in response to such inner states are intentional agents. Simply that an abnormal biological condition played a causal role—and neuroscientific evidence frequently confirms this—does not mean the person was compelled. In some cases, however, the persistence or intensity of the desire or craving makes it supremely difficult for the person to access reason. They still might not be excused, however, if they recognize, as many do, that they are subject to such desires frequently and behave badly when this happens. In such cases, they may have an obligation when they are more rational to take steps to prevent themselves from being in a position to harm themselves or others when they are in the throes of their intense desires.

Neuroscience will surely discover much more about conditions that can compromise rationality and may broaden current legal excusing doctrines or widen the class of people who can raise a plausible claim under current law. Neuroscience may help to adjudicate excusing and mitigating claims more accurately. But unless neuroscience demonstrates that no one is capable of minimal rationality (or that everyone is always responding to supremely intense and persistent cravings)—an implausible scenario discussed below— fundamental criteria for responsibility will be intact.

## Confusions About Responsibility and Excuse

Responsibility has nothing to do with "free will" even though legal cases and commentary concerning responsibility are replete with talk about it. Nor is the truth of a fully physically caused universe (sometimes referred to as "determinism") part of the criteria for any legal doctrine that holds some people nonresponsible. Thinking that causation itself excuses, including causation by abnormal variables, is an analytic error that I have termed "the fundamental psycholegal error." All behavior may be caused in a physical universe, but not all behavior is excused, because causation per se has nothing to do with responsibility. For example, many variables caused you to be reading this article now, but you are perfectly responsible for intentionally reading it. Reading it is presumably not evidence of incapacity for rationality, and presumably no one is forcing you to read it. Causation is not the equivalent of either lack of capacity for rationality or compulsion. If causation negated responsibility, no one would be morally responsible and holding people legally responsible would be extremely difficult.

The fundamental psycholegal error wrongly leads people to try to create a new excuse every time an allegedly valid new "syndrome" or any other cause is discovered that plays a role in behavior. But syndromes and other causes, including those of brain structure and function, do not have excusing force unless they sufficiently diminish rationality in the context in question. In that case, it is diminished rationality that is the excusing condition, not the presence of any particular type of cause. An assertion about "free will" based on causation is simply a conclusion about responsibility that must have been reached based on criteria such as the presence of rationality or absence of coercion.

*Causation is not the excuse*

*Lack of capacity*

## Assessing Responsibility and Excuse

The criteria for excuse—lack of capacity for rationality and the presence of coercion—concern components of human action, such as desires and beliefs, that must in the first instance be assessed behaviorally, including by the use of behavioral tests devised for this purpose. It is human action that is at issue, not the state of the brain. If the person's rational capacities, which we infer from her behavior, seem unimpaired, she will be held responsible, whatever the neuroscience might show, and vice versa. We knew that young children were not fully responsible long before we understood the neuroscience.

Although neuroscientific evidence may surely provide assistance in performing responsibility evaluations, neuroscience could never tell us how much rationality is required for responsibility. The question is social, moral, political, and, ultimately, legal. Moreover, it is unlikely, except in extreme cases in which we wouldn't need brain evidence, that brain states will map legal criteria precisely. For the foreseeable future, neuroscience as a tool cannot replace behavioral investigation and common sense when we assess responsibility.

The issue in deciding if teen killers should be executed, for example, is whether they suffer from sufficiently less rationality than adults, and that must be evaluated by examining adolescents' reasoning and judgment. Brains do not have defective judgment; conscious, intentional agents—people—do. I am an opponent of capital punishment, but if our society decides morally and legally that the capacity for rationality in normal late adolescents is sufficient for capital punishment, even if their brains are less developed than those of adults, then the brain science alone cannot demonstrate that capital punishment is unjustified.

## The Neuroscience Challenge to Personhood and Responsibility

Advances in neuroscience and related fields have revealed hitherto unimagined biological causes of behavior, including abnormal neurotransmitters, that may increase the risk of antisocial or otherwise undesirable behavior,[2] but we have no convincing conceptual or empirical reason to abandon our view of ourselves as creatures whose desires, beliefs, and intentions cause and explain our behavior.

Scientific discoveries might indicate that mental causation does not exist as we think it does, but a brain correlate or cause does not mean that the action is not an action. If actions exist, they have causes, including those arising in the brain. The real question is whether studies have shown that intentional behavior is rare or nonexistent. Despite our intuition and experience that action is ubiquitous and genuinely explainable, increasing numbers of scientists and philosophers claim that intention is an illusion. They cite two kinds of empirical evidence: first, demonstrations that most of our behavior is caused by variables we are unaware of, and second, studies indicating that much behavior occurs when our consciousness is diminished.

A person may not be aware of all the causes of forming and acting on an intention, but this is not to say she did not form an intention and was not a fully conscious agent. Even if we were *never* aware of the causes of our actions, it would not mean that we do not act intentionally and consciously.

Knowledge that a variety of causes can diminish human consciousness long predates contemporary neuroscience. Demonstrating that partial consciousness is more common than it seems extends the range of cases in which people are not responsible or have diminished responsibility. Nevertheless, such studies do not demonstrate that most apparently intentional human actions occur in states of unintegrated consciousness. One cannot generalize from abnormal cases. No scientific study has produced a general demonstration that causal intentionality is an illusion, and I suspect that none ever will. As the eminent philosopher of mind Jerry Fodor has written, the only thing we can be as sure of as the existence of "mid-sized" objects is that we are creatures who act for reasons.[3]

Let us suppose, however, that you were convinced by the mechanistic view of persons that you were not an intentional, rational agent after all. (Of course, the notion of being "convinced" would be an illusion. Being convinced means that you were persuaded by evidence or argument, but a mechanism is not persuaded by anything. It is simply neurophysically transformed or some such.) What should you do now? You know it's an illusion to believe that your deliberation and intention have any causal efficacy in the world. (Again, what does it mean mechanistically to "know" something? But enough.) Nonetheless, you also know that you care about what happens to you and to the world. You cannot just sit quietly and wait for your neurotransmitters to fire. You will deliberate and act. Even if pure mechanism is true, human beings will find it impossible not to treat themselves as rational, intentional agents.

People think that the discovery of causes of behavior over which people have no control, including brain states, suggests that determinism is true and undermines "free will." This concept is what terrifies people about scientific understanding of human behavior, which relentlessly exposes the numerous causal variables that seem to toss us about like small boats in a raging sea. They fear that scientific explanations, biological or otherwise, will demonstrate that we are only mechanisms. But the new neuroscience casts little doubt on responsibility generally; there is no reason to doubt that we are conscious, intentional, and rational creatures. Causation alone does not undermine that knowledge or provide an excusing condition. Western theories of morality and the law properly hold some people responsible and excuse others, and when we do excuse, it is not because a little local determinism has been at work. Determinism cannot tell us which goals we should pursue, and it cannot explain or justify our present practices.

## Neuroscience and Legal Reform

Under what conditions will the new neuroscience suggest reform of existing legal rules? The law is in many respects a conservative enterprise and will always resist supposed reforms that other disciplines suggest. For example, despite the extraordinary advances in the understanding of mental disorder in the past half century and consistent calls for reform based on such understanding, the dominant version of the insanity defense—which excuses if the defendant with a disorder did not know what he was doing or did not know that it was wrong—is scarcely changed from the form adopted in 1843 by the English, the M'Naghten rule. This is unsurprising. Mental health science can teach us much about why some people lack rationality and can help identify and treat those people, but it cannot tell society which rationality defects are sufficient to excuse a wrongdoer. Deciding who is blameworthy and deserves to be punished depends on established social norms and practices about which mental health science must fall silent.

Legal rules do, of course, change in response to evolving principles and new scientific discoveries. Racial discrimination was banned by civil rights legislation simply because it was wrong. It is now unlawful to dump toxic waste, in large part because science demonstrated the health hazards. But before legislators and judges will be rationally justified in changing existing legal rules in response to the discoveries of any scientific discipline, at the least they should be convinced, first, that the data are valid; second, that the data are genuinely relevant to particular rules; third, that the data convincingly imply that specific changes to those rules would have desirable effects; and, fourth, that those changes will not infringe upon other values that may be more important. All this is a tall order.

I predict that neuroscience will not have widespread, profound influence on doctrine in most areas unless its discoveries radically alter our conception of ourselves. On the other hand, one can easily imagine substantial changes in discrete doctrines. For example, neuroscience may teach us much about cognitive processing under stress that would influence our doctrines of informed consent to medical care. For another example, neuroscience may be able to identify when people are consciously lying or consciously or unconsciously discriminating on the basis of objectionable factors such as race. Such discoveries could have profound effects on evidentiary practices. But even an exceptionally sensitive technique to detect lying or discrimination might not be used, because we fear state invasion of our innermost thoughts, even for purposes such as discovering truth or uncovering conscious or unconscious discrimination. In cases involving discrete doctrinal change, we already have the tools to weigh the desirability of the change.

## Potential Threats to Civil Liberties

Neuroscientific discoveries might well raise profound challenges to civil liberties. Other sciences, too, might make discoveries that would do likewise, so the following discussion surely generalizes. The potential of neuroscience to invade our privacy by revealing various aspects of our private, subjective experience may produce the strongest reaction against its use and lead to substantial regulation. Still, the techniques that permit valuable ends such as accurate lie detection may be so alluring that the temptation to use them will be great.

The question is, What constitutional or legislative limits may be placed on such techniques? The government will not be able to use neuroscientific investigative techniques to go on "mental fishing expeditions" generally, but various state interests may permit infringing hitherto protected interests. For example, the Supreme Court recently held that under limited conditions the state has the right to force psychotropic medication on a psychotic criminal defendant solely for the purpose of restoring the defendant's competence to stand trial.[4]

Neuroscientific techniques might also increase the ability to make accurate predictions about various forms of future behavior. If socially trou-blesome behaviors can be accurately predicted, use of such techniques for screening and intervention will be tempting. For example, neuroscientific techniques may well enhance the ability to predict antisocial conduct. It would be less difficult to justify screening prisoners and others under crimi-nal justice control, but widespread screening of others, such as apparently at-risk children—even if the risk status were identified by objective, valid measures—would be legally and politically fraught with civil liberties im-plications. The widespread use of psychotropic medications such as methyl-phenidate (Ritalin) among schoolchildren suggests, however, that a screen-ing/intervention scenario would not be unthinkable under some conditions. It is difficult to envision how society would respond to techniques that iden-tified risk-creating abnormalities highly accurately and to effective interven-tions that would prevent undoubtedly serious social and personal harms. Traditional attitudes toward privacy and liberty might change considerably.

Many consider the potential for direct, biological intervention in the working of the brain and nervous system to change thoughts, feelings, and actions—often polemically characterized as "mind control"—a greater threat to liberty than genetic intervention. The government already has the constitutional authority to compel the use of psychotropic medications un-der limited circumstances, but the potential for widespread intervention to change behavior is apparent. The biological and behavioral definitions of abnormality and disorder can be controversial. At the extremes, little prob-lem exists, but the criteria for abnormal brain structure or function are not self-defining. The criteria for behavioral abnormality are even more fluid, and there is a tendency to pathologize troubling behaviors. Thus, a relative-ly value-neutral criterion of abnormality cannot impose strict limits on the ability of the state to compel behavior-altering interventions. ▴

The current science and political will to accomplish effective, wide-spread behavior control are lacking. Nonetheless, as screening and interven-tion methods become more precise and effective, pressure will build to use them, and proponents will defend their constitutional legitimacy. If neuro-science or other sciences ever reach the levels of understanding and efficacy necessary to make the civil liberties concerns a realistic possibility, it is dif-ficult to predict what legislatures and courts will do. If pressing social prob-lems seem amenable to a technological fix, current political and constitu-tional constraints may weaken.

## New Neuroscience, Old Problems

The new neuroscience poses familiar moral, social, political, and legal challenges that can be addressed using equally familiar conceptual and theoretical tools. Discoveries that increase our understanding and control of human behavior may raise the stakes, but they don't change the game. Future discoveries may so radically alter the way we think about ourselves as persons and about the nature of human existence that massive shifts in all our societal practices and institutions may ensue. For now, however, neuroscience poses no threat to ordinary notions of personhood and responsibility that undergird our civic life and the law.

REFERENCES

1. S. J. Morse, Uncontrollable urges and irrational people, *Virginia Law Review* 88 (2002): 1025.

2. See, e.g., A. Caspi et al, Role of genotype in the cycle of violence in maltreated children, *Science* 297 (2002): 851; R. Z. Goldstein and N. D. Volkow, Drug addiction and its underlying neurobiological basis: Neuroimaging evidence for the involvement of the frontal cortex, *American Journal of Psychiatry* 159 (2002): 1542; M. N. Potenza et al., Gambling urges in pathological gambling: A functional magnetic resonance imaging study, *Archives of General Psychiatry* 60 (2003): 828; M. B. Stein et al., Genetic and environmental influences on trauma exposure and posttraumatic stress disorder symptoms: A twin study, *American Journal of Psychiatry* 159 (2002): 675.

3. J. Fodor, *Psychosemantics: The problem of meaning in the philosophy of mind* (Cambridge, MA: MIT Press, 1987), xii.

4. *Sell v. United States*, 533 U.S. 166 (2003).

STEPHEN J. MORSE, J.D., PH.D., is Ferdinand Wakeman Hubbell Professor of Law at the University of Pennsylvania Law School and professor of psychology and law in psychiatry at Penn's School of Medicine. He is an expert in criminal and mental health law whose work emphasizes individual responsibility in criminal and civil law.

*Chapter 19*

# Moral Cognition and Its Neural Constituents*

*William D. Casebeer*

Identifying the neural mechanisms of moral cognition is especially difficult. In part, this is because moral cognition taps multiple cognitive subprocesses, being a highly distributed, whole-brain affair. The assumptions required to make progress in identifying the neural constituents of moral cognition might simplify morally salient stimuli to the point that they no longer activate the requisite neural architectures, but the right experiments can overcome this difficulty. The current evidence allows us to draw a tentative conclusion: the moral psychology required by virtue theory is the most neurobiologically plausible.

Good moral reasoning is extremely important for *Homo sapiens*. Our lives are more fruitful if we recognize salient ethical norms and reason effectively about their application to our own situations. We are social creatures, and if we are to flourish in our social environments, we must learn how to reason well about what we should do. Despite its importance for our proper functioning, until recently the neural mechanisms of moral cognition were not well studied. This is unfortunate, as co-evolution between the neural constituents of moral cognition and the moral psychologies that are required by the main ethical theories is necessary if we are to make progress in understanding how effective ethical reasoning is embodied in the brain.

*From *Nature Reviews Neuroscience* 4 (2003): 840–847.

To make such progress requires us to probe the nature of moral judgment and its relationship to the experimental regimens that are used to explore such constituents.

Here, I briefly review the neural mechanisms of moral cognition, discuss methodological pitfalls, and consider issues that might inform future experimental work. Ultimately, the current situation makes the moral psychology that is required by virtue theory the most neurobiologically plausible, although this is a tentative, defeasible conclusion, and more work is needed to confirm it.

## Moral Theories and Moral Cognition

To study the neural mechanisms of moral cognition, one must delimit the field of inquiry. What does "moral cognition" encompass? This depends on how we construe the domain of moral theory. Although all moral theories claim to speak to what an agent should do (this is what makes them distinctively moral), they disagree about the substance of such recommendations and the moral psychologies that are required for effective reasoning and action. The three main classic moral theories in the Western tradition are utilitarianism, deontology, and virtue theory.

The typical utilitarian, such as the British philosopher John Stuart Mill (1806–1873), thinks that one should take that action (or follow that "rule") that, if taken (or followed), would produce the greatest amount of happiness for the largest number of sentient beings, where happiness is the presence of pleasure or the absence of pain (and where pleasure and pain are given more sophisticated readings than mere affective satisfaction). The second flavor of utility, "rule utilitarianism," is probably the most popular.

Deontologists, exemplified by the Prussian philosopher Immanuel Kant (1724–1804), do not emphasize the consequences of actions, as utilitarians do. Instead, they focus on the maxim of the action—the intent-based principle that plays itself out in an agent's mind. We must do our duty, as derived from the dictates of pure reason and the "categorical imperative," for duty's sake alone. Deontologists are particularly concerned to highlight the duties that are owed to each other by free and reasonable creatures (paradigmatically, humans). Maximizing happiness is not the goal; instead, ensuring that we do not violate another's rights is paramount.[2]

Virtue theorists, such as the Greek philosophers Plato (427–347 B.C.) and Aristotle (384–322 B.C.), make paramount the concept of "human flourishing"[3,4]; to be maximally moral is to function as well as one can given one's nature. This involves the cultivation of virtues (such as wisdom) and the avoidance of vices (such as intemperance) and is a practical affair.

Each approach asks different things of us cognitively. What follows is an abbreviated discussion of each theory's moral psychology. To make the appropriate judgment about what one should do, the utilitarian would, at least in morally problematic cases, require that a moral agent could recognize and compute salient utility functions. We would then be moved to act on such judgments by cultivation of appropriate altruistic fellow-feeling or, in many cases, merely by self-concern (as utility will often be maximized by having each of us focus on our own happiness as well as the happiness of others). So, in terms of raw computations, a utilitarian moral psychology would require some mechanism for learning what actions or rules would eventually produce happiness. Either implicitly or explicitly, utilitarian computations would constitute the bulk of our moral cognitive capacity. Whether we act on the outcome of those judgments might require some derivative character development (such as the cultivation of concern for the happiness of others), and this would require appropriate training of the emotions.

A Kantian moral psychology would be different. The ability to "reason purely" about the demands of the categorical imperative (the heuristic that is used by Kant to capture our respect for those things that make morality possible—autonomy and rationality) would be the most important part of our cognitive equipment. The best-known formulation of the categorical imperative requires that we act only on maxims that we can will to become a universal law; other maxims are morally impermissible. For example, you could not universalize the maxim that allows you to lie to achieve some end; such a maxim requires that others act with a different intention, of delivering and receiving only true utterances (or else the lie would not be effective). The maxim cannot be made universal; this conceptual truth does not require experimentation to confirm it. We therefore have a perfect (that is, exceptionless) duty not to form the intention to lie. Of note, Kant requires that we be moved to do our duty by the demands of duty alone; if something else (say, the desire to be liked) is moving us, our action is not morally praiseworthy, as it plays more to our animal nature than to our rational (and so human) nature. Kant is thought to give short shrift to character development and related issues, although recent work has "softened up" this position.[5] What exactly the cognitive capacity to reason purely in the Kantian manner would look like has not been a subject of extensive investigation; it would require at least the ability to check universalized maxims for logical consistency in a manner that is separable from the taint of affect and emotion. *[handwritten marginalia: Virtue moral psychol]*

Finally, virtue-theoretic moral psychology is often thought to be the richest of the three. A virtuous person must be able to reason well about what states of being would be most conducive to the best life. What type of person must I become if I am to experience *eudaimonia* (variously translated from the Greek as "flourishing," "proper functioning," or "happiness")? To act on the outcomes of my judgments, I must train my character so that my appetites and "spirit" are coordinated smoothly with the demands of good reason. Virtuous people are moved to do the appropriate thing at the appropriate time; they become angry at unjust events, are sympathetic to recipients of wrongdoing, and so on. Virtue theorists focus on the appropriate coordination of properly functioning cognitive subentities. Moral reasoning and action are therefore "whole-psychology, whole-brain" affairs. Jokingly, then, it could be said that these approaches emphasize different brain regions: frontal (Kant); pre-frontal, limbic, and sensory (Mill); and the properly coordinated action of all (Aristotle).

Kant would say that moral reasoning is a robustly rational affair, where "rational" is given a strict interpretation. With Aristotle, however, I think it is more useful to treat moral judgment in a deflationary manner. Given that the domain of what constitutes a moral judgment is itself in contention, we would be best served by casting our nets wide, narrowing them appropriately as the neurobiological, psychological, and normative aspects of morality co-evolve (admittedly, casting our net so wide might bias us initially toward a virtue-theoretic moral psychology; the give-and-take required by co-evolution of theories at all three levels of analysis would, hopefully, correct any such bias eventually). As a first cut, then, moral cognition comprises any cognitive act that is related to helping us ascertain and act on what we should do. Nonhuman animals (for example, primates and other social animals) might also engage in robust moral reasoning (see, for example, De Waal's work[6]).

My push for this deflationary conception of moral judgment is driven by the recognition (but also has as the upshot) that moral cognition might not be a tightly defined "natural kind" in the sense that other cognitive phenomena might be. For example, the domain of the neural mechanisms of visual cognition, owing to the relatively restricted range of information that is processed by the visual modality, might be more tightly constrained than the domain of the neural mechanisms of basket-weaving. In that sense, the former is a more robust, natural kind than the latter, and is therefore an easier target for neurobiological study. Moral reasoning probably falls somewhere between these two extremes and is still worthy of study by neurobiologists, although this fact might make it more difficult to progress experimentally.[7]

Critics might argue that such a coevolutionary strategy commits the naturalistic fallacy of inferring what should be from what is. The exact nature and status of the naturalistic fallacy is subject to debate (for a summary, see chapter 2 of my *Natural Ethical Facts*[8]). Note, however, that the two most famous arguments against naturalism about ethics, Hume's law[9] and G. E. Moore's open question argument,[10] do not stand up against some contemporary naturalized ethical theories. Both of these arguments rely on an analytic/synthetic distinction that many philosophers agree collapsed in the 20th century. In addition, Hume's argument rules out deductive relationships between facts and norms but not necessarily abductive ("inference to the best explanation" style) relationships. And Moore himself admitted, in the second version of his *Principia Ethica*, that his argument best applies to the two forms of naturalized ethics that he attacks in the book: Spencer's evolutionary ethic and hedonism. I agree that both of these naturalized ethics are poor moral theories, but I disagree that Moore has offered an argument that is general to all attempts to naturalize ethics.

Keep in mind that no good naturalized ethical theory will say that all facts are normative facts, nor that all existing states of affairs—merely because they are "natural" in the sense that they were produced by natural processes—are good. Unlike Greene in his companion piece in this issue,[11] [included here as Chapter 20—*Ed.*]. I think that the neurobiological facts support a version of relational moral realism, but this discussion is beyond the scope of my paper. Ultimately, even the most ardent anti-naturalist would admit that, at the very least, our moral theories must require us to carry out cognitive acts that are also possible for us to implement. The goal of naturalized ethics is to show that norms are natural and that they arise from and are justified by purely natural processes. If this can be done, then the naturalistic fallacy is not actually a fallacy (it merely amounts to saying that you don't have a good naturalized ethical theory yet).

## Emotion and Affect

The rich and diverse literature on the neural mechanisms of moral cognition can be usefully divided into three branches: the moral emotions, theory of mind, and abstract moral reasoning. I will briefly discuss the connections between cortical areas and the limbic system that are necessary for good moral judgment, the neural correlates of theory of mind (TOM) and how they manifest themselves in moral judgment, and useful conceptual tools for thinking about abstract moral reasoning.

The moral emotions are crucial for effective moral cognition. They motivate action, serve as markers of value, are vital for coordinating group activity, and help to filter out and highlight certain aspects of the moral calculus. In mammals and reptiles, the brain's regulatory core is situated in the brain stem/limbic axis, and it subserves important activities such as breathing, arousal, and the coordination of drives (for food, sex, oxygen, and so on) with perceptions (turkey sandwich there, attractive mate here). The basic emotions associated with hunger, thirst, sexual desire, and the like are powerful motivators; as our brain's cortical capacity expanded during evolution, "newer" frontal brain regions remained connected to and were innervated by this regulatory core. No wonder then that these basic, survival-laden emotions serve as the platform on which the moral emotions (and effective moral reasoning) are built.

Exploring the role of the prefrontal regions in connecting limbic areas to frontal areas is difficult; studies of humans with focal brain damage and experimental lesions in monkeys have indicated a relationship between the prefrontal cortex (PFC) and planning, decision making, emotion, attention, memory for spatiotemporal patterns, and recognition of a mismatch between intention and execution. The precise nature of the relationship between these functions and the contribution of the PFC remains unclear, in part because the temporal and spatial resolution of functional magnetic resonance imaging (fMRI) is limited, and because we are still ascertaining the network-level properties of this area.

Despite these difficulties, the link between moral decision making, social cognition, and the emotions is becoming clearer. Converging results from lesion and imaging studies indicate that damage to the ventral and medial PFC is consistently associated with impairments in practical and moral decision making.[12-15] Patients with focal ventromedial lesions show abnormally flat (emotionless) responses when shown emotional pictures and perform poorly on tasks where feelings are needed to guide complex self-directed choices.

Electroencephalogram (EEG) studies of children with self-control disorders[16] also support a link between the ventromedial PFC and moral emotions, as do fMRIs of normal subjects. For example, viewing scenes that evoke moral emotions produces activation in the ventromedial PFC and the superior temporal sulcus.[17,18]

Within the ventral PFC, the orbitofrontal cortex (OFC) is crucial for cueing morally appropriate behavior in adulthood and acquiring moral knowledge during childhood; although patients with adult-onset and childhood-onset OFC damage showed similar abnormal socio-moral behavior, their scores on standardized tests of moral reasoning differed. Those with early damage performed poorly on the tests, exhibiting the egoistic reasoning that is typical of a ten-year-old, whereas adult-onset subjects performed normally despite their abnormal behavior.[15]

The PFC receives important inputs from both sensory and limbic areas. The limbic system is a highly interconnected set of subcortical regions (including hippocampus, amygdala, hypothalamus, and basal forebrain) and the cingulate cortex. The activity of this system is modulated by the neurotransmitters dopamine, serotonin, noradrenaline, and acetylcholine, and changes in the levels of these substances can greatly affect sex drive, moods, emotions, and aggressiveness. The proper operation of the system as a whole is crucial for effective moral judgement.

The amygdala, for example, is part of the complex reward circuitry involving the positive emotions.[19] It is likely that the amygdaloid complex modulates the storage of emotionally important and arousing memories; events that are important to survival provoke specific emotions and, with amygdala activity, are more likely to be permanently stored than neutral events. The amygdala is also crucial for aiding retrieval of socially relevant knowledge about facial appearance; three subjects with total bilateral amygdala damage were asked to judge the trustworthiness of unfamiliar people, and all three judged unfamiliar people to be more approachable and trustworthy than did control subjects.[20]

Hippocampal structures are essential for learning and remembering specific events or episodes, although permanent memory storage lies elsewhere in the cortex. The hippocampus, parahippocampal cortex, entorhinal cortex, and perirhinal cortex all seem to be important for the processing and retrieval of salient "me-relevant" memories. In moral judgment, the hippocampus might facilitate conscious recollection of schemas and memories that allow past events to affect current decisions.

The cingulate cortex has a number of subregions with different functions: regulation of selective attention, regulation of motivation, and detection of malcoordinated intention and execution are associated with anterior regions (anterior cingulate cortex, ACC). Rostral ACC activation (along with the nucleus accumbens, the caudate nucleus, and the ventromedial (VM)/OFC) is needed for cooperative behavior among subjects playing a version of the "prisoner's dilemma"; the hyperscanning methodology used to obtain these results is especially promising[21] (see later in text). Other work supports the conclusion that the ACC is crucial for identifying times when the organism needs to be more strongly engaged in controlling its behavior.[22–25]

## Theory of Mind and Moral Judgment

This brief exploration of limbic areas and their connections undervalues the role of PFC in the second important research area in moral cognition: social judgment and TOM. Our ability to know what others are thinking so that we can interact fruitfully with them is vital—it underlies our ability to empathize with others, to judge how they might react in response to our actions, and to predict the subjective consequences of our actions for con-specifics. Studies of children with autism indicate that TOM might be subserved by the aggregate neural activity of the OFC, the medial structures of the amygdala, and the superior temporal sulcus (STS). The circuit that is formed by the last two structures might mediate direction-of-gaze detection (a crucial component of our ability to infer what others might be thinking about), all three locations are probably involved in mediating shared attention, and the specially coordinated action of all three might therefore constitute TOM processing.[26]

Relatedly, "mirror neurons" in the PFC of the macaque monkey respond either when the monkey makes a specific movement, such as grasping with the index finger and thumb, or when it sees another making the same movement.[27] This indicates that when the animal sees another make the movement, the premotor cortex generates incipient motor commands to match the movement. These signals might be detected as off-line intentions that are used to interpret what is seen (for example, "the dominant male intends to attack me"). Mirror neurons might therefore bootstrap full-blown TOM into existence through an inner simulation of the behavior of others. TOM is probably what allows a chimpanzee to know whether a high-ranking male can see the food she is grabbing or whether the food is occluded from view and can be taken without fear.[28] Robust TOM is necessary for healthy moral judgment; it is also associated with our ability to lie, but is nonetheless probably necessary if a whole host of morally important cognitive abilities are to be realized. For an excellent review of TOM mechanisms, see Adolphs' work.[29]

## Abstract Moral Reasoning

Probably the most difficult aspect of the neural mechanisms of moral cognition is the constituents of abstract moral reasoning. The most important forms of moral reasoning that we rely on daily involve background social skills, tacit use of TOM, ready-at-hand action patterns and interpretive schema, and the like. Much of our day-to-day moral reasoning does not involve highly convoluted moral modeling; mostly we can rely on skills and habits of character as informed by conditioned emotion and affect (indeed, Haidt claims that abstract moral reasoning is a completely post-hoc affair and is almost never the direct cause of moral judgments).[30] Nonetheless, abstract moral reasoning is sometimes necessary. It probably depends on brain structures that subserve morally neutral abstract thought (such as a capacity to model the consequences of an action) and practical reasoning about how to accomplish things. For example, in a classic moral dilemma, such as the trolley problem (in which one has to decide whether to allow an out-of-control trolley to continue down a track where it will strike five people or to throw a switch diverting it onto a track where it will strike only one person, explored in detail in fMRI work by Greene et al.[31]), higher-order cognitive abilities such as planning, executive flexibility, and strategy application are needed.[32] These capacities might be realized in the cerebral cortex by transient cortical networks that Fuster calls "cognits."[33]

The difficulty we have in understanding the neural basis of moral reasoning is indicative of two things: first, that we still need both better theoretical frameworks to understand higher-order cognitive capacities and better network-level tools for probing activity, and second, that such capacities might be overvalued relative to the work that they perform in our cognitive economy (in some respects, then, eliminativism might be called for[34]).

One device that might be useful for helping us to organize abstract moral reasoning is a moral state-space (a concept first articulated by P. M. Churchland).[35] We can think of much of the activity of the frontal cortex and the brain stem limbic axis as consisting of a moving point in an $n$-dimensional space, where $n$ could (in complex cases) be determined by making the activity of every neuron that is involved in the system an axis of that space (in some cases, an axis might be constituted by a single neuron, which might explain results about how single neurons in PFC can seem to encode "rules"[36]). Reducing the dimensions of this space enables us to capture its principal components, which might themselves correspond to traditional moral concepts that have been explored by ethicists for the last 2,500 years. The idea of a moral state-space allows us to aggregate various cortical regions involved in the processing of moral concepts: if we identify neurons or relevant populations of neurons, and tag each of them as being a dimension of the space, using the right statistical tools (principal or independent components analysis primarily), we can reduce the dimensionality of the space to something that is more manageable.

To behave morally would be to have this state-space allocated appropriately (presumably by the conjunction of experience and the ontogeny of native neurobiological equipment) so that one is maximally moral (which, in the case of virtue theory, means being maximally functional). The axes of this reduced state-space would correspond to functionally salient groups of neurons, and regions of the state-space might correspond to the "big three" traditional moral theories that were discussed earlier, or they might help us to identify undiscovered moral concepts. The idea of a moral state-space is one way to conceptually unify disparate brain activity that is related to moral cognition.

## Consilience with Virtue Theory

The evidence, albeit tentative, that we have discussed lends more credence to the moral psychology that is required by virtue theory. Empirically successful moral cognition on the part of an organism requires the appropriate coordination of multi-modal signals conjoined with appropriately cued executive systems that share rich connections with affective and conative brain structures that draw on conditioned memories and insight into the minds of others, so as to think about and actually behave in a maximally functional manner. There is clear consilience between contemporary neuroethics and Aristotelian moral psychology. A co-evolutionary strategy, then, would suggest that some version of pragmatic Aristotelian virtue theory is most compatible with the neurobiological sciences.

The localization work that is mentioned in this paper uses various techniques and experimental regimens. The stimuli normally range from sentences to small photographs. The constraints of rigorous experimental design mean that tests are sometimes conducted in highly artificial situations;

ecological validity for moral reasoning is difficult. Moral cognition exhibits several characteristics that make it difficult to capture in the fMRI chamber.[37] A list of these characterisitics follows.

### Moral Cognition is "Hot."

Owing to evolutionary history, affective and conative states are part and parcel of effective moral judgment. This is endemic to moral reasoning (consider the Damasios' patient E.V.R., whose PFC damage, like that of Phineas Gage, disrupted the connections between limbic "somatic markers" and frontal cortex, resulting in poor moral judgment[14]).Unfortunately, hot cognition is difficult to capture in artificial settings.

### Moral Cognition is Social.

Several crucial components of the neural constituents of moral cognition aim to achieve appropriate behavior in social and group settings. This is no accident: animal and human groups are social groups. Social environments are difficult to simulate in the scanner. A notable methodological improvement in this area is Montague's use of multi-scanner "hyperscanning" methodology, in which several subjects can interact simultaneously while being scanned. Although the technology is currently used to link up subjects from multiple states concomitantly, it could be used to monitor the interactions of subjects who are within sight of each other. Minimally, this technology offers added efficiency in studying social interactions; maximally, however, it adds another dimension to the study of the neural mechanisms of social reasoning. As Montague et al. note, "Studying social interactions by scanning the brain of just one person is analogous to studying synapses while observing either the presynaptic neuron or the postsynaptic neuron, but never both simultaneously. . . . Synapses, like socially interacting people, are best understood by simultaneously studying the interacting components."[21]

### Moral Cognition is Distributed.

Evolution does not build from scratch, but instead tends to work with what is present. Socio-moral behavior is rooted in the brain stem/limbic axis and the PFC, with input and recurrent connections to and from sensory and multimodal cortices and frontal lobe areas: so it involves more or less the entire brain. The reduced stimuli conditions that are necessary to do work in an fMRI chamber might not robustly engage our entire suite of neuro-ethical equipment.

*Moral Cognition is Context-Dependent.*

In one case I might praise you for stealing ("nice work removing that weapon from the terrorist headquarters"), and in another I might condemn you ("please return the candy bar that you stole to its owner"). Experimental set-ups need to take this context sensitivity into account. Experiments with the "trolley problem" do a nice job of teasing apart context-sensitive strands of moral judgment.[31]

*Moral Cognition is Genuine.*

Emotion, reason, and action are bundled together. Selection forces operate on actual behavior, not on hypothetical behavior. Our moral cognitive equipment has evolved to effectively coordinate all aspects of our mind/brain so as to take action that allows us to function properly. Experimental regimens that isolate "dry" thinking-about-things-moral from "wet" here-I-am-doing-moral-things can unnecessarily restrict the scope of the neural mechanisms that are activated.

*Moral Cognition is Directed.*

Moral cognition is about things, broadly construed: How do we interact with the world in a fecund manner? What must I do to function properly? Effective moral cognition is a developmental issue; our socio-moral cognitive system becomes more skilled at navigating a complex physical-social world as time passes. Isolating the act of moral judgment (a knowing "that") from the idea of knowing how to act in the world can be misleading. When in the scanner, I will push the man onto the railroad track to stop the oncoming train; what I would actually do in the real world is more difficult to predict.

This review, and these general observations, lead to several pieces of methodological advice for researchers studying the neural bases of moral cognition.

*Make Things Explicit.*

Ensure that explicit consideration is given to the background moral theory that is affecting the research question being answered and the stimuli domain being used to probe that question. When you use the word "moral" in your research report, ask yourself which moral theory you have in mind, and whether implicit but unexamined background assumptions are causing you to ignore salient data or to choose irrelevant problems.

## Adjust Experimental Regimens Accordingly.

Confirm that you have a theoretically rich (but nonetheless fallible) foundation for the moral intuitions informing the experiment. This requires reviewing the few big-picture survey articles available in the field[7, 29, 32, 37] (and other useful sources [35, 38, 39–41]). Consider that the term "moral" is a theoretical term, and so the link between the stimulus regimen, the problem set, and the theory being tested is more complex than one might think. For example, to assume that certain stimuli sentences are empty of moral content merely because they are purely "factual" is to load the dice against moral realism, while to assume that certain social pictures are morally neutral merely because they aren't threatening is to load the dice against a robustly social conception of morality.

## Keep Ecological Validity in Mind.

Is there any way that you can make the problem and stimulus set more closely resemble an actual socio-moral problem and its accompanying embedded environmental stimuli? Methodologies that are socially robust and that involve interacting with more than just sentential input will be more likely to meet this requirement. Doing all of this intelligently is difficult. There are several problems on the frontiers of brain science, and tackling the neural constituents of moral cognition is surely in the "top ten" in terms of both difficulty and importance. I applaud the researchers who are accomplishing this groundbreaking work. Only by doing it can we seek consilience between normative moral theory, moral psychology, and moral neurobiology; and only by doing these things can we hope to improve our ability to develop and instill good moral judgment in ourselves.

REFERENCES

1. *Ideal code, real world: A rule–consequentialist theory of morality* (New York: Oxford University Press, 2000).

2. T. Hill, *Human welfare and moral worth: Kantian perspectives* (New York: Oxford University Press, 2002).

3. S. Broadie, S. *Ethics with Aristotle* (New York: Oxford University Press, 1991).

4. R. Hursthouse, *On virtue ethics* (New York: Oxford University Press, 1999).

5. G. F. Munzel, *Kant's conception of moral character: The "critical" link of morality, anthropology, and reflective judgment* (Chicago: University of Chicago Press, 1999).

6. R. De Waal, F. *Good natured: The origins of right and wrong in humans and other animals* (Cambridge, MA: Harvard University Press, 1996).

7. J. D. Greene and J. Haidt, How (and where) does moral judgment work? *Trends in Cognitive Science* 6, (2002): 517–523.

8. W. D. Casebeer, *Natural ethical facts: Evolution, connectionism, and moral cognition* (Cambridge, MA: MIT Press, 2003).

9. D. Hume, *A treatise of human nature* (Oxford: Clarendon, 1739/1985).

10. G. E. Moore, *Principia ethica*, rev. ed. (New York: Cambridge University Press, 1994).

11. J. Greene, From neural "is" to moral "ought": What are the moral implications of neuroscientific moral psychology? *Nature Reviews Neuroscience* 4, (2003): 847–850.

12. J. L. Saver and A. R. Damasio, Preserved access and processing of social kowledge in a patient with acquired sociopathy due to ventromedial frontal damage. *Neuropsychologia* 29, (1991): 1241–1249.

13. A. Bechara, A. R. Damasio, H. Damasio, and S. W. Anderson, Insensitivity to future consequences following damage to human prefrontal cortex, *Cognition* 50 (1994): 7–15.

14. A. R. Damasio, *Descartes' error: Emotion, reason, and the human brain* (New York: Putnam, 1994).

15. S. W. Anderson, A. Bechara, H. Damasio, D. Tranel, and A. R. Damasio, Impairment of social and moral behavior related to early damage in human prefrontal cortex, *Nature Neuroscience* 2 (1999): 1032–1037.

16. L. O. Bauer, and V. M. Hesselbrock, P300 decrements in teenagers with conduct problems: Implications for substance abuse risk and brain development. *Biological Psychiatry* 46 (1999): 263–272.

17. J. Moll, J. et al., The neural correlates of moral sensitivity: A functional magnetic resonance imaging investigation of basic and moral emotions, *Journal of Neuroscience* 22 (2002): 2730–2736.

18. J. Moll, R. de Oliveira-Souza, I. E. Bramati, and J. Grafman, Functional networks in emotional moral and nonmoral social judgments, *Neuroimage* 16 (2002): 696–703.

19. S. B. Hamann, T. D. Ely, J. M. Hoffman, and C. D. Kilts, Ecstasy and agony: activation of the human amygdala in positive and negative emotion, *Psychological Science* 13 (2002): 135–141.

20. R. Adolphs, D. Tranel, and A. R. Damasio, The human amygdala in social judgment, *Nature* 393 (1998): 470–474.

21. P. R. Montague et al., Hyperscanning: simultaneous fMRI during linked social interactions, *Neuroimage* 16 (2002): 1159–1164.

22. V. Van Veen, et al., Anterior cingulate cortex, conflict monitoring, and levels of processing. *Neuroimage* 14 (2001): 1302–1308.

23. V. Van Veen and C. S. Carter, The timing of action-monitoring processes in the anterior cingulate cortex, *Journal of Cognitive Neuroscience* 14 (2002): 593–602.

24. S. A. Bunge et al., Prefrontal regions involved in keeping information in and out of mind, *Brain* 124 (2001): 2074–2086.

25. A. W. MacDonald III, J. D. Cohen, V. A. Stenger, and C. S. Carter, Dissociating the role of the dorsolateral prefrontal and anterior cingulate cortex in cognitive control, *Science* 288 (2000): 1835–1838.

26. S. Baron-Cohen, *Mindblindness: An essay on autism and theory of mind* (Cambridge, MA: MIT Press, 1995).

27. G. Rizzolatti, L. Fogassi, and V. Gallese, Neurophysiological mechanisms underlying the understanding and imitation of action, *Nature Reviews Neuroscience* 2 (2001): 661–670.

28. J. Call, Chimpanzee social cognition, *Trends in Cognitive Science* 5 (2001): 388–393.

29. R. Adolphs, Cognitive neuroscience of human social behaviour. *Nature Reviews Neuroscience* 4 (2003): 165–178.

30. J. Haidt, The emotional dog and its rational tail: A social intuitionist approach to moral judgment, *Psychology Review* 108 (2001): 814–834.

31. J. D. Greene, et al., An fMRI investigation of emotional engagement in moral judgment, *Science* 293 (2001): 2105–2108.

32. J. Moll, R. de Oliveira-Souza, and P. J. Eslinger, Morals and the human brain: A working model, *Neuroreport* 14 (2003): 299–305.

33. J. M. Fuster, *Cortex and mind: Unifying cognition* (New York: Oxford University Press, 2002).

34. P.S. Churchland, *Brain-wise: Studies in Neurophilosophy* (Cambridge, Massachusetts: MIT Press, 2002).

35. P.S. Churchland, Towards a cognitive neurobiology of the moral virtues, *Topoi* 17 (1998) 83–96.

36. J.D. Wallis, K.C. Anderson and E.K. Miller, Single neurons in prefrontal cortex encode abstract rules, *Nature* (2001): 411, 953–956.

37. W.D. Casebeer and P.S. Churchland, The neural mechanisms of moral cognition: a multiple-aspect approach to moral judgment and decision-making, *Biology and Philosophy* 18 (2003): 169–194.

38. Moreno, J. D. Neuroethics: an agenda for neuroscience and society. *Nature Reviews Neuroscience*. 4 (2003): 149–153.

39. J.T. Cacioppo, *et al.* (eds), *Foundations in Social Neuroscience* (Cambridge, Massachusetts: MIT Press, 2002).

40. J. Rachels, *The Elements of Moral Philosophy 4th Edition* (New York: McGraw-Hill, 2002).

41. D. Lapsley, *Moral Psychology* (Boulder: Westview, 1996).

WILLIAM CASEBEER, PH.D., is assistant professor of philosophy at the U.S. Air Force Academy.

*Chapter 20*

# From Neural "Is" to Moral "Ought":

*What are the Moral Implications of Neuroscientific*

*Moral Psychology?*\*

*Joshua Greene*

Many moral philosophers regard scientific research as irrelevant to their work because science deals with what is the case, whereas ethics deals with what ought to be. Some ethicists question this is/ought distinction, arguing that science and normative ethics are continuous and that ethics might someday be regarded as a natural social science. I agree with traditional ethicists that there is a sharp and crucial distinction between the "is" of science and the "ought" of ethics, but maintain nonetheless that science, and neuroscience in particular, can have profound ethical implications by providing us with information that will prompt us to reevaluate our moral values and our conceptions of morality.

Many moral philosophers boast a well-cultivated indifference to research in moral psychology. This is regrettable, but not entirely groundless.[1] Philosophers have long recognized that facts concerning how people actually think or act do not imply facts about how people ought to think or act, at least not in any straightforward way. This principle is summarized by the Humean[2] dictum that one can't derive an "ought" from an "is." In a similar vein, moral philosophers since Moore[3] have taken pains to avoid the "nat-

---

\*From *Nature Reviews Neuroscience* 4 (2003): 847–850.

uralistic fallacy," the mistake of identifying that which is natural with that which is right or good (or, more broadly, the mistake of identifying moral properties with natural properties). Prominent among those accused by Moore of committing this fallacy was Herbert Spencer, the father of "social Darwinism," who aimed to ground moral and political philosophy in evolutionary principles.[4] Spencer coined the phrase "survival of the fittest," giving Darwin's purely biological notion of fitness a socio-moral twist: for the good of the species, the government ought not to interfere with nature's tendency to let the strong dominate the weak.

Spencerian social Darwinism is long gone, but the idea that principles of natural science might provide a foundation for normative ethics has won renewed favor in recent years. Some friends of "naturalized ethics" argue, contra Hume and Moore, that the doctrine of the naturalistic fallacy is itself a fallacy, and that facts about right and wrong are, in principle at least, as amenable to scientific discovery as any others. Most of the arguments in favor of ethics as continuous with natural science have been rather abstract, with no attempt to support particular moral theories on the basis of particular scientific research.[5, 6] Casebeer's neuroscientific defense of Aristotelian virtue theory (this issue) [included here as Chapter 19—*Ed.*] is a notable exception in this regard.[7]

A critical survey of recent attempts to naturalize ethics is beyond the scope of this article. Instead I will simply state that I am skeptical of naturalized ethics for the usual Humean and Moorean reasons. Contemporary proponents of naturalized ethics are aware of these objections, but in my opinion their theories do not adequately meet them. Casebeer, for example, examines recent work in neuroscientific moral psychology and finds that actual moral decision making looks more like what Aristotle recommends[8] and less like what Kant[9] and Mill[10] recommended. From this he concludes that the available neuroscientific evidence counts against the moral theories of Kant and Mill, and in favor of Aristotle's. This strikes me as a non sequitur. How do we go from "This is how we think" to "This is how we ought to think"? Kant argued that our actions should exhibit a kind of universalizability that is grounded in respect for other people as autonomous rational agents.[9] Mill argued that we should act so as to produce the greatest sum of happiness.[10] So long as people are capable of taking Kant's or Mill's advice, how does it follow from neuroscientific data—indeed, how could it follow from such data—that people ought to ignore Kant's and Mill's recommendations in favor of Aristotle's? In other words, how does it follow from the proposition that Aristotelian moral thought is more natural than Kant's or Mill's that Aristotle's is better?

Whereas I am skeptical of attempts to derive moral principles from scientific facts, I agree with the proponents of naturalized ethics that scientific facts can have profound moral implications and that moral philosophers have paid too little attention to relevant work in the natural sciences. My understanding of the relationship between science and normative ethics is, however, different from that of naturalized ethicists. Casebeer and others view science and normative ethics as continuous and are therefore interested in normative moral theories that resemble or are "consilient" with theories of moral psychology. Their aim is to find theories of right and wrong that in some sense match natural human practice. By contrast, I view science as offering a "behind the scenes" look at human morality. Just as a well-researched biography can, depending on what it reveals, boost or deflate one's esteem for its subject, the scientific investigation of human morality can help us to understand human moral nature, and in so doing change our opinion of it.

## Neuroscience and Normative Ethics

There is a growing consensus that moral judgments are based largely on intuition—"gut feelings" about what is right or wrong in particular cases.[11] Sometimes these intuitions conflict, both within and between individuals. Are all moral intuitions equally worthy of our allegiance, or are some more reliable than others? Our answers to this question will probably be affected by an improved understanding of where our intuitions come from, both in terms of their proximate psychological/neural bases and their evolutionary histories.

Consider the following moral dilemma (adapted from Unger[12]). You are driving along a country road when you hear a plea for help coming from some roadside bushes. You pull over and encounter a man whose legs are covered with blood. The man explains that he has had an accident while hiking and asks you to take him to a nearby hospital. Your initial inclination is to help this man, who will probably lose his leg if he does not get to the hospital soon. However, if you give this man a lift, his blood will ruin the leather upholstery of your car. Is it appropriate for you to leave this man by the side of the road in order to preserve your leather upholstery?

Most people say that it would be seriously wrong to abandon this man out of concern for one's car seats. Now consider a different case (also adapted from Unger[12]), which nearly all of us have faced. You are at home one day when the mail arrives. You receive a letter from a reputable international aid organization. The letter asks you to make a donation of two hundred dollars to the organization. The letter explains that a two-hundred-dollar donation will allow this organization to provide needed medical attention to some poor people in another part of the world. Is it appropriate for you to not make a donation to this organization in order to save money?

Most people say that it would not be wrong to refrain from making a donation in this case. And yet this case and the previous one are similar. In both cases, one has the option to give someone much-needed medical attention at a relatively modest financial cost. And yet, the person who fails to help in the first case is a moral monster, whereas the person who fails to help in the second case is morally unexceptional. Why is there this difference?

About thirty years ago, the utilitarian philosopher Singer argued that there is no real moral difference between cases such as these two, and that we in the affluent world ought to be giving far more than we do to help the world's most unfortunate people.[13] (Singer currently gives about 20% of his annual income to charity.) Many people, when confronted with this issue, assume or insist that there must be "some good reason" for why it is all right to ignore the severe needs of unfortunate people in far-off countries but deeply wrong to ignore the needs of someone like the unfortunate hiker in the first story. (Indeed, you might be coming up with reasons of your own right now.)

Maybe there is "some good reason" for why it is okay to spend money on sushi and power windows while millions who could be saved die of hunger and treatable illnesses. But maybe this pair of moral intuitions has nothing to do with "some good reason" and everything to do with the way our brains happen to be built.

To explore this and related issues, my colleagues and I conducted a brain imaging study in which participants responded to the above moral dilemmas as well as many others.[14] The dilemma with the bleeding hiker is a "personal" moral dilemma, in which the moral violation in question occurs in an "up close and personal" manner. The donation dilemma is an "impersonal" moral dilemma, in which the moral violation in question does not have this feature. To make a long story short, we found that judgments in response to "personal" moral dilemmas, compared with "impersonal" ones, involved greater activity in brain areas that are associated with emotion and social cognition. Why should this be?

An evolutionary perspective is useful here. Over the last four decades, it has become clear that natural selection can favor altruistic instincts under the right conditions, and many believe that this is how human altruism came to be.[15] If that is right, then our altruistic instincts will reflect the environment in which they evolved rather than our present environment. With this in mind, consider that our ancestors did not evolve in an environment in which total strangers on opposite sides of the world could save each others' lives by making relatively modest material sacrifices. Consider also that our ancestors did evolve in an environment in which individuals standing face-to-face could save each others' lives, sometimes only through considerable personal sacrifice. Given all of this, it makes sense that we would have evolved altruistic instincts that direct us to help others in dire need, but mostly when the ones in need are presented in an "up close and personal" way.

What does this mean for ethics? Again, we are tempted to assume that there must be "some good reason" why it is monstrous to ignore the needs of someone like the bleeding hiker but perfectly acceptable to spend our money on unnecessary luxuries while millions starve and die of preventable diseases. Maybe there is "some good reason" for this pair of attitudes, but the evolutionary account given above suggests otherwise: we ignore the plight of the world's poorest people not because we implicitly appreciate the nuanced structure of moral obligation, but because, the way our brains are wired up, needy people who are "up close and personal" push our emotional buttons, whereas those who are out of sight languish out of mind.

This is just a hypothesis. I do not wish to pretend that this case is closed or, more generally, that science has all the moral answers. Nor do I believe that normative ethics is on its way to becoming a branch of the natural sciences, with the "is" of science and the "ought" of morality gradually melding together. Instead, I think that we can respect the distinction between how things are and how things ought to be while acknowledging, as the preceding discussion illustrates, that scientific facts have the potential to influence our moral thinking in a deep way.

## Neuroscience and Meta-Ethics

Philosophers routinely distinguish between ethics and "meta-ethics." Ethics concerns particular moral issues (such as our obligations to the poor) and theories that attempt to resolve such issues (such as utilitarianism or Aristotelian virtue ethics). Meta-ethics, by contrast, is concerned with more foundational issues, with the status of ethics as a whole. What do we mean when we say something like "Capital punishment is wrong"? Are we stating a putative fact or merely expressing an opinion? According to "moral realism" there are genuine moral facts, whereas moral anti-realists or moral subjectivists maintain that there are no such facts. Although this debate is unlikely to be resolved anytime soon, I believe that neuroscience and related disciplines have the potential to shed light on these matters by helping us to understand our commonsense conceptions of morality.

I begin with the assumption (lamentably, not well tested) that many people, probably most people, are moral realists. That is, they believe that some things really are right or wrong, independent of what any particular person or group thinks about it. For example, if you were to turn the corner and find a group of wayward youths torturing a stray cat,[16] you might say to yourself something like, "That's wrong!" and in saying this you would mean not merely that you are opposed to such behavior, or that some group to which you belong is opposed to it, but rather that such behavior is wrong in and of itself, regardless of what anyone happens to think about it. In other words, you take it that there is a wrongness inherent in such acts that you can perceive, but that exists independently of your moral beliefs and values or those of any particular culture.

This realist conception of morality contrasts with familiar anti-realist conceptions of beauty and other experiential qualities. When gazing upon a dazzling sunset, we might feel as if we are experiencing a beauty that is inherent in the evening sky, but many people acknowledge that such beauty, rather than being in the sky, is ultimately "in the eye of the beholder." Likewise for matters of sexual attraction. You find your favorite movie star sexy, but take no such interest in baboons. Baboons, on the other hand, probably find each other very sexy and take very little interest in the likes of Tom Cruise and Nicole Kidman. Who is right, us or the baboons? Many of us would plausibly insist that there is simply no fact of the matter. Although sexiness might seem to be a mind-independent property of certain individuals, it is ultimately in the eye (that is, the mind) of the beholder.

The big meta-ethical question, then, might be posed as follows: are the moral truths to which we subscribe really full-blown truths, mind-independent facts about the nature of moral reality, or are they, like sexiness, in the mind of the beholder? One way to try to answer this question is to examine what is in the minds of the relevant beholders. Understanding how we make moral judgments might help us to determine whether our judgments are perceptions of external truths or projections of internal attitudes. More specifically, we might ask whether the appearance of moral truth can be explained in a way that does not require the reality of moral truth.

As noted above, recent evidence from neuroscience and neighboring disciplines indicates that moral judgment is often an intuitive, emotional matter. Although many moral judgments are difficult, much moral judgment is accomplished in an intuitive, effortless way. An interesting feature of many intuitive, effortless cognitive processes is that they are accompanied by a perceptual phenomenology. For example, humans can effortlessly determine whether a given face is male or female without any knowledge of how such judgments are made. When you look at someone, you have no experience of working out whether that person is male or female. You just see that person's maleness or femaleness. By contrast, you do not look at a star in the sky and see that it is receding. One can imagine creatures that automatically process spectroscopic redshifts, but as humans we do not. All of this makes sense from an evolutionary point of view.

We have evolved mechanisms for making quick, emotion-based social judgments, for "seeing" rightness and wrongness, because our intensely social lives favor such capacities, but there was little selective pressure on our ancestors to know about the movements of distant stars. We have here the beginnings of a debunking explanation of moral realism: we believe in moral realism because moral experience has a perceptual phenomenology, and moral experience has a perceptual phenomenology because natural selection has outfitted us with mechanisms for making intuitive, emotion-based moral judgments, much as it has outfitted us with mechanisms for making intuitive, emotion-based judgments about who among us are the most suitable mates. Therefore, we can understand our inclination toward moral realism not as an insight into the nature of moral truth but as a by-product of the efficient cognitive processes we use to make moral decisions. According to this view, moral realism is akin to naive realism about sexiness, like making the understandable mistake of thinking that Tom Cruise is objectively sexier than his baboon counterparts. (Note that according to this view moral judgment is importantly different from gender perception. Both involve efficient cognitive processes that give rise to a perceptual phenomenology, but in the case of gender perception the phenomenology is veridical: there really are mind-independent facts about who is male or female.)

Admittedly, this argument requires more elaboration and support, and some philosophers might object to the way I have framed the issue surrounding moral realism. Others might wonder how one can speak on behalf of moral anti-realism after sketching an argument in favor of increasing aid to the poor. (Brief reply: giving up on moral realism does not mean giving up on moral values. It is one thing to care about the plight of the poor, and another to think that one's caring is objectively correct.) However, the point of this brief sketch is not to make a conclusive scientific case against moral realism but simply to explain how neuroscientific evidence, and scientific evidence more broadly, have the potential to influence the way we understand morality. (Elsewhere I attempt to make this case more thoroughly.[17])

Understanding where our moral instincts come from and how they work can, I argue, lead us to doubt that our moral convictions stem from perceptions of moral truth rather than projections of moral attitudes. Some might worry that this conclusion, if true, would be very unfortunate. First, it is important to bear in mind that a conclusion's being unfortunate does not make it false. Second, this conclusion might not be unfortunate at all. A world full of people who regard their moral convictions as reflections of personal values rather than reflections of "the objective moral truth" might be a happier and more peaceful place than the world we currently inhabit.[17] The maturation of human morality will, in many ways, resemble the maturation of an individual person. As we come to understand ourselves better—who we are, and why we are the way we are—we will inevitably change ourselves in the process. Some of our beliefs and values will survive this process of self-discovery and reflection whereas others will not. The course of our moral maturation will not be entirely predictable, but I am confident that the scientific study of human nature will have an increasingly important role in nature's grand experiment with moral animals.

REFERENCES

1. J. M. Doris and S. P. Stich, in *The Oxford Handbook of Contemporary Analytic Philosophy*, ed. F. Jackson and M. Smith (New York: Oxford University Press, 2003).

2. D. Hume, *A treatise of human nature*, ed. L. A. Selby-Bigge and P. H. Nidditch (Oxford: Clarendon, 1739/1978.)

3. G. E. Moore, *Principia ethica* (Cambridge University Press, 1903/1959).

4. H. Spencer, *Data of ethics* (Belle Fourch, Kessinger, 1883/1998).

5. N. L. Sturgeon, in *Essays on moral realism*, ed. G. Sayre-McCord, (Ithaca: Cornell University Press, 1988) 229–255.

6. S. Darwall, A. Gibbard, and P. Railton, Toward a fin de siecle ethics: Some trends, *Philosophical Review* 101 (1992): 115–189.

7. W. D. Casebeer, Moral cognition and its neural constituents, *Nature Reviews Neuroscience* 4 (2003): 841–847.

8. Aristotle, *Nicomachean ethics*, trans. T. Irwin (Indianapolis: Hackett, 1985).

9. I. Kant, *Groundwork of the metaphysic of morals*, trans. H. J. Patton (New York: Harper and Row, 1785/1964).

10. J. S. Mill, *Utilitarianism*, ed. R. Crisp (New York: Oxford University Press, 1861/1998).

11. J. Haidt, The emotional dog and its rational tail: A social intuitionist approach to moral judgment, *Psychology Review* 108 (2001): 814–834.

12. P. Unger, P. *Living high and letting die: Our illusion of innocence* (New York: Oxford University Press, 1996).

13. P. Singer, Famine, affluence, and morality, *Philosophy and Public Affairs* 1 (1972): 229–243.

14. J. D. Greene, R. B. Sommerville, L. E. Nystrom, J. M. Darley, J. D. Cohen, An fMRI investigation of emotional engagement in moral judgment, *Science* 293 (2001): 2105–2108.

15. E. Sober and D. S. Wilson, *Unto others: The evolution and psychology of unselfish behavior* (Cambridge, MA: Harvard University Press, 1998).

16. G. Harman, *The nature of morality* (New York: Oxford University Press, 1977).

17. J. D. Greene, *The terrible, horrible, no good, very bad truth about morality and what to do about it* (Ph.D. diss., Princeton University, 2002).

ACKNOWLEDGMENTS

Many thanks to W. Casebeer, A. Herberlein and L. Nystrom for their valuable comments.

JOSHUA D. GREENE, PH.D., is assistant professor of psychology at Harvard University.

# Part V

# Psychopharmacology

ODERN PSYCHOPHARMACOLOGY confronts us with a growing
variety of difficult ethical questions regarding the ability of drugs
to modify individual brains. The oldest and most difficult of
these questions is how to weigh the potential benefits of psychotropic drugs
against the risks. Given that the brain is the most complex and least understood organ in the body, there may be unforeseen adverse effects of altering
neurons and neural systems. The general aim of pharmacological intervention in the brain is to restore dysfunctional systems responsible for psychiatric or neurological disorders. Yet many psychotropic drugs can have both
positive and negative effects on the brain and mind. This requires a nuanced
discussion of whether or when these drugs can and should be used. We may
agree on treatments that can give life back to a Vietnam or an Iraq war vet-
eran suffering from post-traumatic stress disorder (PTSD). At the same time,
though, we may disagree on whether these treatments should be prescribed
more broadly because of their effects on states of mind that define who we
are. The unifying concern of all these issues is that altering how a person's
brain works may be altering *who* that person is. The articles in this section
address this basic concern about psychopharmacology with respect to therapy and enhancement.

Many psychotropic drugs increasingly are being prescribed for off-label
purposes. These are purposes for which the drugs were not originally designed and for which they did not initially receive FDA approval. One example is gabapentin, an anti-convulsive drug that is now used to treat chronic pain. The beta-adrenergic antagonist propranolol is another. Prescribed
as a first-line antihypertensive and antiarrhythmic drug, it is now being used
experimentally to treat post-traumatic stress and anxiety disorders. More
controversial is the off-label use of pharmacological agents to enhance normal cognition and mood. Whether they are used for therapy or enhancement, all psychotropic drugs can alter brain function and the cognitive and
affective properties that are mediated by the brain.

The President's Council on Bioethics staff working paper "Better Memories? The Promise and Perils of Pharmacological Interventions" examines
the different ways in which psychopharmacology can alter memory. After

distinguishing different types of memory, the council asks whether there may be natural reasons for limits on our ability to remember only so many facts and events. While acknowledging that some emotionally charged memories are too horrible to remember, the council warns that altering memories, even for therapeutic reasons, could adversely affect capacities that are fundamental to our psyche and humanity. The power to blunt or eliminate memories pharmacologically may diminish our capacity for empathy, shame, and remorse and might make us more willing to do things we might not otherwise do. The council concludes by noting that the potential positive and negative consequences of this power oblige us to use it with wisdom and humility.

In "Psychopharmacology and Memory," I discuss the therapeutic use of drugs to treat memory-related psychiatric disorders, as well as the use of drugs to enhance memory. I describe a study conducted by Roger Pitman et al. using propranolol to prevent memories of traumatic events associated with post-traumatic stress disorder. I take issue with the President's Council, arguing that in cases of severe PTSD, the pathological nature of the disorder provides compelling reasons for preventing or erasing the memories that are responsible for it. These reasons outweigh concerns about shame, remorse, and other mental capacities. In the second part of the paper, I consider the prospect of memory enhancement. Like the council, I ask whether there may be natural reasons for the limits on our capacity for memory. I consider whether enhancing our ability to form and store memories might interfere with our ability to retrieve them. In addition, I raise the question of what effects memory enhancement might have on our emotions.

Arthur Caplan and Paul McHugh debate the prospect of enhancing the brain and mind in "Shall We Enhance? A Debate" Responding to the President's Council on Bioethics report *Beyond Therapy: Biotechnology and the Pursuit of Happiness*, Caplan rejects the claim that all forms of enhancement are in principle morally objectionable. He asserts that nothing about enhancement itself will undermine authenticity and virtue in what we achieve, deform our character and spirit, or debase the role of parents toward their children. Many practices in our lives may undermine these traits, but they have nothing to do with enhancement. Caplan argues that while not all enhancements are good or desirable, it does not follow that all enhancements are wrong. A member of the President's Council, McHugh argues that we should resist the temptation to use pharmacological agents to enhance mental capacities. Engaging in this practice would interfere with our ability to respect diversity of excellence and talents among people. It would also alter our expectations of others and thereby undermine emotional aspects of human relations. And it would diminish the value we ascribe to persistence and effort in overcoming obstacles. Such devaluation of effort would consequently distort our appreciation of the importance of these traits in different stages of human development. McHugh cites the misuse of steroids in Major League Baseball as an example of artificial alteration that debases the attributes that ordinarily give meaning to achievement. The costs of psychopharmacological enhancement may be too high for any of its presumed benefits.

In "Neurocognitive Enhancement: What Can We Do and What Should We Do?" Martha Farah et al. raise descriptive and normative questions regarding the use of psychopharmacology to improve cognitive capacities. They discuss safety issues, noting that the long-term effects of intervening in the brain are unknown. This leads one to emphasize that it is one thing to take certain risks with a psychotropic medication to treat a diagnosed mental disorder. It is quite another thing to take these risks for the sake of enhancing normal mental functions. In addition, Farah et al. raise the possibility that some people may feel coerced into taking drugs for neurocognitive enhancement just to "keep up" with their colleagues or peers. If employers see that certain drugs can enhance cognitive function and increase productivity among some employees, then they may come to expect this from all employees. This could deny individuals the free choice of deciding whether or not to take cognition-enhancing drugs for self-improvement. Farah et al. also ask whether neurocognitive enhancement would be fairly distributed and accessible to all people who desired it. While the cost of these drugs might suggest that it would result in unequal access and opportunity, Farah et al. note that in principle neurocognitive enhancement could help to equalize opportunity in society. Like McHugh, though, they question whether taking a pill would undermine the value and dignity of effort in our achievements.

In "The Promise and Predicament of Cosmetic Neurology," Anjan Chatterjee imagines a scenario in which he is a physician prescribing drugs on request for his patients to enhance their cognition and mood. Chatterjee asks whether it would be worth taking risks with these drugs in order to enhance normal mental capacities. He also echoes a concern expressed by McHugh and by Farah et al., questioning whether enhancement of certain mental traits would debase our appreciation of effort and struggle that are essential to the human experience. Responding to the concern that access to enhancement drugs would be unequal, Chatterjee points out that other types of enhancement are already common in society. Cognitive enhancement may not be so different from them. Like Farah et al., Chatterjee considers the possibility that an increase in the use of drugs to enhance cognition would raise social expectations about what we could and should do in work and other aspects of life. This could result in a coercive environment that could limit or even undermine the autonomy of individuals in deciding to take or refrain from taking these drugs. Perhaps the most interesting and significant question that Chatterjee raises is one that McHugh and Farah et al. raise as well. Would prescribing drugs to enhance cognition or mood fundamentally alter the doctor-patient relationship? Whether or how psychopharmacological enhancement might do this will depend on physicians' views of what it means to have a therapeutic relationship with patients. It will depend on whether this is a fiduciary relationship, what it means to act in a patient's best interests, and how physicians conceive of and exercise their professional autonomy in discharging their duty of care to patients.

*Chapter 21*

# Better Memories? The Promise and Perils of Pharmacological Interventions*

*President's Council on Bioethics (Staff Working Paper)*

This working paper seeks to provide background for considering the ethical questions raised by our growing biotechnical abilities to improve or alter human memory. The reasons for seeking such abilities are, at first glance, easy to understand. Because of the centrality of memory in all that we do and are, memory loss has far-reaching and potentially devastating consequences. Perhaps no disease elicits as much horror, sympathy, or biomedical urgency as Alzheimer's disease. We rightly shudder at the prospect of forgetting our own past or not recognizing our loved ones, of being forgotten by an ailing Alzheimer's patient, or of having our own identity "die" to ourselves while we are still living. A massive research effort is under way to understand Alzheimer's disease, and much public support for memory research in particular and brain research more generally is focused on curing this and other memory-destroying diseases.

But the human desire to "improve," "control," or "fix" our memory is not merely medical and therapeutic, and memory loss is not the only memory problem. Recognizing the desire of most people for quicker, sharper, and more reliable memories, many researchers are explicitly pursuing drugs or

*From the President's Council on Bioethics, http://www.bioethics.gov/background/better_memories.html, discussed at the council's Mar. 6, 2003 meeting, session 4.

other pharmacological agents that might improve our "normal" capacity to remember, that might enhance the cognitive performance of both under-achievers (with below-normal "memory IQs") and overachievers (who cannot bear simply to be "normal"), and that might prevent, halt, or reverse age-related memory decline.

In addition to efforts aimed at increasing our power to remember, the goal of producing "better memories" fosters pharmacological efforts aimed at decreasing the necessity of remembering bad things or at reducing the emotional sting of our worst memories. This new class of drugs has the great potential to help those who suffer from traumatic and life-disordering memories, increasing their chances of living, at least partially, a normal life. Yet these drugs also raise new possibilities for abuse and misuse. And even in their most welcome uses, they raise profound questions about the relationship between our subjective experience of memory and the true nature of what we remember.

Before we can begin to make sense of the new science of memory and the biotechnical powers it might set before us, we need to consider the human meaning of memory itself. In particular, we need to consider what memory is, what it would mean to improve it, and how it goes wrong.[1] The analysis that follows aims to provide a survey of some of the major issues and questions, not a comprehensive account. Some important matters are briefly noted without being fully considered, and other important issues are left out altogether. Yet we hope it provides sufficient background for considering the human significance of new pharmacological efforts to produce "better memories," which include interventions aimed at improving our capacity to remember new things, but especially interventions to dull or selectively block our most painful memories.

## The Different Types of Memory

Any effort to understand "human memory," let alone improve "it," must confront a simple, if not always obvious, fact: memory is not a singular phenomenon. Neither is it mediated by a single biological or psychological "system." There are many types of remembering and forgetting.[2] We remember phone numbers that we need only once and the phone numbers that we use every day; we remember the names of old classmates we barely knew and the experiences we shared with our oldest and closest friends; we remember the day we were mugged and the day we were married; we remember how to ride a bicycle and how to speak a foreign language; we remember the soldiers who died in World War II and the names and dates for tomorrow's American history exam; we remember how to drive home from work and what we look like in the mirror. All of the above are surely acts of memory, but each of them involves a different way of remembering, and each of them has a different significance and meaning.

Human memory also looks different when viewed from various human perspectives. There is the vision of the novelist or artist, who attempts to capture descriptively and imaginatively the lived experience of memory; the vision of the philosopher or theologian, who seeks wisdom about the nature of memory and its relationship to human experience and the good life; the vision of the psychologist or clinician, who attempts to research, test, and discover how memory works and how to keep it intact; and the vision of the neuroscientist, who studies the workings of the brain itself, by dissecting and studying the brains of nonhuman animals, by conducting chemical tests on human patients, or by taking pictures of the human brain at work.[3] Much of modern neuroscience has attempted to integrate the study of the "mind" and the "brain," and at least one prominent neuroscientist sees the study of memory as the "Rosetta stone" — that is, as a way to translate between the biological workings of the brain itself and the subjective experiences of those whose brain is at work or malfunctioning.

One of the goals of modern memory research is to organize and describe the different types of memory and "memory systems." These include "short-term memory" and "long-term memory"; "explicit memory" and "implicit memory"; "eidetic memory" and "time-bound memory"; "voluntary memory" and "involuntary memory"; "semantic memory" and "episodic memory." These models of memory are neither fixed nor mutually exclusive, and much of memory research is an effort to refine and perfect how we understand memory's many faces. But building proper models of memory is, as philosopher Eva Brann argues, no easy task, relying as it must on metaphors and images:

> The mechanism of memory itself is conceived of in terms of various models. It may be like a date-stamping machine that time-tags each perceptual event, or like a filing cabinet that is already predated and organized sequentially, or like a fading photograph that indicates time by waning vividness, or like an archaeological dig where dating inferences are made from context, or like a book with cross-references from which one can reconstruct the order of publication, that is, of perceptual occurrence. Sometimes neurophysiologically based systems

are distinguished: "procedural memory" for motor skills, "semantic memory" for languages and facts, "working memory" for temporal order, "episodic memory" for personal experience. Often the memory is analyzed in terms of its structures and the depth as well as the capacity of their levels: immediate and fast-decaying sensory or "iconic" memory, short-term memory where all that is needed for present working purposes is stored, and long-term memory, our deep storage. These models are, of course and of necessity, figurative. Metaphors structure much of our experience in any case. But when we represent interiority we have practically no means except metaphor.[4]

And yet, as alluded to above, the goal of modern neuroscience over the last few decades has been to move beyond metaphor by uncovering scientifically how these different types of memory correspond to the specific functions, sections, or activities of the brain itself. As Steven Rose describes in his book *The Making of Memory: From Molecules to Mind*:

> Brain language has many dialects, spoken by many sorts of biologists—physiologists, biochemists, anatomists—and handles its claims to objectivity with confidence. Mind language can be—generally is—subjective, the language of everyday life, or of the poet or novelist. But in the hands of psychologists, it too aspires towards objectivity. One of the tasks of the new breed of neuroscientist . . . is to learn how to translate between the two objective languages of mind and brain. To help that translation we need a Rosetta stone, some inscription in which the two languages, the Greek of mind and the hieroglyphs of brain, can be read in parallel and the interpretation rules deciphered. Deciphering translation rules is not the same as reducing one language to the other. The Greek is never replaced by the Egyptian; the mind is never replaced by the brain. Instead, we have two distinct and legitimate languages, each describing the same unitary phenomena of the material world. The separate histories of these languages as they have developed over the past century have hitherto made them sometimes rivals, sometimes allies. But . . . the prospect of unity, of healing old divisions and of learning the translation rules, has never seemed brighter.[5]

The first goal of modern memory research, in other words, is to give an objective or scientific account of subjective experience; it is to leave metaphor behind, or at least to verify in the brain which metaphors of the mind are true and which are false. But there is a second, more practical goal as well. Researchers seek to use our knowledge of how memory works—how the specific functions of the brain shape our lived experience of memory— to fix, alter, or manipulate its workings: specifically, to cure dreaded memory disorders, to improve the quality of everyday memory, or to achieve a more perfect control of what we remember and forget. Before moving to describe these potential new powers, it is worth considering both what a "better memory" might mean and the different ways that memory fails.

## What is a "Better" Memory?

To speak about "better memory" is to imply some notion of "best" or "perfect" memory. But it is not easy to specify what having a "perfect" memory in fact means. Perhaps the most obvious reply is that an individual with a perfect memory would never forget anything; he would remember every fact, every face, every encounter, every piece of information, every transgression that he commits himself or suffers at the hands of others. But it does not take much reflection to see that such indiscriminate and perfect recall would be not a blessing but a curse. We would suffer like the Jorge Luis Borges character Funes, the Memorious, who describes his "all-too-perfect" memory as "a garbage disposal"; or like the famous memory patient Shereshevskii, whose photographic memory prevented him from forming normal human relationships. Such total and indiscriminate recall may also make it more difficult to distinguish between big and small, important and trivial, needed and unnecessary, mere facts and their significance.

Perhaps a "perfect" memory means remembering only what we desire, or what we find desirable when we experience it. But this, too, does not seem quite right. For much that is most worth remembering is not by nature desirable; and much that seems undesirable when we first experience it only reveals its true significance, meaning, or value in our lives much later. To remember only what is desirable is to imagine ourselves as more autonomous than we really are—i.e., in full control of our memories—and thus to presume more wisdom than we really have at any given moment. It is also to diminish our horizons. Our memory would become little more than a sequence of seemingly desirable "presents," with no way to relate past, present, and future into a maturing identity.

Perhaps a "perfect" memory means remembering things "as they really are" or "as they actually happen." This seems closer to the truth—though giving an account of what this actually means is rather difficult. It is also only partially the truth. To remember things as they are offers no guidance about what is worth remembering. It provides no insight into the difference between simply cataloging events (the "brain" as camera) and discerning their meaning (the "mind" as photographer and editor). This dilemma raises further questions: Is memory more like an artistic vision (interpretive, creative, contingent) or a static reproduction of past events (objective, given, fixed)? Is the way we remember shaped more by the brain structure of our birth or by the experiences and character of our life? And to what extent do our experiences in the world alter the "memory hardware" of the brain itself?

In the end, there is probably no such thing as a "perfect" memory, just as there is no such thing as a perfect human life. To be imperfect beings means, among other things, having a memory that imperfectly renders many imperfect things. To be creatures of space and time means having a memory that is by nature incomplete. And to be mortal beings means having a memory that must ultimately fail. What we seek, in other words, is not a perfect memory but a good or excellent memory. We seek a memory that honestly helps account for the world as it is and human life as it is lived; a memory that recalls the facts we most need when we most need them; a memory that honors those who came before us and prepares those who will come after us; a memory that allows us to understand other people not simply for what they have done but for what they are; and thus a memory that works not simply chronologically but mosaically, and not simply historically but philosophically.

To remember well, in other words, is to remember at the "right pitch," and it requires both a working instrument (the brain) and a learned capacity to remember with discernment (a well-ordered psyche). This means neither remembering too much, such as trivial facts, minor offenses, or the shames and horrors of life in such a way that we live only in the past, nor remembering too little, such as forgetting the defining moments of life, the information that allows for everyday functioning, or one's own greatest sins and misdeeds. And it means remembering with neither too much emotion, so that we become so haunted by past terrors that our memories control us, nor too little emotion, so that we remember what is joyful, horrible, and inconsequential with the same monotone memory. It also requires an acceptance of the human fact that not all memory is chosen; sometimes memory simply happens to us—both for better and for worse.

## The Ways Memory Fails

Those interested in bettering our memories take their bearings less from an idea of "excellent memory" than from the manifest facts of memory failure.

We should therefore consider the different ways that memory "fails," or fails to satisfy us, and describe those phenomena of remembering and forgetting that individuals experience as "memory problems." But we must do so with caution. As Daniel Schacter has described, some of the apparent "vices" of memory are inextricably linked to its "virtues." "Sometimes we forget the past and at other times we distort it; some disturbing memories haunt us for years," he writes. But the "seven sins of memory," as he calls them, are "by-products of otherwise desirable and adaptive features of the human mind."[6] Put differently, to isolate (and seek to "cure") memory's individual failures risks distorting the way memory works as a whole. It risks disrupting memory's "fragile power," which allows us to weave past, present, and future together in a meaningful way.[7]

Yet some problems of memory are not adaptive but destructive; life is often diminished, not improved, when memory fails, and many memory problems rightfully deserve our best effort to heal them. Consider the following six experiences of memory—all of them memory failures, each of them humanly (and biologically) distinct.

*Group A: Lost Memories*

1. *Alzheimer's Disease:* A condition of declining and ultimately destroyed personal memory; a condition that begins with a self-conscious sense of what is happening and what is coming, and ends with the total loss of self-consciousness itself—or at least consciousness of the life one has lived, the people one has loved, the things one has done, and the world one has known. To cure this disease would mean restoring both a lost memory capacity and the possession of lost memories.

2. *Age-Related Decline ("Mild Cognitive Impairment"):* The decline of memory's power from its peak; it involves the slowing down that comes with human aging, if more quickly or severely than normal. This decline often begins with the reduced ability to remember present names and facts, only to work its way forward, so to speak, by reducing the capacity to remember past experiences. This form of memory loss is described clinically as "mild cognitive impairment," and it has official status in the FDA as a treatable disease. Such treatments would involve a prevention of memory decline or restoration of lost memory capacity.

3. *Head Trauma ("Retrograde Amnesia"):* A condition that results from a physical injury to the brain, resulting in the partial or total loss of one's memory of the past. Such a trauma erases, for the subject, what has already

happened; it shrouds the personal past in mystery, so that this past remains known only (partially) by others. It leaves intact the capacity to learn new things, and yet makes one a stranger to the world—thrown into a life and human relationships that one has no memory of forming. Curing this disease would mean restoring the possession of lost memories.

4. *Head Trauma ("Anterograde Amnesia")*: A condition that results from a physical injury to the brain, resulting in the partial or total inability to remember new things, new events, or new experiences. The known past remains intact as memory, but one is unable to move beyond it. The "new" leaves as quickly as it comes, and our body ages without remembering the experience of being in the world as it ages. Curing this disease would mean restoring a lost memory capacity.

### Group B: Weak Memories

5. *Low Memory IQ*: To have a low "memory IQ"[8] is not to experience a sudden or gradual loss of memory, but rather to be born with a lowered capacity to remember all along, or to be so adversely affected by various environmental factors at an early age that certain memory powers never develop. It is to be slow, not slowing down. Curing this disease, if it is a disease, would mean enhancing a limited memory capacity.

### Group C: Bad Memories

6. *Experiential Trauma ("Post-Traumatic Stress Disorder")*: A condition that results not from direct physical damage to the brain but from the personal experience of something terrible and the haunting effect of how we remember it. To the extent that this phenomenon is understood neurologically—as a problem of the memory system—treating it would require transforming the way we encode and consolidate the memory of emotionally powerful experiences.

The purpose of describing these different memory problems is twofold. First, it is to signal the distinct experiential and biological nature of different memory problems—slow and slowing down, damaged brains and haunted memories, losing the past and losing the capacity to remember what will happen to us in the future. Just as memory has many faces, so do the failure, destruction, and limitations of memory. The second (and hereafter more significant) purpose is to consider the connection between biotechnical efforts to heal these different memory problems and the prospect of biotechnical interventions that go "beyond therapy."

In some of the above cases (Alzheimer's, amnesia, age-related memory decline), the desire for a better memory involves the restoration of something lost, and thus the treatment of an existing affliction.[9] In one case (low memory IQ), it involves the enhancement of a memory that is not "broken" but might have been better made in the first place. In the final case (post-traumatic stress disorder), it involves transforming the way a new memory is made—that is, by intervening after a traumatic experience has occurred to alter the way the experience is "encoded" into memory. It potentially involves treating a "disease" before it happens, and treating a disease that is caused not by a virus or physical trauma but an experience. And it involves intervening in the actual emotional content of our memories. This last case, as we shall see, may be the most profound—both because the technology is close at hand and because the questions it raises lay before us with great clarity the moral dilemmas that come with our expanding control over how we remember.

## Biotechnology and Better Memory

Our focus until now has been on examining the nature of memory itself: what it is, what it might mean to improve it, and the different ways that it fails. We now turn to consider the possibility of actually and actively improving human memory through pharmacology. So far, the effort to treat the major diseases of memory has achieved only limited biomedical success; the problem of memory, in its various guises, is apparently not easy to fix. But this does not mean that we are simply powerless when it comes to trying to better human memory. For while we may not yet be able to remedy unchosen afflictions, we may soon be able to intervene in the workings of memory to achieve our own chosen effects. In what follows, we consider some of the scientific underpinnings and prospects of new biotechnical powers to alter the workings of human memory; we provide a narrative account of where the science stands and where it might be heading, not a comprehensive review of the scientific literature. We focus in particular on two kinds of "enhancements": those that enhance our capacity to remember what is coming, and those that alter the way we remember what has already come. In both cases, we begin to explore the continuity and distinction between therapeutic interventions and interventions that go "beyond therapy."

A. *"Anterograde Enhancement": Preparing for an Unknown Future*

1. While we have not yet cured Alzheimer's disease, which is the focus of much memory research, our understanding of the underlying biology has increased, resulting in at least limited treatments that improve the memory capacity of some early-stage Alzheimer's patients. For example, we have discovered that cholinergic cells are "among the first to die in Alzheimer's patients and that cholinergic mechanisms may be involved in memory formation."[10] This has led to therapeutic interventions with a class of drugs called acetylcholinesterase inhibitors, which inhibit the enzyme that destroys acetylcholine (a neurotransmitter that scientists believe is crucial to forming memories) when it is released. If this enzyme is inhibited, acetylcholine remains at the synapse for a longer period of time. This class of drugs has had a real but limited effect on improving memory in some patients; it can slow down or moderate the effects of the disease, but does not reverse the progressive destruction of the brain.[11]

2. But Alzheimer's treatment is not the only use that has been made of acetylcholinesterase inhibitors, and curing disease is not the only ambition of many memory researchers, who see the prevention of typical age-related decline and the enhancement of everyday memory as a major new market and an exciting new field. For example, a recent study tested the effect of donepezil, one of the major acetylcholinesterase inhibitors, on the performance of middle-aged pilots. As the American Academy of Neurology reports: "The study involved 18 pilots with an average age of 52. First, the pilots conducted seven practice flights on a flight simulator to train them to perform a complex series of instructions. Then half of them took the drug donepezil for 30 days and half took a placebo. They then took the flight simulator test twice more to see if they had retained the training. The pilots who had taken the drug retained the training better than those who had taken the placebo."[12] There is also a large body of research, mostly in animals, demonstrating the effect of "opiate receptor antagonists" on memory formation by stimulating the hormones that are typically released in response to emotionally arousing experiences. This work, as we discuss below, is closely related to recent experiments aimed at dulling the emotional power of certain memories.[13]

3. At the same time, the remarkable complexity of the brain in particular and the human body as a whole makes it very difficult to isolate the functions of memory from other physiological and neurological processes (perception, attention, arousal, etc.) with which it is interconnected. Many

"non-memory drugs" or stimulants have a significant effect on memory; and many "memory drugs" have a significant effect on other bodily functions. So, for example, amphetamines, Ritalin, and dunking one's hand in freezing water have a "positive effect" on the capacity to remember new information, at least over the short term. But these drugs or experiences have their effect not so much by intervening directly in specific memory systems as by affecting other systems of the body that affect how the different memory systems function. They act not directly but indirectly.[14]

4. We should also not assume that bio-technical interventions that address or countervail the biological causes of specific memory diseases will improve the memory capacity of the "worried well," or even prevent the onset of the given diseases themselves. As Steven Rose explains: "The deficits in Alzheimer's Disease and other conditions relate to specific biochemical or physiological lesions, and there is no a priori reason, irrespective of any ethical or other arguments, to suppose that, in the absence of pathology, pharmacological enhancement of such processes will necessarily enhance memory or cognition, which may already be 'set' at psychologically optimal levels."[15] Moreover, even if such drugs or stimulants did improve certain types of memory—such as the speedy retention of new information for a limited period of time—there is little reason to assume that they will improve our memory as a whole, if we understand an excellent memory as remembering at the "right pitch" and with "proper discernment." The cost of "speed" may be missing or misunderstanding what is most memorable, and of course the most powerful stimulants often have other undesirable side effects.

5. Nevertheless, it is indeed possible that we will soon discover a drug that will enhance memory in the ways we desire: by enabling us to retain more new information with less effort, by allowing us to make richer connections between our remembered past and our soon-to-be-remembered future, or by refining the way we remember future experiences. These enhancements, we should note, all involve our capacity to remember experiences and information in the unknown future; they are all forward-looking. Surely there would be a great demand for such drugs if they were to be developed and proved effective, as the craze over lecithin, multivitamins, ginseng, ginkgo biloba, and a variety of other supplements and herbal extracts suggests.

6. We also cannot ignore the profound significance of recent animal studies on the molecular and genetic "switches" that control memory. For example, in 1990, Eric Kandel discovered that blocking the molecule CREB (c-AMP, or cyclic adenosine monophosphate, Response Element Binding

protein) in sea slug nerve cells blocked new long-term memory without affecting short-term memory.[16] A few years later, Tim Tully and Jerry Yin genetically engineered fruit flies with the CREB molecule turned "on"; the resulting flies learned basic tasks in one try, whereas for normal flies it often took ten tries or more. The hypothesis is that "CREB helps turn on the genes needed to produce new proteins that etch permanent connections between nerve cells," and that it is "in these links that long-term memories are stored."[17] Two companies—Memory Pharmaceuticals and Helicon Pharmaceuticals—have been formed to develop potential drugs based on this research. In 1999, Joe Tsien succeeded in genetically engineering mice that learn tasks much more readily. He inserted an additional NR2B gene into a mouse embryo, which caused over-expression of the mice's NMDA receptor 2B: a biological mechanism "embedded in the outer wall of certain brain cells" and "long suspected to be one of the basic mechanisms of memory formation" because it allows the "brain to make an association between two events."[18] Such work, of course, is all very preliminary, and its significance for producing biotechnologies that alter or enhance human memory is uncertain. So far, there seems to be no "silver pill" or "golden gene" for producing better memories with no countervailing biological cost.

## B. "Retrograde Enhancement": Altering Our Remembrance of Things Past

Perhaps closer at hand and more profound are capacities to alter the way we remember emotionally arousing experiences. Recent research on the formation of long-term memories has elaborated two crucial facts: First, there is a period of time after a new experience or new exposure to information during which biotechnical interventions in various memory systems can affect what kind of memories are formed. Second, emotionally arousing experiences activate particular memory systems. These two findings may seem like common sense: After all, the most memorable experiences are typically so complex or so dramatic that "encoding" them into memories cannot happen instantaneously; and precisely because different human experiences have a different meaning, we should expect the brain to encode them differently. And yet, by breaking memory formation down to its component parts—especially the different systems involved in encoding emotionally arousing and emotionally neutral experiences—we are gaining novel forms of control over how we remember.

The desire for such control is of course a very old one. In Shakespeare's

*Macbeth*, Macbeth begs his doctor to free Lady Macbeth from the haunting memory of her own guilty acts:

> Doct. Not so sick, my lord,
>    As she is troubled with thick-coming fancies,
>    That keep her from her rest.
> Macb. Cure her of that:
>    Canst thou not minister to a mind diseas'd,
>    Pluck from the memory a rooted sorrow,
>    Raze out the written troubles of the brain,
>    And with some sweet oblivious antidote
>    Cleanse the stuff'd bosom of that perilous stuff
>    Which weighs upon the heart?
> Doct. Therein the patient
>    Must minister to himself.

Today, the doctor may soon have just the "sweet oblivious antidote" that Macbeth so desired: a class of drugs (beta-adrenergic blockers) that numb the emotional sting typically associated with our memory of intense (and intensely bad) experiences. Seeing how this new antidote was developed—and how it might be used—is a telling tale about the way biological science works and the way the biotechnological fruits of science set before us profound questions about the character of human life. This research also cuts to the heart of memory itself, especially the prospect for controlling our remembrance of things past.

1. By the early 1990s, a body of animal research had established that "newly acquired information can be modulated by drugs or hormones administered shortly after training," focusing in particular on the effect of "opiate receptor agonists" (which impair memory) and "opiate receptor antagonists" (which enhance memory). Additional research in amnesia patients suggested that the amygdala, a "tiny almond-shaped" section of the brain near the hippocampus, "can help to influence or modulate explicit memory for emotionally significant events." James McGaugh, Joseph LeDoux, and many others have done key work in this area,[19] which is summarized by Schacter as follows:

> This modulatory role of the amygdala is linked to its role in determining how various hormones affect memory. Studies of rats and other animals have shown that injecting a stress-related hormone such as epinephrine (which produces high arousal) immediately after an animal learns a

task enhances subsequent memory for that task. This strongly implies that some of the beneficial effects of emotional arousal on memory are due to the release of stress-related hormones by a highly emotional experience. The amygdala plays a key role in this process. When the amygdala is damaged, injecting stress-related hormones no longer enhances memory. The amygdala, then, helps to regulate release of the stress-related hormones that underlie the memory-enhancing effects of emotional arousal.[20]

According to this view, our memory system has specific capacities, involving the amygdala, that match the intensity of an experience with the intensity of our memory of that experience. One adaptive benefit of such a system is that it enables both animals and human beings to attach fearful memories to fearful things, and so to avoid similar experiences in the future. The problem, however, is that the benefits of learning to avoid fearful experiences may be outweighed (or seem to some individuals to be outweighed) by the traumatic and life-disordering effect of the memories themselves. Some experiences for some people are simply too horrible to remember, or induce memories that are too horrible to live with.

2. In 1994, Larry Cahill et. al. used the insights of this animal research to test whether the stress hormone system that modulated memory in human beings could be modified in such a way that the emotional power of certain experiences did not affect how we remember them. The study is summarized as follows:

> Substantial evidence from animal studies suggests that enhanced memory associated with emotional arousal results from an activation of beta-adrenergic stress hormone systems during and after an emotional experience. To examine this implication in human subjects, we investigated the effect of the beta-adrenergic receptor antagonist propranolol hydrochloride on long-term memory for an emotionally arousing short story, or a closely matched but more emotionally neutral story. We report here that propranolol significantly impaired memory of the emotionally arousing story but did not affect memory of the emotionally neutral story. The impairing effect of propranolol on memory of the emotional story was not due either to reduced emotional responsiveness or to nonspecific sedative or attentional effects. The results support the hypothesis that enhanced memory associated with emotional experiences involves activation of the beta-adrenergic system.[21]

More specifically, subjects received either propranolol or a placebo one hour before experiencing either the emotionally neutral or the emotionally arousing version of a slide show. The four different test groups (propranolol/emotion, propranolol/neutral, placebo/emotion, placebo/neutral) were tested for their memory of the different stories one week later. Those taking propranolol and those taking the placebo did not differ in their memory of the neutral story; however, they differed significantly in their memory of the emotionally arousing story, but not in their "subjective emotional reactions to the story assessed immediately after story viewing." In other words, propranolol had little to no effect on how individuals remember everyday or emotionally neutral information but a significant effect on how they remembered emotionally powerful experiences—not on how they reacted to them in the moment but on how they remembered them once the moment had passed. The goal of this study, which moved the underlying research from animals to humans, was to increase our understanding of how we remember, and how we remember emotionally intense and emotionally neutral experiences in different ways. But it established the groundwork for research that aims at clinical (or non-clinical) applications of beta-blockers, and thus for the move from biological science to biotechnology.

3. In 2002, Roger K. Pitman et. al. published a pilot study[22] describing the experimental use of propranolol administered to emergency room patients within six hours after a traumatic experience (mostly car accidents) and for an additional ten days afterward. The patients—both those taking the drug and those taking placebos—were tested for their psychological and physiological response to a retelling (with related images) of the traumatic event. One month after the event, those taking propranolol showed measurably lower incidence of post-traumatic stress disorder (PTSD) symptoms than the control group; and three months later, while the PTSD symptoms of both groups had returned to comparable levels, the propranolol group showed measurably lower psycho-physiological response to "internal cues that symbolized or resembled the initial traumatic event." This study, while preliminary, raises a series of questions: How are we to judge an intervention in human memory so soon after an event, before the ultimate significance of the experience has fully revealed itself and before it is even possible to know whether the individuals will suffer from PTSD? Do such drugs "fix" a "broken" encoding system that attaches too much emotional power to a given memory? Or do they alter a well-working encoding system, one that matches the intensity of one's memory to the intensity of the experience, so that the individual as a whole

can better function? To what extent is the encoding system—and thus the incidence of PTSD—shaped by what is given genetically and to what extent by prior experience or condition of soul? And what lasting effect, if any, do such traumatic experiences have on the brain itself?[23]

The prospect of such "memory numbing" drugs has already elicited considerable public interest in and concern about their potential uses in non-clinical settings: to prepare a soldier to kill (or kill again) on the battlefield; to dull the sting of one's own shameful acts; to allow a criminal to numb the memory of his or her victims.[24] Some of these scenarios are perhaps far-fetched. But the significance of this potential new power—which allows us to separate the subjective experience of memory from the true nature of the experience that is remembered—cannot be underestimated. It surely returns us to the large ethical and anthropological questions with which we began—about the place of memory in shaping the character of human life and about the meaning of remembering things that we would rather forget.

## Philosophical and Ethical Reflections

In this final section of the paper, we explore a few aspects of the possible human significance of new pharmacological powers to intervene in human memory. Most but not all of the issues are especially linked to the matter of memory blunting. Our mode is wisdom-seeking and reflective, preferring to raise questions rather than to presume answers. We are interested in how memory interventions might change human life as a whole, both for individuals and for society.

### 1. The Experience of Memory and the Nature of What We Remember

The power to block or dull the emotional power of certain memories sets before us the "beyond therapy" dilemma in a very clear way. Clearly, some memories are so traumatic that they destroy the lives of those who suffer them. And clearly, many of us desire—at certain moments, if not always—to escape the sting of shameful, embarrassing, or painful memories. But we must consider not only how to draw the line between good and bad, medical and non-medical, uses of this power, but also the meaning of the power itself. For it seems to put truth (remembering events as they really happened and for what they are) in opposition to compassion or well-being (remembering events in a modified way, so that they seem less horrible than they really were). And it potentially gives us new powers of control over how and what we remember, and therefore who and what we are. (In reality, of course, such control is limited, since we cannot escape the memory of others.)

For example, these new powers might make us willing to do things we might not otherwise do, or allow us to do the things we desire without shame, hesitation, or remorse — either by changing our psyche before the act or by giving us the power (known in advance) to numb the sting or shame of the act after the fact. At the same time, the power to numb our memories might make us more "accepting" creatures, by altering our perception of the things we must accept. This forces us to consider the difference between forgiving and forgetting, or between forgiveness that requires soberly facing what needs to be forgiven and that which depends on chemically altering our perception of what needs to be forgiven.[25]

Most profoundly, the power to numb the emotional significance of certain memories alters the inherent connection between how we perceive certain human phenomena and the phenomena themselves and their true nature. Imagine, for example, a witness to a horrible murder. Fearing that he will be haunted by the memory of this event, he immediately takes propranolol to make his memory of the murder "more tolerable." But in doing so, does he risk coming to understand the murder itself as tolerable — that is, as an event that does not sting those who witness it? Does dulling our memory of terrible things make us more "whole" and more "at home in the world"? Or does the experience of terror — the experience of the un-chosen, the inexplicable, the tragic — remind us that we can never be fully whole in this world, especially if we are to take the reality of human evil seriously?

### 2. *Individual Good and Common Good*

One can imagine cases where the good of the individual would be served by numbing the emotional impact of certain memories, for example, those with firsthand experience of the Holocaust.[26] And yet, would the good of society as a whole — or the good of history — be served by a mass numbing of memory? Do those who suffer evil have a duty to remember and bear witness, lest we forget the very horrors that haunt them? The examples of this dilemma need not be so dramatic: the memory of being embarrassed is a source of empathy for others who suffer embarrassment; the memory of loss is a source of empathy for others who experience loss. And yet, can we force those who have lived through a great trauma to endure its memory for the benefit of the rest of us? What kind of people would we be if we did not "want" to remember such things, if we sought simply to make the sting of the Holocaust go away?

And yet, what kind of people are we, especially those who face such horrors firsthand, that we can endure such memories?

### 3. Memory and Human Nature

Memory research raises large questions regarding human nature and human self-understanding. Among other things, it invites reflection on the relation between human beings and animals. Do animals remember in the same way that human beings remember, especially given that human beings have language and perceive the world through lenses colored by opinion and articulable beliefs?[27] Do animals experience the "seven sins of memory"—especially the sins that involve distorting our memory of the past ("bias") in light of our present values and concerns? Or are these sins distinctly human phenomena? Is there an animal equivalent to the word/category test discussed by Dr. Schacter, or the memory process known as "semantic encoding"? Such an exploration of the differences between human beings and other animals might suggest that while there are important biological similarities between the animal brain and the human brain, the differences are perhaps most important. Are these differences explained best or simply in terms of how the different "memory systems" are put together? Or do we need other—perhaps richer—categories for understanding the distinctively human phenomenon of remembering and forgetting?

### 4. Memory and Moral Responsibility

The new science of memory, by describing and seeking to understand memory as a "system," raises questions about the difference between voluntary and involuntary action, and especially about the extent of our responsibility for what we remember and what we forget. For example, to what extent should a woman who forgets her child in a car be held "morally accountable" for her forgetting? Is forgetting an "action" or a "non-action"? Is it something we do or something that happens to us? Can we separate memory failure as a "systems failure" and memory failure as a failure of character (or the result of a disordered psyche)?

### 5. Memory and Coercion

Finally, we cannot be blind to the potentially coercive uses of drugs that alter how we remember and what we forget. Just as drugs that dull the emotional sting of certain memories might be desired by the victim to ease his trauma, they might be useful to the assailant to dull the victim's sense of being wronged. Perhaps no one has a greater interest in blocking the painful memory of evil than the evildoer. And while the use of chemical enhancements of our memory powers may be justifiable or necessary in certain extreme situations (e.g., military), we cannot ignore the potentially coercive nature of normalizing the use of such drugs in certain occupations. Nor can we forget the central place of manipulating memory in totalitarian societies, both real and imagined.[28]

## 6. Conclusion: The (Eternal) Puzzle of Memory

Perhaps Jane Austen captured the mysterious nature of memory and its human significance best: "If any one faculty of our nature may be called more wonderful than the rest, I do think it is memory. There seems something more speakingly incomprehensible in the powers, the failures, the inequalities of memory, than in any other of our intelligences. The memory is sometimes so retentive, so serviceable, so obedient—at others, so bewildered and so weak—and at others again, so tyrannical, so beyond control!—We are to be sure a miracle every way—but our powers of recollecting and of forgetting, do seem peculiarly past finding out." Perhaps it is fitting, as we begin to evaluate the human significance of intervening in the workings of human memory, that we have more questions than answers, more dilemmas than solutions. It is an open question whether we will ever fully understand the nature of memory, a fact that should awaken at least some humility about our capacity to make memory "better." And it is likely that we will be remembered, for better or for worse, by those who follow us, a fact that should inspire at least some sense of responsibility to use our new biotechnical powers wisely.

NOTES

1. A more complete analysis of the human meaning of human memory might begin by asking the following kinds of questions: What does it mean to be the creature that remembers and forgets, that studies and wonders about memory, and that seeks to manipulate and control the way we remember? How does human memory differ from the memory of other animals? Is memory decline actually "normal" for particular age groups? Are remembering and forgetting "activities we engage in" or "experiences that happen to us"? What would it mean to have a "perfect" memory? Why do we so often remember what we would like to forget, and forget what we would like to remember? To what extent is the way we remember shaped by our given genetic or neurological "equipment" and to what extent by our choices, experiences, and upbringing? This paper will touch on many of these issues but adequately address none of them. Keeping them in mind, however, is crucial to considering the ways in which we might alter human memory that go "beyond therapy" and coming to some judgment about the significance or wisdom of doing so.

2. As Daniel Schacter explains: "We have now come to believe that memory is not a single or unitary faculty of the mind, as was long assumed. Instead, it is composed of a

variety of distinct and dissociable processes and systems. Each system depends on a particular constellation of networks in the brain that involve different neural structures, each of which plays a highly specialized role within the system. New breakthroughs in brain imaging allow us to see, for the first time, how these specific parts of the brain contribute to different memory processes." See Schacter, *Searching for memory: The brain, the mind, and the past*, p. 5.

3. Of course, these different perspectives are not mutually exclusive, and much of the best writing about memory draws on all of them. But these perspectives are distinct enough in themselves to be worth noting.

4. Eva Brann, *What, then, is time?* 170–171. Brann also describes additional memory distinctions—including the difference between "access to memory, the memory itself and the memories that it contains—the key, the container, and the contents." And she notes: "There are multitudinous roads of remembrance, ways into memory: spontaneous remembrance, directed recollecting, musing reminiscence, reminding memoranda, directed recollecting, mindful recognition."

5. Steven Rose, *The making of memory: From molecules to mind*, 5–6.

6. For a complete discussion, see Daniel Schacter, *The seven sins of memory: How the mind forgets and remembers*. Schacter describes these sins as: transience, absentmindedness, blocking, misattribution, suggestibility, bias, and persistence.

7. For example: To what extent is the slowing down of memory that comes with growing old a "re-tuning" of memory that allows individuals to make sense of a long life? To what extent are the "memory vices" of old age inextricably linked to the "memory virtues" of coming to know what is most truly memorable or significant?

8. The concept of "memory IQ" is discussed at the October 17, 2002, meeting of the President's Council on Bioethics.

9. Treatment of age-related memory decline might also involve not restoration but prevention: that is, taking drugs before the actual onset of age-related memory decline in an effort to prevent it. This raises important questions about treating diseases that may never arrive, or trying to preemptively stop diseases that might never come.

10. Steven Rose, "Smart drugs": Do they work, are they ethical, will they be legal?, as included in the October 17–18, 2002, briefing book of the President's Council on Bioethics, 6. The above discussion also draws on James McGaugh's testimony before the President's Council on Bioethics, October 17, 2002.

11. For a journalistic account of efforts to produce memory-enhancing drugs, see Stephen S. Hall, Our memories, our selves, *New York Times Magazine*, February 15, 1998, and Robert Langreth, Viagra for the brain, *Forbes*, February 4, 2002.

12. See Donepezil and flight simulator performance: Effects on retention of complex skills, *Neurology* 59 (July 9, 2002).

13. See James L. McGaugh, Significance and remembrance: The role of neuromodulatory systems, *Psychological Science* 1, no. 1 (January 1990): 15–23.

14. The above description draws heavily on Steven Rose, *The making of memory*, 4–5. As Rose explains: "Memory formation requires, amongst other cerebral processes: perception, attention, arousal. All engage both peripheral (hormonal) and central mechanisms. Although the processes involved in recall are less well studied it may be assumed that it makes similar demands. Thus agents that affect any of these concomitant processes may also function to enhance (or inhibit) cognitive performance. Memory formation in simple learning tasks is affected by plasma steroid levels, by adrenaline and even by glucose. At least one agent claimed to function as a nootropic and once widely touted as a smart drug, piracetam, seems to act at least in part via modulation of peripheral steroid levels. Central processes too can affect performance by reducing anxiety, enhancing attention or increasing the salience of the experience to be learned and remembered. Amphetamines, methylphenidate (Ritalin)

antidepressants and anxiolytics, probably act in this way. Other agents regularly cited as potential smart drugs, such as ACTH and vasopressin, may function similarly. Finally, there is evidence from animal studies that endogenous cerebral neuromodulators such as the neurosteroids (e.g., DHEA) and growth factors like BDNF will enhance long-term memory for weakly acquired stimuli." See original for complete list of citations.

15. Rose, *The making of memory*, 3.

16. See Langreth, Viagra for the brain.

17. Ibid.

18. Nicholas Wade, Of smart mice and an even smarter man, *New York Times*, September 7, 1999. See also Joe Z. Tsien et al., Genetic enhancement of learning and memory in mice, *Nature* 401 (September 2, 1999): 63–69.

19. See, for example, J. E. LeDoux, Emotion, memory, and the brain, *Scientific American* 270 (1994): 32–39; James McGaugh, Emotional activation, neuromodulatory systems, and memory, in *Memory distortion: How minds, brains, and societies reconstruct the past*, ed. Schacter et al., 255–273 (1995); and James McGaugh, Memory consolidation and the amygdala: A system perspective, *Trends in Neuroscience* 25, no. 9 ( September 2002).

20. Schacter, *Searching for memory*, 215

21. Larry Cahill et al., Beta-adrenergic activation and memory for emotional events, *Nature* 371 (October 20, 1994): 702–704.

22. Roger K. Pitman et al., Pilot study of secondary prevention of posttraumatic stress disorder with propranolol, *Biological Psychiatry* 51 (2002): 189–142.

23. For a discussion of the effect of traumatic experiences on the brain, see, for example, Amy F. T. Arnsten, The biology of being frazzled: Neurobiological research on response to stress, *Science* (June 12, 1998), and Robert Sapolsky, Stress and your shrinking brain: Posttraumatic stress disorder's effect on the brain, *Discover* (March 1999).

24. See, for example, Ellen Goodman, Matter over mind? *Washington Post*, November 16, 2002, and Erik Baard, The guilt-free soldier, *Village Voice*, January 28, 2003. It is interesting to note the dual appeal of such drugs to both the traumatized victim seeking escape from the horror of his or her experience and the traumatizing assailant looking to escape the inconvenience of his guilty memory.

25. It also forces us to consider the difference between an ethic of just retribution (which requires always remembering) and an ethic of forgiveness (which subordinates remembering the guilty act to the "rebirth" of the guilty).

26. This is not to suggest that drugs would be the only, or even best, way to cope with monumental horror. Many survivors of the Holocaust, through a wide variety of other means, managed without actually forgetting, to make a new life for themselves.

27. In addition to language as a distinguishing characteristic of human beings and human memory, we might also consider the following: man as the being who mourns those who die; man as the being who seeks to be remembered after death; man as the being who celebrates days of remembrance; man as the being who seeks to manipulate memory.

28. George Orwell's *1984* offers just one literary account of how memory control might be central to social coercion.

This staff working paper was prepared by President's Council on Bioethics staff solely to aid discussion and discussed at the PCBE's March 2003 meeting. It does not represent the official views of the council or of the U.S. government.

*Chapter 22*

# Psychopharmacology and Memory[*]

## *Walter Glannon*

Psychotropic and other drugs can alter brain mechanisms regulating the formation, storage, and retrieval of different types of memory. These include "off label" uses of existing drugs and new drugs designed specifically to target the neural bases of memory. This paper discusses the use of beta-adrenergic antagonists to prevent or erase nonconscious pathological emotional memories in the amygdala. It also discusses the use of novel psychopharmacological agents to enhance long-term semantic and short-term working memory by altering storage and retrieval mechanisms in the hippocampus and prefrontal cortex. Although intervention in the brain to alter memory as therapy or enhancement holds considerable promise, the long-term effects of experimental drugs on the brain and memory are not known. More studies are needed to adequately assess the potential benefits and risks of these interventions.

Memory is critical to both human survival and personal identity. Nonconscious emotional memory of fearful or threatening events enables us to recognize and respond appropriately to real threats in the natural and social environment. Episodic memory of events involving personal experience is necessary for the psychological connectedness and continuity that gives one the feeling of persisting through time as the same person. Other forms of memory include nonconscious procedural memory, which enables us to perform basic motor skills and tasks of daily life, and semantic memory,

[*]From *Journal of Medical Ethics* 32 (2006): 74–78.

which enables us to recall and use concepts and facts. Semantic and episodic memory are two forms of declarative memory, which enables us to consciously recall facts and events. Working memory is a short-term version of declarative memory and is involved in such complex cognitive tasks as reasoning and decision making.

Recent advances in psychopharmacology are enabling researchers to intervene in emotional, semantic, and working memory systems. In the first type of intervention, existing drugs are being used to block or reverse the process through which nonconscious fearful memories of traumatic events become pathological and cause post-traumatic stress disorder (PTSD) and similar debilitating mental illnesses. It is because they are used to treat mental disorders that these drugs are a form of therapy. In the second type of intervention, drugs are being designed to enhance the formation, storage, and retrieval capacity of long-term semantic and short-term working memory. It is because they are used to strengthen a capacity considered normal that these drugs are a form of enhancement. These interventions hold considerable promise, but they may also involve pitfalls. Both classes of drugs are experimental in the sense that they are being developed and used for "off label" purposes for which they were not originally designed. Because they are experimental, the long-term effects of these novel forms of psychopharmacology are unknown. I will discuss some of the potential beneficial and harmful effects of psychopharmacology on memory.

## Therapeutic Psychopharmacology

Adrenaline is a hormone released by the adrenal medulla in situations requiring effort to fight against or flee from a perceived or real threat. It is closely related to the other stress hormone, cortisol. Adrenaline is released when the adrenal gland receives a signal from the amygdala in the brain when this structure senses an external threat to a human organism. The amygdala is part of the brain's limbic system, which regulates emotions.[1,2] It is one of the most primitive parts of the brain and plays a critical role in our capacity to avoid threats and survive. One effect of adrenaline is to embed nonconscious emotional memories of fearful or threatening events in the amygdala. If emotional memory is embedded too strongly in the amygdala, however, it can produce a heightened fear response to external events that is out of proportion to the actual nature of the problems at hand. Because emotional memories stored in the amygdala are out of our conscious control, they can be difficult to eradicate or modulate and can adversely influence the nature and content of our beliefs, feelings, and other conscious mental states. Events perceived as stressors or threats can trigger a chronic fear response that puts the brain, body, and mind in a constant state of alert. This describes the pathology and pathophysiology of some forms of depression, anxiety, and the emotionally disturbing flashback memories of traumatic events that characterize PTSD.

Some researchers have raised the possibility of using a beta-adrenergic antagonist, specifically the drug propranolol, to treat PTSD. This drug blocks the effects of the hormone norepinephrine, levels of which rise in the brain in response to adrenaline. The aim of using propranolol would be to prevent the embedding of pathological nonconscious emotional memories of fearful events in the amygdala. Bryan Strange and colleagues have conducted experiments, the results of which appear to support this hypothesis.[3] The key is to intervene in a way that blocks the mechanisms through which these memories are formed and stored. This involves the process of consolidation, whereby an event that one has experienced is first registered by certain neural correlates and is then translated into a permanently stored memory in the brain. Adrenaline appears to play a role in consolidation, ensuring that an emotional memory is strengthened and becomes firmly embedded in the amygdala. If it takes days or weeks for a memory to consolidate and become embedded and stored in the amygdala, and if this process depends on a certain level of adrenaline, then conceivably a sufficient dose of propranolol could block adrenaline and prevent the memory from forming.

This was the hypothesis of Harvard University psychiatrist Roger Pitman, who conducted a study of 41 patients admitted to the emergency room of Massachusetts General Hospital for various traumatic injuries.[4, 5] Subjects in the study took propranolol or a placebo for 10 days immediately after the traumatic event. The idea was to test whether the drug could prevent the consolidation of negative emotional memories of the event by blocking the action of adrenaline and other stress hormones and their effects in the brain. Pitman's hypothesis also rested on the fact that memories that have already formed are vulnerable to erasure over time and need to be reconsolidated. Since many of the same biological mechanisms involved in consolidation are also involved in reconsolidation, it is conceivable that the use of propranolol or other beta-adrenergic antagonists could reverse the mechanisms of consolidation and erase a pathological memory that had already formed. Alternatively, these drugs could be given to people going into harm's way as a form of prevention. They could prevent an excessive release of adrenaline and thereby prevent the formation of heightened fear-inducing memories in the amygdala. Such drugs could be given to soldiers before going into battle, or to paramedics just before responding to a medical disaster. In the light of a recent study showing that about one in six soldiers returning from the war in Iraq have had symptoms of PTSD, anxiety, and major depression, this type of intervention could prevent significant harm to many people.[6]

One possible consequence of using these drugs for this purpose would be the blunting of the natural fear response. This response is adaptive because it offers us a survival advantage in protecting us from external threats. It becomes maladaptive when it puts our bodies and minds on a constant state of high alert and is out of proportion to the real nature of external events. Could we ensure that a beta-blocker designed to prevent a pathological state of fear did not have the extreme opposite effect, weakening our natural fear response to the point of making us vulnerable to real threats? Could we ensure that erasing some harmful memories would not result in the erasure of beneficial memories as well? Even if these drugs were given in carefully calibrated doses, could we accurately weigh the potential benefit against the potential harm in using them for the therapeutic or preventive purposes that I have described?

## Propranolol and PTSD

Suppose that soldiers in the Iraq war were given propranolol as a form of prevention before combat. The aim would be to ensure that no traumatic experience would become embedded in the amygdala as nonconscious emotional memory. This memory could result in a chronic hyperactive fear response when triggered by certain stimuli long after combat. Administering the drug could modulate the fear response. Soldiers would respond appropriately to threatening events but would not form pathological emotional memories of them. Yet if the drug blunted this response too much, soldiers could end up being wounded or killed because they would have lost their normal fight-or-flight response. What was intended as a prophylactic intervention to prevent harm could unwittingly result in harm.

Alternatively, suppose that some of these soldiers already had these pathological memories and were diagnosed with PTSD. The drug would be administered shortly after they returned from combat and ideally would weaken or erase the memories from their storage site in the amygdala. This would be a way of treating the veterans of the war who returned home with the disorder. Here too it is possible that the people treated would develop inappropriate responses to fear-inducing stimuli and would become vulnerable to threats in everyday life. Furthermore, although the amygdala regulates nonconscious emotional memory and the hippocampus regulates conscious episodic memory, it is unclear whether a drug aimed at altering the first type of memory would have any effect on the second. The amygdala and hippocampus are both parts of the limbic system, which regulates general emotional processing, and therefore they are not entirely independent of each other. The action of the drug would have to be very specific, and it would be difficult to predict that the drug would not have any adverse effects on other memory systems. There is no guarantee that targeting negative emotional memories in the amygdala would not result in collateral damage to episodic memories in the hippocampus. Indeed, the results from the experiments by Strange and colleagues seem to confirm this fear. Different memory systems are interconnected to some degree through complex neural pathways. These findings suggest that the benefits of beta-adrenergic blockade or modulation of negative emotional memories may entail costs to the encoding of episodic memories, as well as the potential loss of these memories in retrograde amnesia.

The primary use of beta-adrenergic antagonists such as propranolol is to block or diminish the cardiovascular excitatory response to the stress hormones adrenaline and noradrenaline. Propranolol has been used as an antihypertensive and anti-arrhythmic agent. This and other beta-blockers are generally classified as anti-arrhythmic drugs because they can correct heart rhythm abnormalities by blocking beta-adrenergic receptors in the heart. In some cases, these drugs can have the opposite effect and exacerbate arrhythmias. Presumably, propranolol would not have this effect because of the way it acts on autonomic brain functions. Still, it is not known what effects the chronic use of this drug might have on all the systems involved in the stress response. Specifically, it is unclear whether the intended blocking effect to treat PTSD could be limited to the amygdala.

Perhaps the main problem with beta-adrenergic blockade of stress hormones is that not all people who experience trauma are susceptible to PTSD or to related psychiatric disorders. The initial response is not predictive of who will develop a full-blown disorder. Some people who experience a traumatic event may have sleeping difficulties, nightmares, or obsessive thoughts, but these often disappear not long after the event. In these cases, people would be given a putative therapeutic medication they did not need and would be exposed to its potential risks. One of these risks could be the loss of positive episodic and emotional memories, which could occur if these drugs affected multiple memory systems regulated by multiple cortical-limbic pathways in the brain. These brain pathways include not only the regions that compose the limbic system but also the prefrontal cortex and its projections to and from such limbic structures as the amygdala and the hippocampus. On the other hand, failure to intervene in those who are susceptible could mean losing the opportunity to prevent or effectively treat the disorder. Imaging studies indicate that people with PTSD have smaller hippocampi than those without the disorder. It is not known, however, whether this is a marker of susceptibility to the disorder or an effect of it. Even if magnetic resonance imaging (MRI) or other brain scans could identify those who were at risk, it would not be feasible to scan the brain of every patient admitted to the emergency room following an accident.

Nevertheless, if propranolol could dampen or erase the pathological memories symptomatic of PTSD, then its use could be justified on the ground that a life of psychic suffering is worse than the loss of some positive emotional or episodic memory. The potential benefits of the drug therapy would outweigh the potential risks, significant though they may be. A combination of cognitive-behavioral therapy and anti-anxiety or antidepressant medication has been the conventional treatment for PTSD. Unfortunately, in many cases these interventions have not been effective in treating the disorder. When a condition is intractable to other interventions, when it severely affects one's quality of life, and when it poses a significant risk of harm to oneself and to others, considerations of immediate efficacy can override considerations of long-term safety.

The United States President's Council on Bioethics warned against the psychopharmacological manipulation of memories.[7, 8] It expressed concern about the possibility of therapeutic forgetting on the ground that it could subtly reshape who we are. Unpleasant memories are a necessary imperfection in our human nature. Preventing or eliminating memories would be an undesirable and inherently immoral alteration of our humanity. This position fails, however, to draw the critical distinction between conscious episodic memories that are merely unpleasant and nonconscious negative emotional memories that are pathological. The second type is at the core of psychiatric disorders such as PTSD and severe depression and causes considerable disability and suffering in the people affected by them. Treating these disorders with beta-adrenergic antagonists or other psychopharmacological agents is not meant to alter a normal self but rather to restore a sick self to a normal healthy state. Even if these drugs significantly altered the brain and mind, it seems preferable to alter the self by erasing pathological memories than to retain the self associated with these memories. A substantial alteration of psychological properties might preclude a comparison of earlier and later selves and prevent us from saying that the earlier self was made better off and thus benefited from the alteration. Such a comparison might not be possible if the alteration disrupted psychological connectedness and continuity and thus disrupted personal identity through time. Still, the effect of possibly altering personal identity and the self seems better than the alternative, for intuitively it is preferable to eliminate pathological states of mind than to retain them.

The therapeutic use of drugs targeting emotional memory in mental disorders such as PTSD is very different from the therapeutic use of a distinct class of drugs targeting memory in neurodegenerative disorders such as Alzheimer's disease. Drugs used to treat Alzheimer's aim to prevent additional loss of episodic and semantic memory, especially in the early stages of the disease. For example, donepezil and memantine are designed to do this by preventing further neuronal loss and atrophy in the hippocampus, which is the main brain region implicated in the disease. Propranolol for PTSD aims to erase or prevent pathological emotional memories from forming in the amygdala. These are two distinct interventions with distinct aims involving different memory systems.

The long-term effects of propranolol and similarly acting drugs on nonconscious emotional memory are not known. In particular, it is not known what dosage of the drugs, or when to administer them, would be optimal for preventing pathological fear responses while retaining normal responses to fear-inducing stimuli. In addition, it is not known what effects erasure of these memories would have on other memory systems. It could result in the loss of episodic or other memories that are critical to personal identity through time. The potential side effects of preventing pathological memories from forming could be just as harmful to an individual as the potential side effects of erasing existing pathological memories. While erasing episodic memories could disrupt the psychological connectedness between one's present awareness and awareness of one's past, preventing new episodic memories from forming could disrupt the psychological connectedness between one's present awareness and one's anticipation of the future. This could adversely affect one's ability to learn new things and to plan and undertake new projects. Thus the potential side effects of memory erasure and prevention are equally metaphysically and morally significant. All of these considerations indicate that more studies like those of Strange and Pitman are needed to accurately assess the safety and efficacy of these drugs.

## Memory Enhancement

An even more exciting area of psychopharmacology is the use of drugs to enhance different mechanisms of memory. Unlike the use of beta-adrenergic antagonists to inhibit the formation of, or to erase, pathological emotional memories, these drugs would enhance the storage and retrieval of normally functioning semantic memory and its connection to working memory. As noted earlier, working memory can be characterised as a short-term form of declarative memory, which consists of episodic and semantic memory. Working memory is regulated by the prefrontal cortex, which retrieves semantic and episodic memory of facts and events from a long-term storage site in the hippocampus for short-term use. Drugs that are already under development aim to increase memory storage by acting on the transcription factor cyclic adenosine monophosphate (AMP) and the protein that it modulates, CREB (cyclic AMP response element binding protein). This protein is responsible for switching on and off the genes involved in long-term memory formation and storage. Memory-enhancing drugs would activate CREB through a series of molecular interactions. The drugs would stimulate the neurotransmitter glutamate at the synapses connecting neurons, which would then activate the NMDA (N-methyl-D-aspartate) receptor and increase the supply of CREB inside neurons and thereby strengthen memory consolidation in the pathway between the hippocampus and the prefrontal cortex.

Memory enhancement could benefit individuals by enabling them to access a broader base of factual and conceptual information, as well as to process this information more effectively in decision making and other cognitive tasks. It could promote greater opportunity for individuals to have better education and more lucrative employment. This in turn could benefit society as a whole by creating a more informed population with a higher standard of living.

Some famous cases of people with traumatic brain injuries suggest that different memory systems in the brain operate independently and that damage to them is highly selective. In one case, a person experienced retrograde amnesia and lost episodic memory of the past. He did not, however, experience anterograde amnesia, or loss of the ability to form new memories. In a different case, a person experienced both retrograde and anterograde amnesia, yet he retained semantic and procedural memory.[9] These examples suggest that CREB-boosting drugs could enhance the storage and retrieval of semantic or working memory without adversely affecting conscious episodic memory or nonconscious procedural and emotional memory. If memory systems operate independently of each other, then we should be able to use "smart drugs" to target and enhance one system without having to worry about any collateral damage to other systems.[10-12]

There is some disagreement within the research community about the presumed independence of memory systems. Some believe that, despite reported cases of selective damage to different memory systems, these systems are interconnected through various pathways in the brain.[13-15] The hippocampus and its projections to the prefrontal cortex may, for example, play a critical role in both episodic and semantic memory. If memory systems are interconnected, then could enhancing the mechanisms of one system impair the mechanisms of others? In particular, could enhancing the storage capacity of semantic memory impair the capacity of working memory to retrieve information?

These questions are motivated by an evolutionary interpretation of memory. The limits we have in our capacity to remember only so many facts or events may be part of a natural design that is critical for our survival. Each memory system may have optimal levels of formation, storage, and retrieval, in which case trying to increase the storage capacity of memory in a particular system could have deleterious effects on formation or retrieval in that system or in other systems. Ideally, we would want to use drugs that both increased memory formation and storage and made memory retrieval more efficient across all memory systems. A more rapid rate of memory retrieval might, however, affect the brain's ability to form and store memories. Moreover, increased storage would not necessarily mean quicker retrieval. More facts stored in the brain might result in an overloaded short-term working memory, which could impair the ability to execute cognitive tasks and to learn new things, which depends on a certain degree of forgetting. This makes good evolutionary sense. It is advantageous for us to recall facts and events, or to learn new facts, to the extent that they enable us to perform cognitive and physical tasks of daily life that are necessary for our survival. It is not obvious that storing many memories beyond what is useful in helping us to carry out these tasks could benefit us in this regard or any other.

## CREB-Boosting Drugs

These considerations suggest that there may an optimal amount of CREB in our brains for memory. Too much CREB could result in an overproduction and oversupply of memory, which could result in our brains and minds becoming cluttered with memories of facts and events that served no purpose. This memory overload could impair our ability to perform complex cognitive tasks, which ironically would defeat the aim of using these drugs. CREB-boosting drugs could also overstimulate glutamate. Too much glutamate can kill neurons and thus inhibit the formation of new memories or even eliminate memories that already have been formed and stored in the brain. Admittedly, this is speculative. Yet if there is an optimal balance between remembering and forgetting, then it seems plausible to hypothesize that increased semantic memory storage and decreased forgetting could result in impaired semantic memory retrieval. Neuroscientist Martha Farah supports this point: "We understand very little about the design constraints that were being satisfied in the process of creating a human brain. Therefore, we do not know which 'limitations' are there for a good reason. . . . Normal forgetting rates seem to be optimal for information retrieval."[16, 17] Farah further warns of "hidden costs" of trying to enhance memory, and that evolutionary considerations should make us wary of the prospect of general cognitive enhancement as a "free lunch." Remembering the gist of what happened is an economical way of storing experiences without cluttering memory with trivial details. This also enables us to anticipate what may happen in the future. We should be wary of making the inference that if a certain amount of memory is good, then more memory is better. This point was made by James McGaugh in his testimony before the U.S. President's Council on Bioethics.[7]

The potential problem of memory overload that I have described could be avoided by separating retrieval of recent memory from retrieval of remote memory. Presumably, researchers could avoid any adverse effects by using smart drugs enhancing the storage and retrieval of recent memory while allowing normal forgetting rates of remote memory. Nevertheless, even recent memory can involve many trivial details that could clutter the mind, and it is unclear how specially designed drugs could effectively weed out recent memory of trivial facts from recent memory of useful facts. Nor is it clear that quicker retrieval would not have any untoward effects on the formation and storage of useful semantic memory. Again, the idea of an optimal balance within and among memory systems serving an adaptive purpose seems intuitively plausible. It is unclear how artificial manipulation of naturally designed memory systems that have served us so well could improve these systems. More importantly, it is unclear whether this could be done without any short- or long-term risks to these systems. There is no "memory bank" in the brain corresponding to computer memory. Memory systems in the human brain could not easily be upgraded or expanded by activating neurons with certain drugs, unlike replacing or adding silicon chips in computers to upgrade or increase computer memory. Human memories are not encoded in specific neuronal connections, but are distributed across multiple neural pathways. Because of the complexity of these pathways, it would be difficult to design drugs that could effectively and safely target specific functions of specific memory systems without adversely affecting other functions of other memory systems.[18]

A different worry is that altering regions of the brain that control memory and other cognitive functions might disrupt emotional functions. Cognitive and emotional processing are part of an interconnected system in the mind, which is regulated by interconnected cortical-limbic pathways in the brain. Because of these interactions, trying to enhance cognitive processing could impair emotional processing. A drug that made one "smarter" might also make one emotionally flat by blunting one's affective capacities. Even the therapeutic use of psychopharmacological agents to treat cognitive deficits could have this effect. Anecdotal evidence suggests that dextroamphetamine (Adderall), which is in the same class of drugs as methylphenidate (Ritalin), can have adverse effects on mood. One woman taking this drug to improve attention, memory, and other cognitive abilities that had become impaired due to an earlier series of concussions noted: "I worked like a demon, but I found myself disconnected. At the computer I was entirely focused, but off duty, certain pleasures, like wandering around aimlessly in my own mind, were no longer available to me."[19] It is also possible that, given the connection between cognition and emotion, too much of one and too little of the other could impair some forms of reasoning. In psychopathy, for example, there appears to be a correlation between the inability to experience certain emotions and the inability to rationally consider the long-term consequences of action.[20–22] The constellation of psychological effects of cognitive enhancement might not be so desirable.

No one knows what the long-term cognitive, affective, or conative effects of memory-enhancing drugs would be. Chronic use of psychotropic drugs could lead to the remodeling of synapses and changes in neural circuitry, and it is not known whether this would all be salutary or benign. To be sure, this concern is not unique to enhancement drugs but applies to therapeutic drugs as well. Still, it is one thing for a physician to prescribe a drug with potential adverse effects for therapeutic treatment of a mental disorder. It is quite another thing for a physician to prescribe a drug with potential adverse effects to enhance normal mental functions. Until there is a better understanding of any risks of using drugs for cognitive enhancement, the potential harm from long-term use of these drugs justifies limiting them to short-term use in special circumstances and only when there is a compelling reason to use them.

## Conclusion

Psychopharmacology can be used therapeutically to prevent or erase pathological emotional memory. It can also be used non-therapeutically to enhance the normal formation, storage, and retrieval capacity of semantic and working memory. Each of these interventions raises a different set of medical, metaphysical, and ethical questions. Because off-label and other novel uses of psychotropic and other drugs to alter the neurobiological substrate of memory are experimental, there is still considerable uncertainty about their effects on memory. Accordingly, more longitudinal studies involving a significant number of people are needed to accurately assess the risks and benefits of these drugs. These studies will determine how safe and effective the drugs are, which in turn will influence physicians' decisions regarding whether to prescribe them and individuals' decisions regarding whether to take them. Memory is essential to our experience of persisting through time and is in this respect an essential component of personal identity, the self, agency, and responsibility. It is also critical to our survival in enabling us to recognize and respond appropriately to threats in the natural and social environment. There are different memory systems serving different biological and psychological functions. Some of these systems may function independently, whereas others seem to be interdependent. The psychological importance of memory and its neurobiological complexity make it clear that a better understanding of the effects of psychopharmacology on memory is needed before we can argue that this type of intervention in the brain could be justified as a general practice.

REFERENCES

1. J. LeDoux, *The emotional brain* (New York: Simon and Schuster, 1996).
2. J. LeDoux *The synaptic self* (New York: Norton, 2002), ch. 3.
3. B. Strange, R. Hurlemann, R. Dolan, An emotion induced retrograde amnesia in humans is amygdala and beta-adrenergic dependent, *Proceedings of the National Academy of Sciences* 100 (2003): 13626–13631.
4. R. Pitman, K. Sanders, R. Zusman, et al., Pilot study of secondary prevention of post-traumatic stress disorder with propranolol, *Biological Psychiatry* 51 (2002): 189–192.
5. G. Miller, Learning to forget, *Science* 302 (2004): 34–36.
6. C. Hoge, C. Castro, S. Messer, et al., Combat duty in Iraq and Afghanistan, mental health problems, and barriers to care, *New England Journal of Medicine* 351 (2004): 13–22.
7. U.S. President's Council on Bioethics, seventh meeting, October 17, 2002, session 3: Remembering and forgetting: physiological and pharmacological aspects, www.bioethics. gov/transcripts/oct02/session3.html (accessed Jul 22, 2005).
8. L. Kass, *Beyond therapy: Biotechnology and the pursuit of happiness.* (New York: HarperCollins, 2003).
9. D. Schacter, *Searching for memory* (New York: Basic Books, 1996), ch. 5.
10. T. Tully, R. Bourtchouladze, S. Gossweiler, et al., Targeting the CREB pathway for memory enhancers, *Nature Reviews, Drug Discovery* 2 (2003): 267–277.
11. R. Scott, Bourtchouladze, S. Gossweiler, et al., CREB and the discovery of cognitive enhancers. *Journal of Molecular Neuroscience* 19 (2002): 171–177.
12. G. Lynch, Memory enhancement: The search for mechanism-based drugs. *Nature Neuroscience* 5 (2002): 1035–1038.
13. E. Tulving, Where in the brain is the awareness of one's past? in *Memory, brain, and belief,* ed. D. Schacter and E. Scarry, 208–228 (Cambridge, MA: Harvard University Press, 2000).
14. E. Tulving, Episodic memory: From mind to brain, *Annual Review of Psychology* 53 (2002): 27–51.
15. L. Squire, Declarative and non-declarative memory: Multiple brain systems support learning and memory, in *Memory systems,* ed. D. Schacter and E. Tulving, 203–232 (Cambridge, MA: MIT Press, 1994).
16. M. Farah, Emerging ethical issues in neuroscience, *Nature Neuroscience* 2 (2002):1123–1129, at 1125.
17. M. Farah, J. Illes, R. Cook-Deegan, et al., Neurocognitive enhancement: What can we do and what should we do? *Nature Reviews Neuroscience* 5 (2004): 421–425.
18. S. Hyman and W. Fenton, What are the right targets for psychopharmacology? *Science* 299 (2003): 350–351.
19. C. Jakobson Ramin, In search of lost time, *New York Times Magazine,* December 5, 2004, 11–17, at 15.
20. R. J. R. Blair, Neurological basis of psychopathy, *British Journal of Psychiatry* 182 (2003): 5–7.
21. R. D. Hare, *The Hare psychopathy checklist* (Toronto: Multi-Health Systems, 1991).
22. H. Cleckley, *The mask of sanity* (St. Louis: Mosby, 1967).

WALTER GLANNON, PH.D., is Canada Research Chair in Medical Bioethics and Ethical Theory at the University of Calgary, where he is associate professor of philosophy and associate professor of community health sciences.

*Chapter 23*

# Shall We Enhance? A Debate<sup>*</sup>

*Arthur L. Caplan* and *Paul R. McHugh*

*Your kid's schoolwork not up to par? Looking for Mr. or Ms. Right? Any other problems caused by a mind's eye seemingly not quite on the ball? Answers might lie in a brain-enhancing pill. Some argue this is merely better living through chemistry and in line with humanity's self-improving actions throughout history, but others suggest that quick-fix medications could well distort the very things that make us human. Here a leading bioethicist squares off with a member of the President's Council on Bioethics on the controversy about pursuing better brains with a little help from biotechnology.*

*Arthur L. Caplan*

## Straining their Brains: Why the Case Against Enhancement Is Not Persuasive

By some estimates, 1.5 million Americans have undergone laser surgery to improve their vision. The purveyors of this procedure often promise that those who have it will see better than they ever have before, even with the aid of glasses or contact lenses. Laser surgery sometimes can give eyes better than 20-20 vision.

*From *Cerebrum* 6 (2004): 13–29.

So have those who have undergone this type of procedure and achieved enhanced vision done something immoral? If you were to read the recent report of the President's Council on Bioethics, entitled *Beyond Therapy: Biotechnology and the Pursuit of Happiness,*[1] or a recent article by a member of that council, Michael J. Sandel,[2] you might believe they had.

The arguments the Council and its members make against enhancement merit close scrutiny, not only because they have commanded much attention from policy makers and the media but because these arguments are mistaken. And as new knowledge about enhancing another part of the human nervous system—the brain—becomes available, it is all the more important that flawed moral reasoning does not hinder how this knowledge is used.

## Wishing for Better Brains

*Beyond Therapy* is not mainly focused on new knowledge of the human eye or brain per se or how such knowledge might be used to enhance humanity. It is mostly concerned with efforts to improve children through genetic manipulation, gene therapy, and pharmacological agents, as well as with efforts to extend life and control aging through genetic engineering and other means. Nonetheless, it is easy to see that the kinds of concerns fueling the angst that so permeates this report are surely meant to generally rebuke efforts to improve mental performance as well.

For the most part, those who study the brain have very little interest in enhancing or optimizing anything. They seek to know how the brain works. Many scientists and physicians are also keenly interested in determining how, if possible, to repair the devastating impact of injury, disability, and disease that strike the brain.

But potential interest in brain enhancement is enormous. Already, a number of pharmaceutical and nutritional supplement companies are interested in selling drugs that, like modafinil, allow individuals to go without sleep for longer periods of time than they otherwise could or herbal substances that allegedly improve memory or sexual enjoyment. These are, arguably, enhancement drugs. And two things have not escaped some scientists' notice: that a drug capable of helping an Alzheimer's patient retain memory function might also provide some enhancement to those who simply have poor memory skills and that the market possibilities for selling a drug such as a memory enhancer are huge.

Many students, for example, are keenly interested in any drug that might improve their ability on tests or in musical, dramatic, or athletic performances by allowing for increased short-term memory, greater attention span, or reduced anxiety. The military has an interest in seeing mental performance improved so as to increase the combat effectiveness of individuals and units. And not a few of us drink coffee, tea, colas, and other stimulants to try to enhance our cognitive performance. Many people take various drugs, foods, and herbs, or utilize technology such as virtual reality, to try to enhance their mood, emotional state, sexual enjoyment, or range of sensory experience.

While these activities can be, and sometimes are, abused, it would hardly seem morally objectionable to try to improve one's mental abilities. Surely it is the critics of efforts to improve the brain that bear the burden of showing why this is wrong.

## Going Bad by Doing Better

So what are the council's and Professor Sandel's moral concerns about efforts to improve, enhance, or optimize our brains, vision, or any other human organ or trait? Their objections seem to be as follows: 1. The happiness or satisfaction achieved through engineering is seductive and will lead to a deformation of our character and spirit. 2. Engineered improvements in performance are not authentic, not earned, and therefore not morally commendable. 3. To accept enhancement for our children will undermine and deform the role of the parent.

None of these arguments provides a sufficient reason to oppose enhancement or optimization, be it of our vision or our brains, our own or our children's. Each argument carries some emotive force but is not a sound basis for rejecting choices that individuals might make to improve or optimize themselves or their children. This is not to say that every choice for enhancement or optimization is beyond moral criticism or even morally valid. But it is to say that those who would have us turn away in principle from all forms of enhancement or optimization have not made a convincing case.

Consider this question from the President's Council, which suggests that all efforts at enhancement will distort or deform our character: Indeed, why would one need to discipline one's passions, refine one's sentiments, and cultivate one's virtues—in short, to organize one's soul for action in the world—when one's aspiration to happiness could be satisfied by drugs in a quick, consistent, and cost-effective manner?

The concern expressed here is that if we enhanced ourselves and our achievements and enjoyments came easy, why would we continue striving to be good and virtuous people?

## The Wrong Culprit

The problem with this argument is that many people who do not now strive to be good and virtuous are neither enhanced nor optimized in any way. Laying the blame for vice at the foot of enhancement ignores the inconvenient fact that the desires for quick returns, easy money, and instant gratification have nothing at all to do with enhancement. They are traits of many, if not most, human beings. The notion of character development implicit in this account has deeper roots in fictionalized accounts of young men at boarding schools than in anything that accurately describes how human beings actually evolve the character traits that they manifest.

Still, the council broods in *Beyond Therapy*, easy pleasures and cheap thrills will make us weak and spineless. There is nothing like misery to make us stronger. Sorrow, courageously confronted, can make us wiser and more compassionate. By the same kind of logic, the selective serotonin reuptake inhibitor (SSRI) antidepressants, when used to reduce our sorrows, would endanger this aspect of affective life. Because they dull our capacity to feel psychic pain, they would render us less capable of experiencing and learning from misfortune or tragedy or empathizing with the miseries of others. If some virtues can be taught only through very trying circumstances, those virtues might be lost or at least less developed. Putting aside the fact that sorrow can also drive some to suicide and bring others to dysfunction and despair, is it really true that improvement and virtue cannot coexist? The council's argument is a bit like those who worried what the military airplane would do to the virtues of the ground combat soldier—that the improved technology would make obsolete the kind of courage needed for a frontal assault. Oh, really? Tell that to the fighter pilot who needs to evade ground-to-air missiles or to the helicopter pilot evacuating a wounded soldier under a barrage of ground fire.

Improving performance is not necessarily toxic to virtue. It simply shifts how virtue is manifested. It is highly unlikely that those with enhanced vision or muscles or brains would lack for challenges in the real world.

## Satisfaction Not Guaranteed

So the case is not made that improving our brains will destroy our character. What then? The council wrings its collective hands at the prospect that enhancement of the brain or optimization of brain performance will cheapen the value of our experiences. But seldom do those who win by cheating or who love by deceiving cease to long for the joy and fulfillment that come from winning fair and square or being loved for who one truly is. Many stoop to fraud to obtain happiness, but none want their feeling of flourishing itself to be fraudulent. Yet a fraudulent happiness is just what the pharmacological management of our mental lives threatens to confer upon us.

Translation: If you don't really earn your performance, if you do not sweat and toil at it, then it will not be authentic, and it will ultimately prove unsatisfying. One is tempted to ask who is writing this stuff—is the council somehow psychically channeling our Puritan Protestant ancestors?

Certainly it is exciting to achieve satisfaction by testing our limits, by seeing what we can achieve by striving, struggling, and working to overcome innate boundaries. But it is also very satisfying to have benefits that simply come from out of the blue or through good fortune. No people with enhanced vision that I have ever encountered feel the least bit of guilt, shame, or doubt that the improved vision they enjoy is fraudulent because they did nothing to deserve or earn it except pay their money and let a laser do its thing. Life is full of many pleasures that are not earned by testing our limits but that are fully and thoroughly enjoyed. Think of the pleasure in winning the lottery; or in being reassured that your friends like you even though you cheat at cards, cannot stop smoking, eat too much, or are sometimes boring; or in solving problems using computers and any other form of technological assistance you can muster to aid your fallible brain.

We do not always have to "earn" our happiness to be really and truly happy. Nor do we reject as fraudulent those things that make us happy that we have done little or nothing to earn. An enhanced brain or improved cognitive functioning would not in principle undermine the ethos of authenticity that undergirds human satisfaction because that infrastructure is not as the council depicts it. Authentic happiness sometimes results from success in the battle against limits, but authentic happiness can also result from luck, happenstance, serendipity, gifts, indulgence, whimsy, and, although the council seems unable to fathom the possibility, even vice.

## Improving Children

Lastly, consider the concerns of the council's Sandel writing in the *Atlantic Monthly*. He is worried that if we seek to perfect our children—to enhance and optimize them— we will no longer see them as "gifts." In a social world that prizes mastery and control, parenthood is a school for humility. That we care deeply about our children and yet cannot choose the kind we want teaches parents to be open to the unbidden. Such openness is a disposition worth affirming—it invites us to abide the unexpected, to live with dissonance, to rein in the impulse to control.

Put aside the irony of the author, a professor at a school (Harvard University) that inspires parents to devote enormous resources to enhancing their children's abilities so that they may enter there, extolling the idea of accepting your kids as they "are." Ignore the fact that the vision of parenting put forward seems unduly bound by an upper-class American vision of what makes for desirable parenthood—no collective parenting or parent-child estrangement clouds Sandel's vision. Is there value to be found in accepting the random draw of the genetic lottery with respect to one's children? Should a random point mutation that produces a slight change in a trait, or a spontaneous recombination of genetic material, really be seen as the source of value in creating the unexpected in our offspring?

It seems to me that much of what parents traditionally do is try to shape and control their children. Would changing what the accidents of nature produce really result in a child that is any less the object of parental design? And would such change lessen parental affection for the child? It is not self-evident that this must be so. One can accept a gift, embellish, tweak, noodle, and modify it in order to improve it, and still cherish what was given as a gift.

The case against all enhancements is not made. Which, again, is not to say that all enhancements are, of necessity, good or desirable. But it is to say that "in-principle" objections to enhancement should not deter those who seek to improve their own minds or those of their children.

REFERENCES

1. President's Council on Bioethics. *Beyond Therapy: Biotechnology and the Pursuit of Happiness.* Washington, D.C.: Dana Press, 2003.
2. Sandel, MJ. The case against perfection. *Atlantic Monthly.* April 2004, 51–62.

*Paul R. McHugh*

# No Veterinarian to "the Naked Ape" I

As I recount to colleagues our debate within the President's Council on Bioethics leading to the publication of the book *Beyond Therapy: Biotechnology and the Pursuit of Happiness*, many ask, "Why are you guys worrying about the off-label use of medications" such as growth hormones, steroids, stimulants, and antidepressants? By "off-label" they mean the use of these drugs and hormones not, as originally intended, to cure people of conditions such as depression, infection, or hormone deficiency but to enable the healthy to become stronger, quicker, or taller than they would naturally. "After all," they note, "who's to say where sickness ends and health begins—and anyway, why can't folks try stuff as long as it doesn't hurt them?"

These natural questions are relatively easy to answer, as they all in some way turn on concerns over the risks involved in taking medications. But I remind my interlocutors that people do certainly sense other problems in "off-label" medications and express their concerns. Witness the recent outcry in the newspapers, picked up and amplified by the president's State of the Union address, over Major League Baseball players who increased their strength—and disrupted the credibility of their records—by using muscle-enhancing steroids and growth hormones on the advice of their trainers and physicians.

Some critics of this practice were concerned over the risks to health that these professional athletes were prepared (or pressured) to accept. Indeed, these risks are not trivial. But many more were troubled by what biologic enhancements implied about the meaning of achievement in sport and the values expressed in athletic competition.

## A Question of Purpose

Several of my questioners did identify this challenging question from the controversy over sports by asking: "Just what are you trying to preserve or defend when you debate the use of medications to enhance some trait, rather than treat an illness?" I hold that answering this question of purpose is central not only to the sports issue but also to the mission of the council itself. Therefore, I begin by noting how this council was charged by the president to spur public discussion on bioethics in a fashion that would get beyond some simple calculus of risks and benefits to consider what challenges to human values and moral purpose the new discoveries in biomedicine could bring to us as people. Sport is one arena where such challenges would emerge, but hardly the only one. Specifically, in working with our chairman, Leon Kass, to produce *Beyond Therapy*, we council members explored how medications with effects on mood and cognition, so useful in treating certain mental disorders, might alter a doctor's practice with people seeking to enhance desirable traits.

Doctors, after all, do not see themselves as veterinarians to Desmond Morris's *Naked Ape*—workers who tinker with the bodily structure and function of a human as if they were simply beefing up a biologic machine. They hold that, as advisors and teachers, they treat people who need more than technological know-how in order to thrive, who need help to understand what goes into a good human life and how it can go awry. However, as information spreads about medications, some patients—perhaps better called "clients"—are turning up asking for and expecting novel pharmacologic services from their doctors, services that may not extend the patients' best interests. *Beyond Therapy* intends to spur the public to think about these matters.

Case examples help make these ideas about apt and inapt use of medications— especially the newly discovered medications—clear. Here are three, chosen because each depicts a particular aspect of contemporary life in a psychiatric practice and represents a situation where human hopes and fears are in play. In each, medications are an issue even though a "quick fix" with some medication not only would have fallen short of a solution but might well have distracted everyone from the central and deeply human issues at the heart of the problem.

## A Frustrated Young Man

To begin: at least once a year, I am asked to see some young man (seldom a young woman) whose parents worry about his school performance and are wondering whether some medications—either sedatives for his mild test anxiety or stimulants for his mild distractibility—might enhance it. The parents are gifted professionals with long records of academic success and honors (valedictorians, Phi Beta Kappa election, etc.). They worry that their son's present school record and lack of scholastic achievements matching theirs indicate either that something is wrong with him that I might fix with one of these new medications they have read about or that he has some unapparent psychological conflict that I might resolve for him.

The truth is that the son does not have the superior IQ of his parents. The statistical "reversion to the mean" inherent in the genetic roulette of a polygenic feature such as IQ has brought him a somewhat lower capacity than his gifted parents. But often he, and subjects like him, more than balance this aspect of their makeup by displaying—and in fact surpassing their parents in—several other fine human characteristics. He may be handsome, charming, athletic, graceful. These traits are visible and acknowledged by all, even though, on the day I see him, his most prominent feature is his frustration over disappointing his parents.

My task in this situation is to get the parents to forget about adjusting him to their aims with medications or anything else. I want them to appreciate what he brings to them and to all of us in life-affirming ways. I point out that no one can "major in IQ" in life, but anyone can use a whole variety of assets to make life work for him or her. These parents need to understand the young man for what he is and use their talents—and social connections if need be—to guide him toward enterprises that will employ his particular talents and skills to build a life and a career. They should emphasize his strengths, stop trying to make him more like themselves, and give up their notion (common, I've discovered, among the gifted) that the only path to success in life is the one they followed.

I do not immediately succeed in this process with some of these parents, primarily because at the start they assume that my job is to do their bidding and "fix" the young man rather than reinterpret their situation for them. But with time I can usually win them over, thanks mainly to the natural affection all parents have for their offspring, but also because I, an outsider, embarrass them into thinking about the gifts of life by emphasizing what is attractive about their son.

## Right Feelings, Wrong Objects  .

Here is a second prototypic example of how assumptions about life can, in the present era, prompt a search for enhancement medications that misses the point. A young woman arrives in my office depressed and concerned about what she imagines to be some flaw in her psychological makeup that renders her unattractive to others. Her concerns, it turns out, have emerged from several failed romances. Each seems to have followed the same course: She meets an attractive young man and develops a relationship that rather promptly—as is customary with young people now—becomes an intimate one. After some months, and just as she has begun to hope they will marry and start a family together, he tells her he is "not ready" for such a serious commitment and its attendant responsibilities. She concludes he is not sufficiently interested in her, and soon they part.

The repetitiveness of this experience— right down to the stock expression "I'm not ready"—leads her to believe that something about her is to blame. She wonders, as she reflects on her feelings and her behavior, if she's "too intense," "too possessive," or "too needy." She's certainly disheartened and demoralized, and she asks me for medication for her mood and perhaps some other medications that would reduce her anxiety around men—making her perhaps more "relaxed" about these matters.

I notice how she is distressed and concerned about male withdrawal but seeks to explain it as a result of her shortcomings. With these ideas in my mind, I try to show her that, in expecting intimacy to lead to commitment, she is the one who is acting in a natural way, and her boyfriends are not. I tell her that she needs neither a sedative for her thoughts nor an antidepressant to rid her of her low mood but a better assessment of the situation she faces.

When I eventually point out how contemporary sexual mores, supported by easy contraception, tend to emphasize what one receives from an intimate relationship rather than what one brings to it—i.e., taking something from one another rather than making something together—she may wonder, primarily because she has never heard such ideas from a doctor, whether she has come to the right office. Only after figuratively catching her breath does she ask exactly what I think she should do in these situations. I respond to this question by saying she will need some coaching or "cognitive-behavioral" psychotherapy as she approaches affectionate relationships in the future. I suggest several therapists—usually female—who have helped other young women I referred.

She came with the belief that her moods and distress represented some set of pathologic features in herself. I try to help her appreciate that she has been cooperating with a cultural system that permits males to remain perpetual adolescents (and even offers them a standard excuse line—"I'm not ready"), postponing indefinitely their transition into responsible—read "stand-up"—men. Her goal should be to figure out how to stop cooperating with this system and its misuse of her.

## Tempting Thoughts

As a final example of the temptation to use pharmacologic tools for enhancement, I offer an experience and thought experiment from my personal, rather than professional, life. I enjoy periodic, several-day visits from my eight-year-old grandson. We do many things together, but one that we enjoy is playing chess and analyzing situations on the board. He's pretty good for a youngster, and for a period of about half to three-quarters of an hour, we can concentrate together on these problems.

But as the time passes I sense his attention waning and eventually—sooner than I do—he wearies of these "if the opponent makes that move then we should follow with this response" analyses. I've learned to offer him something else to do with me then—best something more physical such as running or throwing a ball—all with the tacit agreement that "maybe later" we could return to chess.

The thought experiment, though, comes as I realize how, with a medication such as Ritalin, I could hold him longer at the chessboard, enjoy the interplay with him for a greater stretch of time, and even, so I might rationalize, make him a better player. The thought is enough to identify the injustice. To use my medical skills to draw something I want from him rather than to accept and support the break from effort his nature seeks is to deny, indeed belittle, his boyhood. "More recess, less Ritalin" I regularly prescribe to people worried about how boys tend to be restless in class. I'm even more confident of the wisdom in that prescription after spending time so happily with a first-rate example of the group.

## Cheating Victory of Its Meaning

With these case examples in mind, let us now return to the sports problem that may be the greatest source of public interest and disquiet over pharmacologic enhancements today. I hold that the expressions of concern brought out in those discussions resemble in many ways the concerns raised in my clinical examples. I also believe that some aspects of the solutions likely to be effective for these athletes will apply to practice with patients such as I've described. Much will depend on attitudes in the community about what is to be admired and what is to be scorned, about what advances and what retards our human pursuits.

William James referred to organized sports as "the moral equivalent of war." And for most of us that's just why we are drawn to the games, as both players and spectators. Nowhere else can we see human beings struggling to be their best, displaying the strenuous, dare one say manly, virtues of courage, tenacity, and self-sacrifice for some collective victory in an arena where blood is not shed and lives are not lost. At its best, organized sport works as a tangible and direct moral educator to us all, by identifying people who have honed wonderful physical gifts and, often, demonstrating how adversity and stress can be overcome through persistence and bravery put into play with a sense of purpose.

Major League Baseball should free itself from the misuse of steroids and other drugs not just by appropriate supervision, rules, and stiff fines but as well by ridicule, contempt, and moral reprobation of the offending athletes by their peers and by the supporters of the game. This reproving stance derives from rejecting the "anything goes" view of athletic competitions and is inspired by respect for the opportunity in sport to witness remarkable combinations of human gifts and virtues, played out in a framework of conventions that give those gifts and virtues a stage. Artificially altering the players—distorting their bodies and making them somehow chemically different from the rest of us—debases this opportunity.

Most of us can see these points immediately and appreciate that unnatural procedures, by severing performance from effort, cheat victory of its meaning. In the same way, I try to encourage my patients to see the real goals embedded in their pursuit of happiness. Thus I do not aim to cover over a painful but natural response to life circumstances or tone down some cosmetic flaw. Rather I seek to help a person find coherence and direction in his or her life so as to resolve some of the difficulties prompting the trip to a psychiatrist. "Man does not live on pharmaceuticals alone," we might say today in updating the Gospels. I apply that lesson repeatedly in my office. Each person in my case examples needed help to recognize just how, like the use of steroids by baseball players, the pharmacologic interventions they wanted would be wrong. Here medicating might not be against some formal rule, but it would in important ways distort the goal of treatment and often turn attention away from the real nature of the situation. In all three cases this goal was to recognize the challenging realities built into human life and how best to meet them. The first case illustrated how one should recognize and honor the diversity of excellence to be found among people, the second how to recognize and honor the natural assumptions of human affection, and the last how to recognize and honor psychological characteristics built into and appropriate for the different stages of human development. To intervene with medications in any of these examples might have helped achieve some narrow aim but would have done so at the price of loss of reverence for the good things that life, outside our command, brings to us and prompts us to fulfill.

## The Moral Bottom Line

As anyone who reads *Beyond Therapy* will quickly appreciate, the council was not calling for laws to deal with these issues of "off-label" treatments. We thought and wrote differently here than we would about matters where life and death—or even physical well-being—are involved. Different members brought different experiences with biologic enhancements to this discussion. All, though, wanted to help the public to appreciate both what great goods these new medicines bring to our treatment of the mentally ill and the many aspects of human life at stake as our knowledge in psychopharmacology expands.

In particular, I wanted to emphasize what psychiatric practice taught me about what to behold, identify, and admire in individual lives. I've learned that, when no psychiatric illness disrupts the picture and calls for medical relief with these new medications, these assets usually offset the challenging blemishes that remain for each of us to overcome. People triumph over these milder handicaps when they are helped to make sense of their circumstances, live up to their gifts, and cultivate those strenuous virtues of self-sufficiency, energy, loyalty, and independence of mind that grow with practice over time.

As a doctor, the moral "bottom line" for me in the use of all medications (not just the new ones) is: Turn to a medication only after you have thought carefully about the patient's symptoms and complaints and decided that these issues represent or express some disruption of brain function or structure in need of medical management. Otherwise, help those who consult you to see what they can do to make better sense of their situations and deal more effectively with them. If this method of assessment is followed, then the new discoveries in pharmacology will work as they were designed and a coherent, effective practice of psychiatry will proceed for the benefit of all.

*Rebuttal by Arthur Caplan*

# Important Treatment, Wrong Diagnosis: Enhancement Is Not to Blame for the Abuse of Autonomy

Every so often a physician will complain to me that it is impossible to get families to consent to the cessation of life-extending medical treatments for their terminally ill loved ones. These doctors are not talking about borderline cases; they are talking about situations in which a patient is in the final throes of cancer, the last stages of AIDS, or at the brink of expiring from congestive heart failure. But, still, the family insists that everything be done.

My response is to ask why the doctors are listing options and alternatives. When they look at me with an expression of confusion, I elaborate: Why are they not presenting their recommendation about what they believe is the appropriate thing to do—in these particular situations, to shift from therapeutic interventions to palliation. The physicians sometimes angrily retort that as someone in the field of bioethics I should realize that their role is not to tell people what to do (as these professionals' role models and mentors once did) but to give patients and families choices consistent with the doctrine of informed consent that bioethicists have supposedly drummed into their heads as key to the moral practice of medicine. This line of argument does not move me, I respond, as informed consent is not simply giving people options and alternatives. It is also sharing with them the wisdom of experience and judgment about what option is the best one to pursue.

Dr. McHugh's comments on his experiences with patients and families who turn to the field of medicine for pills and nostrums to fix the everyday woes of life put me in mind of these conversations. He correctly points out that the right thing to do when confronted with parents who want a pill to make their child smarter or patients requesting a drug to calm their anxieties about an unsatisfactory love life is to offer counsel about the acceptance of limits or the need to learn to cope with the challenges of disrespectful, outlandish, or crude behavior. Not every insult, slight, failure, disappointment, and challenge in life merits the prescription of a pill.

Doctors are not waiters; they do not simply respond to the orders and preferences of their patients. If bioethicists have given medicine the idea that this is what informed consent means, whether in the ICU or in the practice of psychiatry, then bioethics is wrong. That said, the fact that medicine has become beguiled by respect for patient autonomy does not mean that Dr. McHugh is right to commingle concerns about enhancement with concerns about an undue obeisance to patient preferences and demands. While he is on the right path in terms of providing sounder alternatives to patients' requests for "quick fixes" and in not yielding to their anger when he offers direction rather than drugs, the source of the more general problem is the medical community's overindulgence of patients' demands—not the fact that they sometimes demand enhancement.

Nor does the example of steroid use in baseball take Dr. McHugh exactly where he wants to go in cautioning about the destructive influence of enhancement. True, some forms of enhancement seem to undermine fair play in sports. But not all.

I am reminded of a conversation I had with an Olympic official who pointed out that the pleasure Americans and Europeans take in beating basketball teams from Africa is in no way diminished by the fact that the African teams are undernourished, poorly coached, and have almost no access to training facilities. We are so used to these forms of enhancement in our sports that not only do we not protest them, we are downright angry if our favorite athletes do not have access to the best dietitians, masseuses, sports psychologists, sports physiologists, and strength coaches.

The dislike of steroids is the dislike of a dangerous drug. But it may or may not extend to the professional baseball player who has had laser surgery to improve his vision or who pitches even better after reconstructive surgery than he did before.

It is hard to draw the line when it comes to enhancement. Surely medicine should not simply prostitute itself to the whims of its patients. The need to rein in autonomy-run-amok in doctor-patient encounters is real. However, the patient who seeks improvement may, even post-counseling, continue to seek it. The point is not that all enhancement is bad, but that it is bad medicine to assume that if a patient wants the enhancement then it must be bad.

*Rebuttal by Paul McHugh*
# Where's the Wisdom?

The book *Beyond Therapy* here under discussion is a product of the President's Council on Bioethics. That preposition, "on," gets little attention but this debate demands that we note what it implies. The council is not a crew of bioethics experts presenting its judgments to the world. It is an assembly of scientists, physicians, and humanists deliberating on the mandate and portfolio of bioethics itself. By discussing vexatious biomedical matters and publishing treatises drawing from these open discussions, we are ultimately trying to discern whether the relatively new discipline of bioethics is entitled to special standing or authority as a body of organized thought enriching the alliance between physicians and the public. Specifically, does it improve upon the physician's traditional ethics of recognizing good practice in enterprises that function (to quote the Hippocratic oath) for "the benefit of the sick"?

Professor Caplan, an academic and highly regarded bioethicist, is a spokesman of this new discipline, and thus I'm deeply disappointed that his views on the matters dealt with in *Beyond Therapy* are so dismissive. We have heard rumblings of dissatisfaction with our council's thought and leadership from him, so I was eager to see a thoughtful recasting of the specific issues we raised that might amplify our efforts or display alternative ways to come at them. Rather, Professor Caplan disparages the whole enterprise to consider and debate potentially deleterious sides to the promiscuous use— "beyond therapy"—of pharmacological treatments.

The council members certainly knew the value of these medications for the sick but thought it wise to consider how they might be used—or requested—when no sickness was involved. Call our endeavor a needed break in the headlong rush toward nonmedical exploitation of biotechnology, a pause to consider any Huxleyan brave new world implications that might rest within the marvelous medical advances provided by modern drugs that affect the mind and brain.

*Beyond Therapy* boils down to the following questions: Are there any serious problems to discuss here? Do, as Professor Caplan claims, only rugby-school-cold-shower folk worry about biotechnology just as they abjure all creature comforts? Are there things about human nature that we want to preserve when thinking about prescribing these new medicines? Professor

Caplan treats all these questions off-handedly—likening, I have to suppose, any concern for the huge increase in Ritalin prescriptions for children to worries over the American appetite for Starbucks coffee. Not only is his response to the book a rebuke to the council's effort, but his answer to any concern expressed by the council is to tell us to get with the program of modernity and stop bothering everybody with trivia.

But he often seems to miss the point. For example, he sees us as worrying that "seductive" promises of bioengineering "deform the role of parent." We do have concerns over "seduction" but not because these promises "undermine . . . the parent." Rather, they distort everyone's—parent's and child's—attitude toward childhood in exactly the way pushing kids to attend Harvard, a practice apparently deplored by Professor Caplan, does.

I spoke as a practicing psychiatrist about my worries with some of the empty promises carried by the new pharmacology in case examples I won't repeat here. But I'd like to mention something other than the flippancy of Professor Caplan's exercise.

I detect no sense of direction in his commentary. As we doctors strive to help patients separate the real from the false, we can hope that bioethics, for which Professor Caplan speaks, might one day come up with better insights than suggesting we meet all demands for therapy, especially novel ones, by going with the tide. My cautions about enhancement are no resistance to change but recognition of how, for doctors, going with the tide may mean forsaking the responsibility—and opportunity—to consider new technologies in the light of wisdom derived from living and sympathetic contact with real people. Proposing the treatment that best allows a patient to flourish is seldom simple, especially when many treatment programs are available and each carries its own complications. Surely physicians can reasonably expect their professional ethics to offer some principled inclination or direction for their practices, given the weight of responsibilities they carry.

Without thoughts that provide a sense of direction—such as emerges from the reflections in the Hippocratic tradition about the nature of "benefit" and "harm" in their dealings with patients—physicians are at the mercy of technologic illusions and detachments. These tend to promote actions at costs they don't anticipate and exact a penalty, in loss of trust, that they are called to pay when damage cannot be repaired.

Our council published *Beyond Therapy* not to worry the public or to legislate practices but to spur discussion about often unseen features of the new biotherapies. Bioethics was spawned from the philosophical faculties but is now being tied more and more into medical discourse. If Professor Caplan's unconcerned response is representative of expert opinion from the new world of "official" bioethics, the public will be disappointed by the contributions from that quarter and may wonder about the value of that new enterprise in thought.

ARTHUR L. CAPLAN PH.D. is a professor of bioethics, chair of the Department of Medical Ethics, and director of the Center for Bioethics at the University of Pennsylvania.

PAUL R. MCHUGH M.D. is a professor of psychiatry at Johns Hopkins University School of Medicine and professor in the Department of Mental Health in the Bloomberg School of Public Health, Johns Hopkins University.

*Chapter 24*

# Neurocognitive Enhancement: What Can We Do and What Should We Do?*

*Martha J. Farah, Judy Illes, Robert Cook-Deegan,
Howard Gardner, Eric Kandel, Patricia King,
Erik Parens, Barbara Sahakian, and Paul Root Wolpe*

Our growing ability to alter brain function can be used to enhance the mental processes of normal individuals as well as to treat mental dysfunction in people who are ill. The prospect of neurocognitive enhancement raises many issues about what is safe, fair, and otherwise morally acceptable. This article resulted from a meeting on neurocognitive enhancement that was held by the authors. Our goal is to review the state of the art in neurocognitive enhancement, its attendant social and ethical problems, and the ways in which society can address these problems.

Many are predicting that the 21st century will be the century of neuroscience. Humanity's ability to alter its own brain function might well shape history as powerfully as the development of metallurgy in the Iron Age, mechanization in the Industrial Revolution, or genetics in the second half of the twentieth century. This possibility calls for an examination of the benefits and dangers of neuroscience-based technology, or "neurotechnology," and consideration of whether, when, and how society might intervene to limit its uses.

*From *Nature Reviews Neuroscience* 6 (2004): 421–425.

At the turn of the century, neurotechnology spans a wide range of methods and stages of development. Brain-machine interfaces that allow direct two-way interaction between neural tissue and electronic transducers remain in the "proof of concept" stage, but show substantial promise.[1] Neurosurgery is increasingly considered as a treatment for mental illnesses and an array of new procedures are under development, including the implantation of devices and tissue.[2] Noninvasive transcranial magnetic stimulation (TMS) of targeted brain areas is the basis of promising new treatments for depression and other psychopathology.[3]

On the leading edge of neurotechnology is psychopharmacology. Our ability to achieve specific psychological changes by targeted neurochemical interventions,which began through a process of serendipity and trial and error in the mid-20th century, is evolving into the science of rational drug design. The psychopharmacopia of the early 21st century encompasses both familiar, and in some cases highly effective, drugs, and a new generation of more selective drugs that target the specific molecular events that underlie cognition and emotion.[4] For the most part, these drugs are used to treat neurological and psychiatric illnesses, and there is relatively little controversy surrounding this use. However, psychopharmacology is also increasingly used for "enhancement"—that is, for improving the psychological function of individuals who are not ill.

The enhancement of normal neurocognitive function by pharmacological means is already a fact of life for many people in our society, from elementary school children to aging baby boomers. In some school districts in the United States the proportion of boys taking methylphenidate exceeds the highest estimates of the prevalence of attention deficit hyperactivity disorder (ADHD),[5] implying that normal childhood boisterousness and distractibility are being targeted for pharmacological intervention. The use of prescription stimulants (such as methylphenidate and dextroamphetamine) as study aids by high school and college students who do not have ADHD has recently drawn attention, and might include as many as 16% of the students on some campuses.[6] Sales of nutritional supplements that promise improved memory in middle age and beyond have reached a billion dollars annually in the United States alone,[7] despite mixed evidence of effectiveness.[8] In contrast to the other neurotechnologies mentioned earlier, whose potential use for enhancement is still hypothetical, pharmacological enhancement has already begun.

## What Can We Do?

Many aspects of psychological function are potential targets for pharmacological enhancement, including memory, executive function, mood, appetite, libido, and sleep.[9, 10] We will use the first two of these, memory and executive function, as examples to show the state of the art in psychopharmaceutical enhancement, the ethical issues raised by such enhancement, and the policy implications of these ethical issues. A brief review of the state of the art in neurocognitive enhancement is offered here; additional information is freely available to readers of this article at www.nyas.org/ebrief/neuroethics and in recent articles by Rose,[11] Lynch,[12] and Hall.[7]

## Memory Enhancement

Memory enhancement is of interest primarily to older adults. The ability to encode new memories declines measurably from the third decade of life onward, and by the fourth decade the decline can become noticeable and bothersome to normal healthy individuals.[13] Memory difficulties in middle or old age are not necessarily a harbinger of future dementia but can be part of the normal pattern of cognitive aging, which does not make it any less inconvenient when we misplace our glasses or forget the name of a recent acquaintance. What can current and imminent neurotechnologies offer us by way of help?

The changes that underlie normal age-related declines in memory probably differ from those that underlie Alzheimer's disease, indicating that the optimal pharmacological approaches to therapy and enhancement might also differ. Although donepezil, a cholinesterase inhibitor that is used to treat Alzheimer's disease, did enhance performance in one study of healthy middle-aged pilots after flight simulator training,[14] drug companies are looking elsewhere for pharmacological approaches to memory enhancement in normal individuals. Recent advances in the molecular biology of memory have presented drug designers with many entry points through which to influence the specific processes of memory formation, potentially redressing the changes that underlie both normal and pathological declines in memory. Most of the candidate drugs fall into one of two categories: those that target the initial induction of long-term potentiation and those that target the later stages of memory consolidation. In the first category are drugs that modulate AMPA ($\alpha$-amino-3-hydroxy-5-methyl-4-isoxazole propionic acid) receptors to facilitate depolarization, including Cortex Pharmaceuticals'Am pakines.[12] In the second category are drugs that increase CREB (the cAMP response element-binding protein), a molecule that in turn activates genes to produce proteins that strengthen the synapse. One such drug is the molecule MEM1414, which is being tested by Memory Pharmaceuticals[7] (a company cofounded by one of the authors (E.K.)).

The pursuit of mastery over our own memories includes erasing undesirable memories as well as retaining desirable ones. Traumatic events can cause lifelong suffering by the intrusive memories of post-traumatic stress disorder (PTSD), and methods are being sought to prevent the consolidation of such memories by pharmacological intervention immediately after the trauma.[15] Drugs whose primary purpose is to block memories are also being developed by the pharmaceutical industry.[7] Extending these methods beyond the victims of trauma, to anyone who wishes to avoid remembering an unpleasant event, is another way in which the neural bases of memory could be altered to enhance normal function.

## Enhancement of Executive Function.

"Executive function" refers to abilities that enable flexible, task-appropriate responses in the face of irrelevant competing inputs or more habitual but inappropriate response patterns. These include the overlapping constructs of attention, working memory, and inhibitory control. Drugs that target the dopamine and noradrenaline neurotransmitter systems are effective at improving deficient executive function, for example in ADHD, and have recently been shown to improve normal executive function as well.[16, 17]

For example, one of the authors (B.J.S.) found that healthy young volunteers performed the Tower of London problem-solving task more accurately after being given methylphenidate than after being given a placebo when the task was novel.[16] Methylphenidate also increased accuracy in a complex spatial working memory task, and this was accompanied by a reduction in the activation of areas of the brain that are related to working memory, as shown by positron emission tomography (PET).[17] For the latter task, the amount of benefit was inversely proportional to the volunteers' working memory capacity as assessed by a different working memory task, digit span, with little or no benefit to those with the highest digit span performances. This is of interest in discussions of enhancement, because it indicates that, for this medication and this cognitive ability at least, those with lower levels of performance are more likely to benefit from enhancement than those with higher levels.

Indeed, it is possible that some drugs would compress the normal range of performance in both directions. One of the authors (M.J.F.) found that the dopamine agonist bromocriptine improved performance on various executive function tasks for individuals with lower-than-average working memory capacity, but lowered the performance of those with the highest working memory capacities.[18] Whether enhancement can boost the performance of already high-performing individuals must be determined empirically for each drug and for each type of cognitive ability.

Newer drugs might improve executive function in different ways, influencing different underlying processes and interacting in different ways with individual differences (for example, in working memory capacity) and states (such as restedness). The newest potential neurocognitive enhancer is the drug modafinil, which is approved for the treatment of narcolepsy and is increasingly prescribed "off label" for other purposes.[19] One of the authors (B.J.S.) found that it increases performance among healthy young adults on a set of executive function tasks that differs partly from those that are influenced by methylphenidate, with its effects resulting at least in part from an improved ability to inhibit impulsive responses.[20]

## What Should We Do? Ethical Problems and Policy Solutions

Neurocognitive enhancement raises ethical issues for many different constituencies. These include academic and industry scientists who are developing enhancers, and physicians who will be the gatekeepers to them, at least initially. Also included are individuals who must choose to use or not to use neurocognitive enhancers themselves, and parents who must choose to give them or not to give them to their children. With the advent of widespread neurocognitive enhancement, employers and educators will also face new challenges in the management and evaluation of people who might be unenhanced or enhanced (for example, decisions to recommend enhancement, to prefer natural over enhanced performance or vice versa, and to request disclosure of enhancement). Regulatory agencies might find their responsibilities expanding into considerations of "lifestyle" benefits and the definition of acceptable risk in exchange for such benefits. Finally, legislators and the public will need to decide whether current regulatory frameworks are adequate for the regulation of neurocognitive enhancement or whether new laws must be written and new agencies commissioned.

To focus our discussion, we will dispense with some ethical issues that are important but not specific to neurocognitive enhancement. The first such issue is research ethics. Research on neurocognitive enhancement, as opposed to therapy, raises special considerations mainly insofar as the potential benefits can be viewed as smaller, and acceptable levels of risk to research subjects would be accordingly lower. This consideration is largely academic for those neurocognitive enhancers that come to market first as therapies for recognized medical conditions, which includes all of the substances that are now available for enhancement, although this might not be true in the future. Another important ethical issue concerns the use of neurocognitive enhancement in the criminal justice system, in which a large proportion of offenders fall in the lower range of cognitive ability in general[21] and executive inhibitory control in particular.[22] Although neurocognitive enhancement brings with it the potential for subtle coercion in the office or classroom, "neurocorrection" is more explicitly coercive and raises special issues of privacy and liberty that will not be discussed here. Finally, the ethical problems that are involved in parental decision making on behalf of minor children are complex and enter into the ethics of neurocognitive enhancement in schoolchildren, but will not be discussed here.

The remaining issues can be classified and enumerated in various ways. Four general categories will be used here to organize our discussion of the ethical challenges of neurocognitive enhancement and possible societal responses.

## Safety

The idea of neurocognitive enhancement evokes unease in many people, and one source of the unease is concern about safety. Safety is a concern with all medications and procedures, but our tolerance for risk is smallest when the treatment is purely elective. Furthermore, in comparison to other comparably elective treatments such as cosmetic surgery, neurocognitive enhancement involves intervening in a far more complex system, and we are therefore at greater risk of unanticipated problems. Would endowing learners with super-memory interfere with their ability to understand what they have learned and relate it to other knowledge? Might today's Ritalin users face an old age of premature cognitive decline? The possibility of hidden costs of neurocognitive enhancement might be especially salient because of our mistrust of unearned rewards, and the sense that such opportunities can have Faustian results.

With any drug, whether for therapy or enhancement, we can never be absolutely certain about the potential for subtle, rare, or long-term side effects. Instead, our regulatory agencies determine what constitutes a sufficiently careful search for side effects and what side effects are acceptable in view of a drug's benefits. Although consensus will have to be developed on these issues in connection with neurocognitive enhancement, we see no reason that the same approach cannot be applied here.

## Coercion

If neurocognitive enhancement becomes widespread, there will inevitably be situations in which people are pressured to enhance their cognitive abilities. Employers will recognize the benefits of a more attentive and less forgetful workforce; teachers will find enhanced pupils more receptive to learning. What if keeping one's job or remaining in one's school depends on practicing neurocognitive enhancement? Such dilemmas are difficult but are not without useful legal precedent. Many of the relevant issues have been addressed in legislation such as Connecticut's statute Policies Regarding the Recommendation of Psychotropic Drugs by School Personnel[23] and case law such as *Valerie v. Derry Cooperative School District*.[24]

Of course, coercion need not be explicit. Merely competing against enhanced coworkers or students exerts an incentive to use neurocognitive enhancement, and it is harder to identify any existing legal framework for protecting people against such incentives to compete. But would we even want to? The straightforward legislative approach of outlawing or restricting the use of neurocognitive enhancement in the workplace or in school is itself also coercive. It denies people the freedom to practice a safe means of self-improvement, just to eliminate any negative consequences of the (freely taken) choice not to enhance.

## Distributive Justice

It is likely that neurocognitive enhancement, like most other things, will not be fairly distributed. Ritalin use by normal healthy people is highest among college students, an overwhelmingly middle-class and privileged segment of the population. There will undoubtedly be cost barriers to legal neurocognitive enhancement and possibly social barriers as well for certain groups. Such barriers could compound the disadvantages that are already faced by people of low socioeconomic status in education and employment. Of course, our society is already full of such inequities, and few would restrict advances in health or quality of life because of the potential for inequitable distribution. Unequal access is generally not grounds for prohibiting neurocognitive enhancement, any more than it is grounds for prohibiting other types of enhancement, such as private tutoring or cosmetic surgery, that are enjoyed mainly by the wealthy. Indeed, in principle there is no reason that neurocognitive enhancement could not help to equalize opportunity in our society. In comparison with other forms of enhancement that contribute to gaps in socioeconomic achievement, from good nutrition to high-quality schools, neurocognitive enhancement could prove easier to distribute equitably.

## Personhood and Intangible Values

Enhancing psychological function by brain intervention is in some ways like improving a car's performance by making adjustments to the engine. In both cases the goal is to improve function, and to the extent that we succeed without compromising safety, freedom of choice, or fairness we can view the result as good. But in other ways the two are very different, because modifying brains, unlike engines, affects persons. The fourth category of ethical issue encompasses the many ways in which neurocognitive enhancement intersects with our understanding of what it means to be a person, to be healthy and whole, to do meaningful work, and to value human life in all its imperfection. The recent report of the President's Council on Bioethics[25] emphasizes these issues in its discussion of enhancement.

Attempts to derive policies from these considerations must contend with the contradictory ways in which different values are both challenged and affirmed by neurocognitive enhancement. For example, we generally view self-improvement as a laudable goal. At the same time, improving our natural endowments for traits such as attention span runs the risk of commodifying them. We generally encourage innovations that save time and effort, because they enable us to be more productive and to direct our efforts toward potentially more worthy goals. However, when we improve our productivity by taking a pill, we might also be undermining the value and dignity of hard work, medicalizing human effort, and pathologizing a normal attention span. The self-transformation that we effect by neurocognitive intervention can be seen as self-actualizing or as eroding our personal identity. Neither the benefits nor the dangers of neurocognitive enhancement are trivial.

In weighing the dangers of neurocognitive enhancement against its benefits, it is important to note the many ways in which similar trade-offs are already present in our society. For example, the commodification of human talent is not unique to Ritalin-enhanced executive ability. It is probably more baldly on display in books and classes that are designed to prepare preschoolers for precocious reading, music, or foreign language skills, but many loving parents seek out such enrichment for their children. Americans admire the effort that was expended in Abraham Lincoln's legendary four-mile walk to school every day, but no one would do that (or want their child to do that) if a bus ride were available. Medicalization has accompanied many improvements in human life, including improved nutrition and family planning. And if we are not the same person on Ritalin as off, neither are we the same person after a glass of wine as before, or on vacation as before an exam. As these examples show, many of our "lifestyle" decisions end up on the right side of one value and the wrong side of another, but this does not necessarily mean that these decisions are wrong.

## Disentangling Moral Principle and Empirical Fact

Since pre-Socratic times, philosophers have sought ways of systematizing our ethical intuitions to identify a set of guiding principles that could be applied in any situation to dictate the right course of action. All of us have ethical intuitions about most situations; one goal of ethics is to replace case-by-case intuitions with principled decisions. A practical social advantage of ethical principles is that they can provide guidance when intuitions are unclear or inconsistent from person to person. The success of an ethical discussion depends on the discussants' ability to articulate the relevant principles as well as the relevant facts about a situation to which the principles apply.

In the ethics of neurocognitive enhancement we are still feeling our way toward the relevant principles and we still have much to learn about the relevant facts. Is it a matter of principle that "medicalization" is bad, or that hard work confers "dignity"? Or are these moral heuristics, rules of thumb that might be contradicted in some cases? And is it a matter of fact that Ritalin reduces our opportunities to learn self-discipline, or could it in fact have no effect or even help us in some way? Until we have disentangled the *a priori* from the empirical claims, and evaluated the empirical claims more thoroughly, we are at risk of making wrong choices.

## When Not to Decide Is to Decide

Neurocognitive enhancement is already a fact of life for many people. Market demand, as measured by sales of nutritional supplements that promise cognitive enhancement, and ongoing progress in psychopharmacology, portend a growing number of people practicing neurocognitive enhancement in the coming years. In terms of policy, we will soon reach the point where not to decide is to decide. Continuing our current laissez-faire approach, with individuals relying on their physicians or illegal suppliers for neurocognitive enhancement, risks running afoul of public opinion, drug laws, and physicians' codes of ethics. The question is therefore not whether we need policies to govern neurocognitive enhancement, but rather what kind of policies we need. The choices range from minimal measures, such as raising public awareness of the potential practical and moral difficulties of neurocognitive enhancement, to the wholesale enacting of new laws and the creation of new regulatory agencies. Between these extremes lies a host of other options, for example the inclusion of neurocognitive enhancement policies in codes of ethics of the professional organizations of physicians, scientists, human resource managers, and educators, and short-term moratoria on neurocognitive enhancement.

Francis Fukuyama[26] has argued for new legislation to control the use of neurocognitive enhancement, among other biotechnologies. He characterizes the work of groups such as the President's Council on Bioethics in the United States and the European Group on Ethics in Science and New Technology as the "intellectual spade work of thinking through the moral and social implications of biomedical research," and suggests that "it is time to move from thinking to acting, from recommending to legislating. We need institutions with real enforcement powers."

We admit to being less certain about the right course of action. With respect to the first three categories of issue, concerning safety, freedom, and fairness, current laws and customs already go a long way toward protecting society. With respect to the fourth category of issue, we believe that there is much more "spade work" (in Fukuyama's words) to be done in sorting out the moral and social implications of neurocognitive enhancement before we move from recommendations to legislation. We should draw an object lesson from the history of federal stem cell legislation in the USA, which was enacted hastily in the wake of reported attempts at human reproductive cloning with limited public understanding of the issues. That legislation is now viewed by many as a setback for responsible biomedical research, and two states have now enacted their own laws to permit a wider range of research activity.

The need for more discussion of the issues is a predictable conclusion for an article like this one, but nevertheless a valid one. One urgent topic for discussion is the role of physicians in neurocognitive enhancement.[27] Although Western medicine has traditionally focused on therapy rather than enhancement, exceptions are well established. Cosmetic surgery is the most obvious example, but dermatology, sports medicine, and fertility treatments also include enhancement among their goals. Enabling a young woman to bank her eggs to allow later childbearing, for example, is not therapeutic but enhancing. Will neurocognitive enhancement join these practices? If so, will it be provided by specialists or family practitioners? What responsibility will physicians take for the social and psychological impact of the enhancements they prescribe, and by what means (for example, informal or formal psychological screening as used by cosmetic surgeons or fertility specialists)?

Beyond these immediate practical issues, we must clarify the intangible ethical issues that apply to neurocognitive enhancement. This requires interdisciplinary discussion, with neuroscientists available to identify the factual assumptions that are implicit in the arguments for and against different positions, and ethicists available to articulate the fundamental moral principles that apply. As a society we are far from understanding the facts and identifying the relevant principles. With many of our college students already using stimulants to enhance executive function and the pharmaceutical industry soon to be offering an array of new memory-enhancing drugs, the time to begin this discussion is now.

## REFERENCES

1. J. Donoghue, Connecting cortex to machines: Recent advances in brain interfaces, *Nature Neuroscience* (Suppl.) 5 (2002): 1085–1088.

2. G. R. Malhi and P. Sachdev, Novel physical treatments for the management of neuropsychiatric disorders, *Journal of Psychosomatic Research* 53 (2002): 709–719.

3. M. S. George and R. H. Belmaker, *Transcranial magnetic stimulation in neuropsychiatry* (Washington, DC: American Psychiatric Press, 2000).

4. S. Barondes, *Better than Prozac: Creating the next generation of psychiatric drugs* (New York: Oxford University Press, 2003).

5. L. H. Diller, The run on Ritalin: Attention deficit disorder and stimulant treatment in the 1990s, *Hastings Center Report* 26 (1996), 12–18.

6. Q. Babcock and T. Byrne, Student perceptions of methylphenidate abuse at a public liberal arts college, *Journal of American College Health* 49 (2000): 143–145.

7. S. S. Hall, The quest for a smart pill, *Scientific American*, September 2003, 54–65.

8. P. E. Gold, L. Cahill, and G. L. Wenk, Ginkgo biloba: a cognitive enhancer? *Psychological Science in the Public Interest* 3, (2002): 2–11.

9. M. Farah, Emerging ethical issues in neuroscience, *Nature Neuroscience* 5 (2002): 1123–1129.

10. M. Farah and P. R. Wolpe, Monitoring and manipulating brain function: New neuroscience technologies and their ethical implications, *Hastings Center Report* 34 (May-June 2004): 35–45.

11. S. P. R. Rose, "Smart drugs": Do they work? Are they ethical? Will they be legal? *Nature Reviews Neuroscience* 3 (2002): 975–979.

12. G. Lynch, Memory enhancement: The search for mechanism-based drugs, *Nature Neuroscience* 5 (2002): 1035–1038.

13. F. I. M. Craik and T. A. Salthouse, *The handbook of aging and cognition* (Hillsdale, NJ: Lawrence Erlbaum, 1992).

14. J. A. Yesavage et al., Donepezil and flight simulator performance: Effects on retention of complex skills, *Neurology* 59 (2002): 123–125.

15. R. K. Pitman et al., Pilot study of secondary prevention of posttraumatic stress disorder with propranolol, *Biological Psychiatry* 15 (2002): 189–192.

16. M. A. Mehta et al., Methylphenidate enhances working memory by modulating discrete frontal and parietal lobe regions in the human brain, *Journal of Neuroscience* 20 (2000): RC65.

17. R. Elliott et al., Effects of methylphenidate on spatial working memory and planning in healthy young adults, *Psychopharmacology* 131 (1997): 196–206.

18. D. Y. Kimberg, M. D'Esposito, and M. J. Farah, Effects of bromocriptine on human subjects depend on working memory capacity, *Neuroreport* 8 (1997): 3581–3585.

19. E. Teitelman, Off-label uses of modafinil, *American Journal of Psychiatry* 158 (2001): 1341.

20. D. C. Turner, Cognitive enhancing effects of modafinil in healthy volunteers, *Psychopharmacology (Berl.)* 165 (2003): 260–269.

21. T. Holland, I. C. Clare, and T. Mukhopadhyay, Prevalence of criminal offending by men and women with intellectual disability and the characteristics of offenders. *Journal of Intellectual Disability Research* 46 (Suppl.) (2002): 6–20.

22. M. C. Brower and B. H. Price, Neuropsychiatry of frontal lobe dysfunction in violent and criminal behaviour: A critical review, *Journal of Neurology, Neurosurgery and Psychiatry* 71 (2002): 720–726.

23. Legislative Commissioners' Office, General Statutes of Connecticut. Title 10, Ch. 169, Sect. 10-212b (January 1, 2003).

24. United States District Court, New Hampshire, Case No. C-88-412-L (August 1, 1991).

25. L. Kass, *Beyond therapy:Biotechnology and the pursuit of happiness* (New York: HarperCollins, 2003).

26. F. Francis, *Our posthuman future* (New York: Farrar, Strauss and Giroux, 2002).

27. A. Chatterjee, Cosmetic neurology: the controversy over enhancing movement, mentation, and mood, *Neurology* 63 (2004): 968–974.

ACKNOWLEDGMENTS

This paper is based, in part, on a meeting held at the New York Academy of Sciences in June 2003, supported by a grant to J.I. from the National Science Foundation with cosponsorship of a Mushett Family Foundation grant to the Academy. The writing of this paper was supported by NSF and NIH grants to M.J.F. and an NIH grant and a Greenwald Foundation grant to J.I.

*Competing-Interests Statement* The authors declare competing financial interests: see Web version for details.

MARTHA FARAH, PH.D. is at the Center for Cognitive Neuroscience, University of Pennsylvania.

JUDY ILLES, PH.D., is at the Stanford Center for Biomedical Ethics and the Department of Radiology.

ROBERT COOK-DEEGAN, M.D., is at the Center for Genome Ethics, Law, and Policy, Institute for Genome Sciences and Policy, and Department of Public Policy Studies, Duke University.

HOWARD GARDNER, PH.D., is at the Graduate School of Education, Harvard University.

ERIC KANDEL, M.D., is at the Center for Neurobiology and Behavior, Columbia University, and Howard Hughes Medical Institute.

PATRICIA KING, J.D., is at the Georgetown University Law Center.

ERIK PARENS, PH.D., is at the the Hastings Center.

BARBARA SAHAKIAN, F.MED.SCI., is at the Department of Psychiatry, University of Cambridge, Addenbrooke's Hospital.

PAUL ROOT WOLPE, PH.D., is at the Departments of Psychiatry, Medical Ethics, and Sociology, Center for Bioethics, University of Pennsylvania.

# Chapter 25

# The Promise and Predicament of Cosmetic Neurology*

*Anjan Chatterjee*

Advances in cognitive neuroscience make cosmetic neurology in some form inevitable and will give rise to extremely difficult ethical issues.

Consider the following hypothetical case study. A well-heeled executive walks into my cognitive neurology clinic because he is concerned that he is becoming forgetful. It turns out that he is going through a difficult divorce, and my clinical impression is that his memory problems stem from the stress he is experiencing. I place him on a selective serotonin reuptake inhibitor, sertraline, and in a few weeks he feels better. Around this time his thirteen-year-old daughter has difficulty at school and is diagnosed by the school psychologist as having attention deficit disorder. I place her on adderall, a stimulant combination drug, which seems to help with her behavior in school. My patient then comes to me because he is experiencing the "tip of the tongue" phenomenon more frequently. He is concerned that his word-finding difficulty interferes with his ability to function in high-level meetings. I suggest that we try a cholinesterase inhibitor to see if this helps. I am careful to explain that the Food and Drug Administration does not approve such a use for this medication. He wants to try it and is pleased with the results.

A few months later, this patient visits me with his sixteen-year-old son, a talented middle-distance runner. His father thinks if he were just a bit bet-

*From *Journal of Medical Ethics* 32 (2006): 110–113.

ter, among the elite high school runners in the state, he would be far more competitive as an applicant for selective colleges. We discuss various options. Because of a recent report that sildenafil, which is used conventionally for male impotence, may improve oxygen-carrying capacity, I prescribe this medication. The son does not object.

Encouraged by these pharmacologic successes, my patient approaches me with an interesting problem. He is planning a trip to Saudi Arabia in a couple of months to bid on a lucrative contract. He thinks that learning Arabic would give him a decided edge over his competitors and is enrolling in an intensive crash course to learn the language. He wants to know if I can help. Because of data suggesting that amphetamines promote neural plasticity and improve recovery in aphasic patients, I advise him to take a small dose of dextroamphetamine half an hour before each of his classes.

When he is ready to fly to Saudi Arabia I give him my recently patented "travel pack"—a hypnotic, zolpidem, to be taken when he gets on the plane and a stimulant, modafinil, to be taken when he gets off the plane. He goes to Saudi Arabia, impresses the royal family with his Arabic, and wins the contract. Triumphant, he makes a large donation to my research program. And we all live happily ever after.

Or do we? If such a scenario is plausible, is it desirable or is it dystopic? In what follows, I review what is plausible in the practice of pharmacological enhancements and the kinds of ethical issues that would surface from such a practice. While the hypothetical case described may seem extreme now, it might not in the future.

## The Promise

What can be done in cosmetic neurology and what is likely to be possible in the near future? This topic has received some attention in the lay press[1-6] and in the scientific literature,[7-10] but relatively little in clinical circles.[11] The possibilities for enhancement fall into three broad categories: motor abilities, cognition, and affective systems.

The targets for enhancement of motor abilities encompass cardiovascular, peripheral motor, and central nervous systems. For cardiovascular systems, human erythropoietin is used to increase oxygen-carrying capacities for better endurance.[12] New transfusion methods are likely to be used in this way, and, as mentioned already, sildenafil may have similar effects.[13] To enhance motor systems, athletes use anabolic steroids commonly, an issue that has preoccupied even those at the highest level of American politics.[14] Insulin-like growth factor may increase muscle mass and prevent muscular decline associated with aging.[15, 16] Musicians frequently use beta-blockers to dampen physiological tremors in order to improve their performances.[17] Finally, targeting the central nervous system, dopamine agonists may improve the acquisition of motor skills. Such agonists are associated with greater neural plasticity, and the use of dextroamphetamine, when paired with physical therapy, appears to hasten motor learning following stroke.[18, 19]

Intense research efforts in the last few decades are yielding novel treatments for cognitive disorders such as Alzheimer's disease and attention deficit disorder. These medications are also likely to modulate attention, memory, and learning in healthy individuals. Cholinesterase inhibitors may improve normal performance under some circumstances.[20] Modafinil can be used to improve vigilance and reduce impulsive responding,[21] especially in sleep-deprived states, and it is being studied extensively by the armed services.[22] New non-addictive stimulant medications, such as atomoxetine, are also likely to improve levels of arousal in normal subjects. Based on the belief that these drugs improve test performance, the use of stimulant medications among college students in the United States is widespread.[23] Interestingly, the effects of these medications may be influenced by genetic endowments such as which catechol omethyltransferase alleles are inherited.[24] This observation raises the possibility that enhancement cocktails might eventually be tailored to individual genetic profiles.

Particularly intriguing is the development of new classes of drugs, such as ampakines and cyclic AMP response element binding protein (CREB) modulators. They are striking, because they are not being developed with a disease in mind. These medications promote the intracellular cascade of events leading up to the structural neural changes associated with the acquisition of long-term memories.[25–28] Most of the drugs discussed in this paper are developed to treat disorders. As an afterthought, they may also enhance normal abilities. By contrast, ampakines and CREB modulators are developed to augment normal encoding mechanisms. They might then also apply to disease states.

Finally, we continue to refine ways to modify affective systems. Such developments are desirable given that some estimate that up to one in five Americans are depressed,[29] and recent surveys suggest that close to half of adult Americans suffer from affective and substance abuse illnesses.[30] Given that affective illnesses often lie on continua, more people than those who meet checklist criteria might actually benefit from these medications. Beta-blockers, sometimes used for anxiety, appear to help with post-traumatic symptoms in individuals who come to emergency departments after car crashes.[31] Serotonin reuptake inhibitors are used widely and seem to promote affiliative behavior in healthy states.[32, 33] Around the corner are a host of potentially new ways of controlling affective states with the modulation of neuropeptides[34] such as substance P, vasopressin, galanin, and neuropeptide Y. Corticotropin release factor (CRF) seems to mediate the long-term effects of stress,[35, 36] and blocking CRF may blunt these effects.[37] The subtlety with which affective states might be modulated in the future is hard to predict. However, heralding the way in which emotional states might be "fine-tuned," a recent study found that inhaling oxytocin promotes feelings of trust, and that these feelings affect behavior.[38] The general point that I would like to highlight is the following: The armamentarium of drugs that could be used to enhance healthy individuals is growing. We can expect that this growth will continue for the indefinite future. Medications for impotence, hair loss, and obesity are sometimes referred to as "lifestyle" drugs.[39] The medications under consideration here seem to have more pervasive effects—where the altering of substance rather than style is what is at issue. We can expect that drugs will be targeted for specific effects and that they will be targeted for specific genetic profiles.

## The Predicament

Four reasons might give pause to the practice of cosmetic neurology. These concerns have to do with safety, character, justice, and autonomy.

Safety concerns are familiar. Most medications can have unpleasant side effects. Are the risks of these effects worth the expected benefits? The use of drugs in various combinations could complicate the safety concern in unpredictable ways. Physiological and psychological addictions might occur. Since most clinical trials are designed to test safety over relatively short periods, potential long-term toxicities are not known when drugs are introduced into the market. In disease states, one weighs the potential benefits versus the potential risks in making decisions. Thus, one might tolerate significant risk when the alternative is a relentlessly progressive disease like Creutzfeldt-Jakob disease. Are any risks tolerable when the alternative is normality?

In my view, safety is more of a pragmatic than an ethical concern. The nature of drug development is that some problematic effects will occur that could not be predicted. However, all the parties involved—patient/consumers, physicians, and pharmaceutical companies—are interested in having drugs that are safe. Since there are no inherent conflicts of interest, and as long as information about side effects is not suppressed, the ethical issues do not cut deeply.

The character concern has to do with undermining our sense of identity and what gives meaning to our lives.[40] The concern is often placed in a "no pain, no gain" framework. Struggling in some situations and experiencing distress and failure are quintessential aspects of human experience. Enhancing cognition is somehow cheating. Sidestepping distress is somehow cheapening. These experiences give rise to desirable personal attributes. Recent studies find that observing someone in pain activates the same neural circuits that are involved when one experiences pain.[41, 42] One infers from such studies that some painful experiences are probably necessary in developing empathy. The character concern is hard to dispense with. While this remains a deep concern, it is hard to see how this concern would precipitate into public policy or even into consistent social norms. Who decides which pains should be suffered to build character and which can be reasonably avoided? The meaning given to pain that women might experience in childbirth has varied in different settings, from atonement for original sin to promotion of mother-infant bonds.[43] Pain and suffering more generally can take on spiritual significance.[44] Yet many would not accept mandates that prohibit the amelioration of specific pains. In cultures with strong libertarian tendencies it is hard to see how individuals will not insist on making decisions about what to do with their own bodies and brains, for better and for worse.

The justice concern is about equitable distribution of resources. Medications used for enhancements are unlikely to be paid for by insurance companies or by socialized health care systems. That means the wealthy will avail themselves of designer drugs, whereas the poor will be confined to coffee, booze, and cigarettes. On the assumption that the enhancement drugs work to improve abilities, unequal access to them will widen disparities at the ends of the economic spectrum.

Concerns about distributive justice are also difficult to dispense with. Again, it is hard to see how these concerns will prevent the use of pharmacological enhancements. In the United States, wide disparities in access to and quality of health care and education are tolerated. Pharmacological enhancement may not be so different from these other "life enhancers."

The autonomy concern is directed at the possibility that what starts out as a matter of choice ends up as a coercive force. These coercive forces may be explicit or implicit. Explicit coercion might be seen with classes of individuals who might be expected to take certain medications for the greater good. Such precedents exist in the military,[45] and they may seep into other specialized professions. One study found that commercial pilots taking a cholinesterase inhibitor performed better in emergency situations on simulation experiments than did pilots taking placebos.[20] If these results were robust and reliable could pilots be encouraged through financial incentives to take these? Could they be required to take such medications? Could individuals with medical contraindications to these medications be banned from the profession?

The implicit coercive pressures are more complicated, and, in some sectors of society, they are likely to be quite forceful. In winner-take-all environments, slight incremental advantages have disproportionate consequences.[46] This point is made most clearly in sports. Thus, the difference between being first or fourth in the 100 meters at the Olympics is huge, even though objectively both athletes are indistinguishable when compared to the population at large. Similar pressures apply to athletes in other professional sports, such as baseball or football. The pressure to take advantage of slight improvements is sufficient to have athletes risk significant side effects of medications as well as public sanctions for their behavior. Also, many athletes are willing to engage in pharmacological enhancements in an environment in which "fairness" is explicitly valued. Many business and professional environments are set up to make the most of competition. It is not unusual for professionals to work 80 or 90 hours a week, while their children enroll in several sports programs and after-school music programs to ensure that they can make competitive applications to colleges. The pressures for such children to take stimulant drugs to help with academic performance are already evident. The worry is that we may encounter the "Red Queen" principle.[47]

When Alice in Wonderland finally catches up with the Red Queen, she finds that they are both running hard but not moving forward. The Red Queen points out to Alice that sometimes one needs to run as fast as possible just to stay in place. In some sectors of our society one might need to make use of every possible advantage, including enhancements, just to stay in place. In my view, the practice of cosmetic neurology is inevitable. This claim is predictive, not prescriptive.[48] While the ethical concerns are real and run deep, the countervailing social pressures seem overwhelming. Pharmaceutical companies have significant economic incentives to expand their markets to healthy individuals. Since 1997, the Food and Drug Administration has allowed direct advertising to consumers. Television advertisements now give permission to indulge in a pepperoni pizza without the fear of heartburn because one could take an $H_2$ blocker prophylactically. One would be surprised if similar advertisements did not recommend getting an edge with cognitive enhancers or a boost with mood manipulators.

While the coming of cosmetic neurology is in my view inevitable, the specific shape it will take may vary in different locations—for example, winner-take-all pressures vary in different cultures and within different sectors of society. The ways that these promises and predicaments will settle into practice is likely to be reflective of cultural norms. Education is an example of an enhancer that is potentially available to everybody and has a huge impact on social well-being. Perhaps current disparities in availability and quality of education in different countries may predict future norms of access to pharmacological enhancements.

## A (Hypothetical) Clinical Scenario

My clinical practice of neurology has changed. Having struggled through a classic winner-take-all environment, the world of National Institutes of Health (NIH) research funding, I have given up my career as a physician/ scientist. In the last few years, NIH funding rates dropped by half of what they were in an already extremely competitive environment. Grant awards are based on increasingly slight and probably unreliable differences in judgments about the merits of an application. These small differences have disproportionate impacts on people's careers. Encouraged by my original patient, I open a cosmetic neurology clinic on elegant Rittenhouse Square in Philadelphia. My patient, who has invested in this clinic, is a great advocate. I soon have a busy and lucrative practice, largely fueled by word of mouth. The patients are wealthy and for the most part grateful. They sign all the necessary waivers, understand that no specific effects are guaranteed and that the medications are being used in ways not specifically approved by the Food and Drug Administration. I no longer bother with bureaucratic burdens imposed by insurance companies. Things go so well that we open another clinic on Madison Avenue in New York. This clinic is also enormously successful. We are now negotiating to open a clinic in London, with a further eye to Paris and Milan. I am invited frequently to give talks at corporations. Motivational speakers routinely include a discussion of pharmacological enhancements in their exhortations. A few other brain spas are opening, but this simply increases the demand for services at my clinics. I work harder to keep ahead.

## Conclusion

My intentions in this paper are threefold. Firstly, I have tried to make the case that advances in cognitive neuroscience and neuropharmacology make cosmetic neurology plausible and in some form inevitable. The issue is not isolated to doping athletes or pharmacologically insomniac students. These examples are simply the nose of a camel that is well on its way into the tent.

Second, I have tried to emphasize that the ethical issues that arise, particularly those centered on character, coercion, and justice, are extremely difficult. My own views on these issues are not settled. I think it makes little sense to have a singular opinion about the prospect of cosmetic neurology. Each possibility would need to be considered on its own merits. Particularly tricky are situations in which individuals' desires to engage or not to engage in enhancements are at odds with societal desires.

Third, I expect that the practice of cosmetic neurology will challenge conventional notions of the role of physicians. In the last century plastic surgery struggled with its identity as demand for services shifted from reconstructive to cosmetic procedures.[49] In the coming century, clinical neurosciences are likely to struggle similarly. The challenge for physicians will be sorting out their relationships with individuals as patients and consumers, especially when fiduciary and commercial interests collide.

## Acknowledgments

I would like to thank Lisa Santer for her enhancing suggestions on an earlier draft of this paper and Mette Hartlev for alerting me to social meanings that get attached to pain associated with childbirth. A version of this paper was presented at a EURECA workshop in Bologna, in May 2005.

REFERENCES

1. J. Groopman, Eyes wide open, *New Yorker* 3, 2001, 52–57.

2. Anonymous, The ethics of brain science: Open your mind, *Economist*, May 2002, 77–79.

3. D. Plotz, The ethics of enhancement, *Slate*, March 12, 2003; www.politics.slate.msn.com/id/ 2079310/ (accessed August 18, 2005).

4. R. Bailey, The battle for your brain, *Reason* online, February 2003; www.reason.com/0302/ fe.rb.the.shtml (accessed August 18, 2005).

5. S. Begley, Memory drugs create new ethical minefield, *Wall Street Journal*, October 2004, B1.

6. D. Fallik, Improve my mind, please, *Philadelphia Inquirer*, March 2005, D1.

7. P. Whitehouse, E. Juengst, M. Mehlman, et al., Enhancing cognition in the intellectually intact, *Hastings Center Report* 27 (1997): 14–22.

8. S. J. Marcus, ed., *Neuroethics: Mapping the field*. New York: Dana Press, 2002.

9. P. Wolpe,. Treatment, enhancement, and the ethics of neurotherapeutics, *Brain and Cognition* 50 (2002): 387–395.

10. M. J. Farah, Emerging ethical issues in neuroscience, *Nature Neuroscience* 5 (2002): 1123–1129.

11. A. Chatterjee, Cosmetic neurology: the controversy over enhancing movement, mentation, and mood, *Neurology* 63 (2004): 968–974.

12. E. Varlet-Marie, A. Gaudard, M. Audran, et al., Pharmacokinetics/pharmacodynamics of recombinant human erythropoietins in doping control, *Sports Medicine* 33 (2003): 301–315.

13. A. Gaudard, E. Varlet-Marie, F. Bressolle, et al., Drugs for increasing oxygen transport and their potential use in doping, *Sports Medicine* 33 (2003): 187–212.

14. T. Heath, Senate warns baseball on steroids testing, *Washington Post*, March 2004, A01.

15. D. Rudman, A. Feller, H. Nagraj, et al., Effects of human growth hormone in men over 60 years old, *New England Journal of Medicine* 323 (1990): 1–6.

16. E. Barton-Davis, D. Shoturma, A. Musaro, et al., Viral mediated expression of insulin-like growth factor I blocks the aging-related loss of skeletal muscle function, *Proceedings of the National Academy of Sciences* 95 (1998): 15603–15607.

17. B. Tindall, Better playing through chemistry, *New York Times*, October 17, 2004; http://www.nytimes.com/2004/10/17/arts/music/17tind.html?ex=1170046800&en=b66b122a3ed43ef7&ei=5070 *(free, registration required, last accessed, January, 2007)*.

18. D. Walker-Batson, P. Smith, S. Curtis, et al., Amphetamine paired with physical therapy accelerates motor recovery after stroke: Further evidence, *Stroke* 26 (1995): 2254–2259.

19. C. Grade, B. Redford, J. Chrostowski, et al., Methylphenidate in early poststroke recovery: A double-blind, placebo-controlled study, *Archives of Physical Medicine and Rehabilitation* 79 (1998): 1047–1050.

20. J. Yesavage, M. Mumenthaler, J. Taylor, et al., Donezepil and flight simulator performance: Effects on retention of complex skills, *Neurology* 59 (2001): 123–125.

21. D. Turner, T. Robbins, L. Clark, et al., Cognitive enhancing effects of modafinil in healthy volunteers, *Psychopharmacology* 165 (2003): 260–269.

22. J. J. Caldwell, J. Caldwell, N. R. Smythe, et al., A double-blind, placebo-controlled investigation of the efficacy of modafinil for sustaining the alertness and performance of aviators: a helicopter simulator study, *Psychopharmacology* 150 (2000): 272–282.

23. Q. Babcock, T. Byrne, Student perceptions of methylphenidate abuse at a public liberal arts college, *Journal of American College of Health* 49 (2000): 143–145.

24. V. Mattay, T. Goldberg, FF, et al., Catechol omethyltranserase val158-met genotype and individual variation in the brain response to amphetamine, *Proceedings of the Natural Academy of Sciences* 100 (2003): 6186–6191.

25. G. Lynch, Memory enhancement: The search for mechanism-based drugs, *Nature Neuroscience* 5 (Supplement) (2003): 1035–1038.

26. M. Ingvar, J. Ambros-Ingerson, M. Davis, et al., Enhancement by an ampakine of memory encoding in humans, *Experimental Neurolgoy* 146 (1997): 553–559.

27. T. Tully, R. Bourtchouladze, R. Scott, et al., Targeting the CREB pathway for memory enhancers, *Nature Reviews Drug Discovery* 2 (2003): 267–277.

28. R. Scott, R. Bourtchouladze, S. Gossweiler, et al., CREB and the discovery of cognitive enhancers, *Journal of Molecular Neuroscience* 19 (2002): 171–177.

29. The National Institute of Mental Health, The numbers count: Mental disorders in America. (Washington, DC: NIH Publication, 2003, No. 01-4584).

30. R. Kessler, W. Chui, O. Demler, et al., Prevalence, severity, and comorbidity of 12-month DSM-IV disorders in the National Comorbidity Survey Replication. *Archives of General Psychiatry* 62 (2005): 617–627.

31. R. Pitman, K. Sanders, R. Zusman, et al., Pilot study of secondary prevention of post-traumatic stress disorder with propanolol, *Biological Psychiatry* 51 (2002): 189–192.

32. B. Knutson, O. Wolkowitz, S. Cole, et al., Selective alteration of personality and social behavior by serotonergic intervention, *American Journal of Psychiatry* 155 (1998) 373–379.

33. W. A. Tse, Bond. Serotonergic intervention affects both social dominance and affiliative behavior, *Psychopharmacology* 161 (2002): 373–379.

34. A. Holmes, M. Heilig, N. Rupniak, et al., Neuropeptide systems as novel therapeutic targets for depression and anxiety disorders, *Trends in Pharmacological Science* 24 (2003): 580–588.

35. D. Walker, D. Toufexis, M. Davis, Role of the bed nucleus of the stria terminalis versus amygdala in fear, stress, and anxiety. *European Journal of Pharmacology* 463 (2003): 199–216.

36. M. Davis, Are different parts of the extended amygdala involved in fear versus anxiety? *Biological Psychiatry* 44 (1998): 1239–1247.

37. J. Salzano, Taming stress, *Scientific American* 289 (2003): 87–95.

38. M. Kosfeld, M. Heinrichs, P. Zak, et al., Oxytocin increases trust in humans, *Nature* 435 (2005): 673–676.

39. R. Flower, Lifestyle drugs: Pharmacology and the social agenda, *Trends in Pharmacological Sciences* 25 (2004): 182–185.

40. President's Council on Bioethics, *Beyond therapy: Biotechnology and the pursuit of happiness,* www.bioethics.gov/reports/beyondtherapy/ index.html (accessed August 18, 2005).

41. T. Singer, B. Seymour, J. O'Doherty, et al., Empathy for pain involves the affective but not sensory components of pain, *Science* 303 (2004): 1157–1162.

42. A. Avenanti, D. Bueti, G. Galati, et al., Transcranial magnetic stimulation highlights the sensorimotor side of empathy for pain, *Nature Neuroscience* 8 (2005); 955–960.

43. D. Caton, *What a blessing she had chloroform: The medical and social response to the pain of childbirth from 1800 to the present* (New Haven, CT: Yale University Press, 1999).

44. J. Cusick,. Spirituality and voluntary pain, *APS Bulletin* (2003); www.ampainsoc. org/pub/ bulletin/sep03/path1.htm *(accessed August 18, 2005).*

45. M. Russo, C. Maher, W. Campbell, Cosmetic neurology: The controversy over enhancing movement, mentation, and mood [letter], *Neurology* 64 (2005): 1320–1321.

46. R. J. Frank, P. J. Cook, *The winner-take-all society* (New York: Free Press, 1995).

47. L. Van Valen, A new evolutionary law, *Evolutionary Theory* 1 (1973): 1–30.

48. R. Dees, Slippery slopes, wonder drugs, and cosmetic neurology, *Neurology* 63 (2004): 951–952.

49. D. A. Sullivan, *Cosmetic surgery. The cutting edge of commercial medicine in America* (New Brunswick, NJ: Rutgers University Press, 2001).

ANJAN CHATTERJEE, M.D., is associate professor of neurology and is in the Center for Cognitive Neuroscience at the University of Pennsylvania.

# Part VI

# Brain Injury and Brain Death

ONE OF THE GREATEST CONCERNS about advances in the science of the brain is what it could mean for our concepts of human life and dignity. Many people fear that electroencephalography (EEG) and structural and functional neuroimaging will lead to a utilitarian view of our brains and to practices that would violate the intrinsic value of our lives. This fear is especially acute with regard to alleged arbitrary and value-laden judgments about brain death. Yet by providing more refined information about brain structure and function, neuroscience can offer a clear picture of the neural correlates of our capacity for consciousness. It can also offer diagnostic clarity in defining human death and determining when it occurs. In these respects, neuroscience can secure our conviction in the intrinsic value and dignity of human life. The articles in this section address different medical and ethical dimensions of brain injury and brain death.

Before the 1960s, death was a relatively straightforward phenomenon. People were declared dead when their heartbeat and breathing stopped. This standard changed with advances in mechanical ventilation. Patients' heartbeat and blood circulation could be sustained and they could be kept alive indefinitely by artificial means. But these advances made it unclear when death occurred and on what clinical basis death could be declared. It was especially problematic for organ transplantation. Transplant surgeons faced the moral dilemma of taking an organ from a patient's body when it was not clear whether the patient was dead. In particular, they faced the dilemma of taking a beating heart, or a heart that could be restarted, from a patient's body without killing the patient.

To resolve these and related problems, in 1968 an ad hoc committee at Harvard University formulated a "whole-brain criterion" of death. The whole-brain criterion said that death occurs when there is permanent cessation of all brain functions, including the brainstem. In 1981, the President's Commission for the Study of Ethical Problems in Medicine and Biomedical and Behavioral Research adopted a similar position. It defined death as the irreversible cessation of cardiorespiratory function, *or* the irreversible cessation of all brain functions. This definition was formalized in the Uniform Determination of Death Act (UDDA) of 1981. In the United Kingdom, death is defined as the permanent cessation of brainstem function. In the United States and Canada, the whole-brain criterion has become the legal and clinical standard for determining death. This criterion does not supplant but complements the cardiorespiratory criterion of death.

In "Brain Death in an Age of Heroic Medicine," Guy McKhann addresses several challenges to the whole-brain criterion of death and proposes a revised version of it. He argues that brain death should be defined as the irreversible cessation of *integrated* brain functions, not the failure of separate brain functions to respond to testing in a brain-death evaluation. The integrated functions include brainstem activity required to support breathing and circulation, as well as higher-brain activity required for alertness, attention, and awareness of one's environment. He describes how advances in mechanical ventilation, cardiopulmonary resuscitation (CPR), and heart transplantation posed problems for the cardioculatory criterion of death and motivated the need for a whole-brain criterion. McKhann argues that a higher-brain criterion of death by itself would be flawed, since many individuals who meet this criterion and have no cognitive activity retain critical integrated brain functions and are still very much alive. He also takes issue with the brainstem criterion of death, noting that even when brainstem functions cease, integrated functions of the cerebral hemispheres may be intact. Most interesting, McKhann says that, while loss of higher-brain function cannot be the biological basis of declaring a person dead, it may be the basis for discontinuing life-sustaining treatment. Withdrawing life-support on grounds of futility can be justified in cases where there is no expected benefit for the patient but only excessive burdens.

In "Constructing an Ethical Stereotaxy for Severe Brain Injury: Balancing Risks, Benefits, and Access," Joseph Fins underscores the need for diagnostic clarity in assessing patients with traumatic brain injuries. He criticizes the "neglect syndrome" and the "therapeutic nihilism" that assume that all individuals with severe brain injury are beyond any hope of recovery and cannot benefit from any intervention. He distinguishes the persistent vegetative state (PVS) from the minimally conscious state (MCS), where there may be preserved networks of integrated brain activity. Patients in an MCS may recover some degree of consciousness, depending on the nature and extent of their structural and functional brain injuries. Fins proposes an "ethical stereotaxy," which involves calculating the right balance between benefits in access to clinical trials using brain stimulation and protection of those who would participate in these trials. In addition to a duty to avoid exposing subjects to unnecessary risks, researchers conducting these trials have a duty to uphold principles of beneficence and justice. They should not withhold interventions that might promote self-determination by restoring some degree of cognitive function to individuals with brain injuries.

Nicholas Schiff and Fins further develop Fins's ethical stereotaxy in "Hope for 'Comatose' Patients." They describe different types of brain injury and the varying states of consciousness, or lack thereof, resulting from traumatic brain injuries. These states include coma, PVS, and MCS. Failure to differentiate these states can lead to misdiagnosis and preclude any prospect of recovery. They cite the case of Terry Wallis to illustrate this problem and to show that some patients in an MCS do not progress to the PVS. For Schiff and Fins, differences in integrated brain function, rather than levels of resting brain activity, are the key to distinguishing these states and assessing the probability of recovering consciousness. They note the importance of neuroimaging techniques in clarifying a diagnosis. These techniques can identify traumatic brain injury patients who may emerge from an MCS and respond to therapeutic deep-brain stimulation. They also comment on the case of Terri Schiavo, who was declared dead on March 30, 2005, following the removal of the feeding tube that was keeping her alive. Schiff and Fins cite unequivocal neurological evidence that she was in a permanent vegetative state. One can infer from this diagnosis that it was ethically justified to withdraw the feeding tube in this case. Once Terri Schiavo fell into this state, she could not benefit from continued feeding or continued life. Schiff and Fins argue that not all brain-injured patients are beyond hope, however. Patients like Terry Wallis deserve just as much protection from errors of omission in not being considered for potential therapeutic interventions, as they deserve protection from any errors of commission.

In "Rethinking Disorders of Consciousness: New Research and Its Implications," Fins refines the distinctions between different types of brain injury and impaired states of consciousness. He points out that some patients can progress from a persistent vegetative state to a minimally conscious state, and they may emerge from an MCS without becoming permanently vegetative. He also notes how often neurologists fail to draw the diagnostic distinction between a PVS and an MCS. In some cases, this failure may reflect value judgments that may preclude careful balancing of burdens and benefits of continuing care for patients in these states. Fins criticizes the media for contributing to the conflation of different states of consciousness, which perpetuates misunderstanding of these states in the popular culture. All of these issues underscore the need for greater diagnostic clarity and knowledge about brain function following brain injury. Fins concludes his chapter on a philosophical note, pointing out how individual narratives can change as a result of brain injuries. Adjusting expectations for altered mental abilities may lead to the discovery or rediscovery of different dimensions of the self.

# Chapter 26

# Brain Death in an Age of Heroic Medicine*

## Guy M. McKhann

Since before biblical times, the accepted criterion of death has been cessation of cardiorespiratory function. The heart has stopped, breathing has stopped, the person is dead. Every paramedic is trained to determine if the accident or heart attack victim is breathing and has a pulse. Even in the modern hospital, cessation of cardiorespiratory function is sufficient to pronounce death in 90 to 95 percent of cases. Only recently has the triumphant advance of medical technology brought our deep-seated assumptions about death into question and given rise to an alterative criterion: the irreversible loss of brain function that we call "brain death."

The first challenge to the traditional criterion came from life-support machines. With their aid, the heart of a person who has suffered irreversible brain damage may continue to beat—either spontaneously or aided by the machine—for months. By the traditional criterion of cardiorespiratory function, the person is "alive." But at some point in the course of care we may ask: "Should this person be maintained in this state, not only with no chance to return to a level of functioning life but with no chance to return to interaction with the external or internal environment?"

The other challenge arising from this century's medical and technological advances is to determine when a person is a potential donor of vital or-

---

*From Cerebrum 1 (1998): 7–16.

gans such as heart, lungs, or liver. It seems obvious that in these circumstances irreversible cessation of heart function cannot be the criterion for death. This same heart is going to be removed and transplanted into a recipient, where it is expected to return to normal functioning—for years, one hopes.

Over the past thirty years, a new criterion— brain death—has come into use. This criterion has worked well in medical practice, and is now widely applied (although with vastly differing interpretations), but it also has spawned a host of medical, legal, ethical, and philosophical debates. Lately, it has come under attack by groups as disparate as the religious right in the United States, the German Green Party, and some physicians and ethicists.

My involvement with the idea of brain death began thirty years ago when, as a neurologist, I was asked to help develop criteria for brain death in connection with the first heart transplants in the United States. I have continued this involvement at the Johns Hopkins Hospital. Most recently, I have been one of a small group of advisors to the Vatican on the biological and ethical validity of using brain death as a criterion for death of a person. This committee consisted of physicians, philosophers, and theologians, some who favored the concept of brain death and others who, on various biological and ethical grounds, did not.

My position is that cessation of brain function, if clearly defined, is an appropriate, defensible criterion of death at our current stage of technology. The alternative criteria being proposed are less coherent, and some raise implications that are troubling indeed.

## Enter the Heart Transplant

In 1967, Dr. Christian Barnard of South Africa performed the first heart transplant in a human being. Shortly thereafter, Dr. Norman Shumway of Stanford University, who did much of the animal research that made heart transplants feasible, performed the first human heart transplant in the United States. I was at Stanford and became the neurologist involved in questions raised by transplantation. Our committee, comprising physicians from Stanford, Stanford trustees, and people from the community, including clergy, met repeatedly to discuss the biological, ethical, and legal implications. Concerns that dominated those discussions persist today.

At that earlier time, the accepted (indeed only) criterion for death was cessation of the beating heart, leading to absence of circulation and thus lack of perfusion of all organs. The traditional criterion implied another essential principle: irreversibility. The assumption was not only that the heart had ceased beating but also that it could not be resuscitated—or it would have been. In this context, death was seen as an irreversible biological process that occurred in all species. But the transplantation of a heart, and subsequent restoration of its function in the recipient, threw these assumptions into doubt.

If irreversible loss of heart function no longer determined death, could failure of another organ determine it? The brain was the obvious candidate. Without brain function, vital functions of respiration and maintenance of blood pressure could not be sustained, and higher functions, such as communication, attention to the environment, and awareness of surroundings, would not exist. Loss of brain function could be irreversible, and it was (and still is) irreplaceable by any machine.

Our committee repeatedly discussed the distinction between two situations: the first is the person who has had severe brain damage resulting in a non-communicative, non-responsive state; the second is the person who meets the criteria for brain death. In the first situation, the medical and ethical issue is whether or not to stop life-supportive measures where the outcome is hopeless. Many of the legal cases about cessation of supportive measures involve persons in this state. In brain death, the whole-brain function (as discussed below) has ceased. This is an important distinction, because withdrawal of life support in a hopelessly brain-damaged person will result in brain death, but that person is not "brain dead" while still being supported by machines.

## Who Is a Donor?

Another focus of our discussion was the patient who might be an organ donor. From the start, it was clear there would be more potential recipients than donors. Demand for donors would be high. There would be pressure to enlarge the donor pool, particularly because, in the 1960s and early 1970s, when little was known about tissue immunology, many patients rapidly rejected the transplanted heart and required another one immediately. Under these circumstances, pressures to "bend the rules" and accept as a donor a patient who had irreversible brain damage but was not brain dead would be enormous. Perhaps the public sensed this, because people were concerned that they might be seen as potential donors, not patients. Our position was clear: The primary responsibility of the potential donor's physician was proper care of that patient (and the patient's family). Until the patient was declared "brain dead," absolutely no measures that differed from the normal standard of care should be introduced to preserve organs that potentially could be donated.

For the patient and family, it was important that the criteria for brain death be precisely definable. When a patient was declared brain dead, the family could be informed, and a decision about transplantation then made. Only after the family had agreed to transplantation would decisions about medical procedures relative to donation be integrated into the physician's planning. Without clear criteria for brain death, the basis for deciding to stop supportive measures could be vague, variable, and subjective.

The Stanford committee saw the need for a patient (donor) advocate. Thus three people reviewed each possible donor: a neurologist chosen from a pool and often not associated with Stanford, a non-medical person (usually a clergyman chosen by the patient's family), and me. This group, independently and together, repeatedly reviewed the patient's examinations and other data before deciding about brain death.

We also considered the rights of the recipient, a person alive but with a heart progressively failing. Considering transplantation implies a medical judgment that this patient's heart will stop despite all medical measures. If he is to receive a heart, he is taken to the operating room and his own beating heart stopped and removed. For a time, therefore, this person is without heart function, and may be without spontaneous circulation or respiration. It is significant that this person is not considered dead or to have been killed. There are two reasons:

1. The recipient's body, including his brain, is protected by machines that maintain blood pressure and oxygen supply. Note that in all cardiac surgery where the heart is stopped and its actions replaced by a machine, the assumption is that supportive procedures will keep the body components not only alive but capable of returning to integrated function after the period of no spontaneous circulation or respiration.

2. The recipient will receive a replacement heart that will return to normal functioning.

## Competing Criteria for Death

Resistance to the concept of brain death arises from several sources. First, there are far fewer organ donors than potential recipients because relatively few patients declared brain dead actually become donors. An increasing number of potential recipients are dying before transplants are available. From the point of view of transplantation needs, then, the criterion of brain death begins to be seen as the limiting factor.

Second, resistance comes from physicians, ethicists, and others who challenge the validity of brain death as a criterion. They argue that from a biological point of view brain death is imprecise. After all, when life support machines are stopped, people do not always immediately die. Ethicists and theologians also object that the death of the brain is not the same as the death of the person. They add, for good measure, that the issue of defining death is too important to leave to physicians. Third, groups fundamentally anti-science and anti-technology resist a definition of death inherently based on science and technology.

These dissatisfactions with the concept of brain death have led to proposals for alternative criteria for defining the end of life. Let me comment briefly on their implications.

The spontaneous, irreversible cessation of respiration and circulation is the traditional criterion for death in all parts of the world. Some organs, such as skin, corneas, and perhaps kidneys, can be transplanted from a person who, by this criterion, has died (in other words, transplanted from a cadaver without a beating heart). Vital organs like the heart and the lungs, however, cannot be transplanted unless specific preservative measures are instituted before or immediately after death of the donor. Thus, a return to use of this criterion would result in a major decrease in availability of these vital organs for transplantation.

Physicians and ethicists who reject the criterion of absence of brain function have suggested that death be defined as some time point after support systems are discontinued, and the heart stops. Recognizing that a return to the criterion of cessation of heart function would eliminate or greatly reduce certain organ transplantations, they propose that when life support systems are discontinued and the heart stops, the person be declared dead after an arbitrary interval without heartbeat, (suggestions range from two to thirty minutes). They assume that this arbitrary elapse of time is the maximum interval during which the heart could be removed, resuscitated, and transplanted.

They also suggest that the heart and lungs might be protected during this period by being perfused with protective solutions.[1] To me, this proposal is a dodge, a way to make transplantation possible without having to consider death by some other criterion than cessation of the heart. The essential issue is still that any heart that will function in a recipient very likely could have been resuscitated in the donor, if the medical team had chosen to do so.

Of particular concern to me are the criteria for withdrawing life support from these patients in the first place. Alan Shewmon suggests that any patient who is ventilator-dependent (whether brain dead or not) and for whom the ventilator is about to be legitimately discontinued as an extraordinary or disproportionate means, and who wants to donate his or her organs, is a potential candidate.[2] Such a patient would be taken to the operating room with the transplant teams ready, perhaps even with catheters placed for perfusing the organs with cooling solutions prior to death. The problem is that this proposal implies no real criteria for discontinuing life-support. Conceivably, a person with a normally functioning brain, such as a person with a high cervical cord transection or a ventilator-dependent person with a medical disease such as Guillain-Barre syndrome, in which full recovery may occur, would be a candidate!

Others, such as Professor Robert Truog of the Boston Children's Hospital, at least recognize the paradox of this approach and indicate that the process of organ procurement would require substantial changes in the law. The process of organ procurement would have to be legitimated as a form of justified killing, rather than the dissection of a corpse.[3]

Another proposed criterion for death is the failure of function of other organs. For example, death could be based on failure of the liver or kidneys. In both instances, death will occur if failure of these organs is not reversed. True, but both the liver and the kidneys can be replaced by transplantation — and kidney failure can be overcome by dialysis.

## Clarifying the Brain Death Criterion

We are forced back, it seems to me, to the brain — the organ that cannot be replaced. What is needed is a clarification of the brain death criterion. This requires understanding three concepts related to brain function:

- higher-brain functions, which confer our properties of humanness such as communication and awareness of our environment and response to it;

- brainstem functions, which support vital processes of breathing, circulatory homeostasis, and maintenance of awareness and attention by activation of the cerebral cortex;
- whole-brain functions, which require both higher-brain functions and brainstem functions, as well as ability to integrate sensory, motor, and homeostatic mechanisms to maintain blood pressure, temperature, and similar functions.

Each of these concepts has given rise to proposals for different criteria for brain death, particularly in situations involving a potential organ donor.

1. The first proposed criterion is "the irreversible loss of higher-brain functions." This definition would apply to persons who still have intact brainstem functions (that is, their breathing and cardiac functions are being maintained) but have irreversibly lost higher-brain functions. We see in such patients that death may not be a spontaneous event; it may require that life-sustaining measures be actively discontinued. Using this definition, for death to occur and organs to become available for transplantation, it would be necessary for life-sustaining measures to be discontinued.

To me, however, equating death with the loss of higher-brain functions is flawed. First, these patients have lost only one of the groups of functions of the brain. Their brain-based supportive functions may be intact; in some instances, breathing and cardiac function may continue for long periods of time, even without external support.

In this group of people, the distinction between withdrawal of life-support and brain death becomes important. This is not the place to discuss the medical, ethical, and legal issues involved in withdrawing life-support, but I will note that there is general (although not universal) acceptance that, when continuing support is futile, this support can be withdrawn. Pope John Paul II states that "euthanasia must be distinguished from the decision to forgo so-called 'aggressive medical treatment,' in other words, medical procedures which no longer correspond to the real situation of the patient, either because they are by now disproportionate to any expected results or because they prove an excessive burden on the patient and his family."[4] Under these circumstances, shutting off support machines causes the death of the person, as defined by the brain being irreversibly damaged or the heart stopping.

Therefore loss of higher-brain function cannot be the biological basis for declaring a person dead. It may at some point be the basis for discontinuing treatments and so result in death, but it is not the criterion of death.

2. Others, chiefly physicians in the United Kingdom, propose "the irreversible loss of brainstem functions" as the definition of brain death. As described above, the brainstem has multiple vital functions, primarily support of respiration and circulation. Even if brainstem functions are lost, however, the integrated activities of the cerebral hemispheres may be intact, and remain functional. Since brainstem functions could be replaced or bypassed by machines, and the person maintained on life-support systems might retain those very higher-brain functions that others promote as the basis of our humanness, I see no biological basis for this criterion of death.

3. In the United States, "brain death" is synonymous with "the irreversible loss of whole-brain functions." This implies that functions essential for the integrated existence of the person are interrupted at all levels of the nervous system: the brainstem (for support of breathing and the control of circulation), the brainstem and midbrain (for integrating sensory, motor, and regulatory mechanisms), and the cerebral hemispheres and cerebral cortex (for the performance of higher-brain functions such as communication and interaction with the environment). Interaction among these levels of brain activity is also interrupted.

In some formulations of this criterion, the words used are "the irreversible cessation of all clinical functions of the entire brain." Inclusion of the word "entire" has caused confusion, however, because in this context it has no anatomical basis unless it refers to cessation of every function of every neuron, including those of the spinal cord. Opponents of the concept of brain death rightly point out that not all nerve cells may be dead. Neurons stop functioning at different rates after injuries. Isolated functions may persist. For brain death, what is significant is that the integrated activities of neuronal populations supporting the vital functions of the brain be irreversibly lost.

I propose, therefore, that the term "brain death" be defined more explicitly as: "the irreversible cessation of functions of the brain involved with the integrated actions required for support of breathing, circulation, and higher brain functions such as alertness, attention, verbal and non-verbal communication, awareness of one's environment, and response to the environment." In addition, I would emphasize that brain death is irreversible, that the mechanism of the acute global destruction of the brain must be known, and that the integrated functions of the brain cannot be replaced by any machine or organ transplantation.

## How Is the Criterion of Brain Death Used?

Perhaps because cardiac transplantation focused attention on the need to change the criterion of death, it is commonly assumed that the concept of brain death is used only in connection with transplantation of vital organs. This is not the case at my hospital, however, or at several other institutions that I have surveyed. I have reviewed data and procedures in the Neurology-Neurosurgery Intensive Care Unit of Johns Hopkins Hospital, the Maryland Shock Trauma Unit (a state facility for accident victims and others acutely injured), and the University of Washington. Of 703 persons who died at Johns Hopkins between January and August 1997, the outcome of brain death was determined in only 26 patients (3.7% of the deaths). Of these 26 patients, only 6 became organ donors. Data from the Maryland Shock-Trauma unit are very similar. Of 748 deaths, 67 patients were identified as having brain death. Of these, only 19 became organ donors. In all three institutions, the practice is to make a determination of brain death before there is any discussion of organ donation with the families involved (unless the family spontaneously brings up the issue). These statistics show that many patients defined as brain dead do not become organ donors.

Let me repeat that discussions of brain death per se are substantially different from discussions of irreversible loss of higher-brain function. The basic differences are the preciseness of the definition of brain death, the irreversibility of brain death, and that when support measures are discontinued in brain death, the vital functions of breathing and circulation will cease immediately .

End-of-life issues have always been with us. They are particularly poignant now with respect to termination of life-support in the irreversibly brain-damaged patient. The concept and precise definition of brain death are important in discussions with families about the futility of continuing therapy. When a patient has had irreversible injury to the brain, and the medical staff has determined that brain death has occurred, continued treatment and support are futile. In talking with family members, some of whom may need reassurance that "everything possible is being done," the ability to state that the patient meets the clinical criteria for brain death is comforting.

## Looking Ahead

The concept of brain death is biologically sound. Brain death can be defined, is irreversible, and represents the cessation of higher-brain functions and vital integrating functions that cannot be replaced. If, in response to current pressures and attacks, the concept of brain death were abandoned, the consequences for patients needing transplants and families making heart-wrenching decisions about ending life-support would be tragic. The specific definition of brain death I have proposed, if more widely understood and put into practice, could help to reduce pressures that threaten its acceptance.

The concept of brain death has been used for thirty years, during which there have been remarkable advances in both mechanical and biological replacement of organs, organ systems, cells, and now even cellular components, by transplantation. The global functioning of the brain, however, has so far resisted being replaced by these advances. If, at some time, the brain can be replaced, as the heart is now routinely replaced, then the concept of brain death will have to be reconsidered, much as we are now reconsidering the traditional concept of cessation of cardiorespiratory function.

Almost certainly, there will be further attempts to develop mechanical hearts or to genetically engineer organs from other species to be immunologically compatible with humans and transplanted into them. Donation of organs by one human to another might be eliminated. Despite these advances, we will need not only to accept and use a consistent definition of death but to continually examine this definition to see how it might need to evolve.

## Acknowledgment

I wish to thank Dr. James Bernat, Professor of Neurology at Dartmouth, for his helpful input. His thoughtful discussion of many of the issues discussed in this paper is in reference 5.

REFERENCES

1. S. J. Younger and R. M. Arnold, Ethical, psychosocial, and public health policy implications of procuring organs from non-beating heart donors, *Journal of the American Medical Association* 269 (1993): 2769–2774.

2. A. Shewmon, Recovery from "brain death": A neurologist's apologia, *Linacre Quarterly* 64 (1997): 30–96.

3. R. D. Truog, Is it time to abandon brain death? *Hastings Center Report* 27 (1997): 29–37.

4. Pope John Paul II, *The encyclical letter on abortion, euthanasia, and the death penalty in today's world. The gospel of life (Evangelicum vitae)* (New York: Random House, 1995).

5. J. L. Bernat, A defense of the whole brain concept of death, *Hastings Center Report.* 28 (1998): 14–23.

GUY M. MCKHANN, M.D., is a professor of neurology and director of the Zanvyl Krieger Mind/Brain Institute at Johns Hopkins University.

*Chapter 27*

# Constructing an Ethical Stereotaxy for Severe Brain Injury: *Balancing Risks, Benefits and Access*[*]

## *Joseph J. Fins*

Imagine a world where severely brain-injured patients were taken seriously. State of the art research facilities would stand next to clinical centers equipped with the latest imaging technologies to elucidate the underlying mechanisms of cognitive disability. Basic neuroscientists and engineers would collaborate on neuroprosthetic devices. Comprehensive psychosocial support would be provided to patients and families compelled to contend with the burdens of cognitive impairment and disability.

In an ideal world, we would devote resources to the treatment, not to mention the prevention, of traumatic brain injury that were commensurate with the magnitude of the problem. We would have centers for brain injury that would not only address the laudable management of acute injury, but the equally crucial questions of cognitive rehabilitation and longitudinal assessment, with an emphasis on neuropsychological mechanisms of recovery. These strategies would be informed by a sophisticated knowledge of neuroscience, neurosurgery, psychiatry and rehabilitation medicine. But this would be in an ideal world.

Here I examine why these aspirations have not been fulfilled. I consid-

[*]From *Nature Reviews Neuroscience* 4 (2003): 323–327.

er the legacies of two traditions—the right-to-die movement and psychosurgery—that undermine our intellectual curiosity about severe brain injury and its potential management in light of recent progress in cognitive neuroscience. A review of these advances, particularly the potential use of neuromodulation to treat chronically impaired cognitive function, makes this analysis more than an exercise in rhetoric. These advances create an ethical imperative to consider our moral obligations to those members of society who have severe cognitive impairment.

## Brain Injury and Public Health

In the real world, traumatic brain injury (TBI) – the leading cause of acquired cognitive disability – has been described as a 'silent epidemic,' underfunded and underappreciated as a threat to public health.[1] The sad reality is that, after aggressive treatment following acute injury, patients are often relegated to custodial care, spending years without careful neurological assessment or imaging studies. Research is woefully underfunded and complicated by the geographical separation of patients with cognitive impairment from academic medical centers.[2] Empirical studies on quality of life after TBI are scarce.[3] Impressions about patient preferences and quality of life are often inferential and made on the basis of the interpretation of observed behaviors in family members.[4, 5] Furthermore, research in cognitive rehabilitation is often perceived as ethically disproportionate and inappropriate.[6]

These barriers to treatment and research are even more paradoxical when we consider the magnitude of the problem. In the United States alone, TBI has an incidence of 1.5–2.0 million people per year, with an overall prevalence of 2.5–6.5 million people with permanent impairment. The yearly cost for new cases is between $9 and $10 billon, and lifetime costs per person have been estimated[7] to be between $600,000 and $1,875,000. Even more crucially, TBI is the leading cause of long-term disability in children and young adults.

In neurological practice, we often speak of a 'neglect syndrome' when a patient does not attend to a part of his visual field. Why have the needs of those with severe cognitive impairment been so far out of our gaze as clinicians, scientists and public policy makers? Why has the study of the fundamental alteration of the self that can accompany severe brain injury not engendered intellectual curiosity and widespread support?

These questions, perhaps rhetorical in the past, are now issues that we can begin to address from both philosophical and scientific perspectives.

## The Right to Die

Society's neglect syndrome becomes understandable if we recall that the right to die and the redefinition of death itself were established in patients with overwhelming brain injury.[8] One of the challenges posed by emerging research on patients with profoundly impaired consciousness is that we are now asking society for a moral warrant to positively intervene in patients who resemble the very patients in whom the right to die was first established.

The right to die evolved in patients with severe cognitive impairment largely because of their grave neurological state and the perceived futility of continued treatment.[9] Indeed, the legal scholar G. Annas has observed that all of the important right-to-die cases involved patients in the persistent vegetative state (PVS). In the celebrated Quinlan case, for example, the New Jersey Supreme Court allowed the removal of life-sustaining therapy when a return to a 'cognitive sapient state' was irretrievably lost.[10]

The question of consciousness was also central to the establishment of brain-death criteria in 1968.[11] The ethical justification for brain death centred on a moral valuation of permanent unconsciousness and the utilitarian possibilities offered by the burgeoning field of transplantation.[12] At the time, the Harvard anaesthesiologist and bioethicist H.K. Beecher urged that death be linked to the permanent loss of consciousness in the 'hopelessly damaged brain' so as to permit organ harvesting. To do otherwise would have a "desperately radical result: the curable, the salvageable, can thus be sacrificed to the hopelessly damaged and unconscious who consume the time and space and money better devoted to those who could be helped. To pretend that no such alternative exists is nonsense—what one gets, the other is deprived of."[13]

Time has added a subtlety to Beecher's analysis because the description of brain states has been refined. Both patients who are brain dead, and patients in the PVS are "hopelessly damaged and permanently unconscious."[14, 15] But patients in the PVS recover the cyclical alteration of arousal patterns that are not present in cases of brain death. Although this feature does not herald the recovery of consciousness, it shows the existence of residual brainstem function in contrast to whole brain death.[16] The recently described minimally conscious state (MCS) is further distinguished from the PVS by the presence of reliable but inconsistent awareness of oneself or the environment.[17] Patients who meet the criteria for the MCS can show a spectrum of behaviors that might range from simple responses to selective environmental stimuli, to following commands or even the production of inconsistent verbal communication. All patients in the MCS across this functional spectrum are inconsistent in their responsiveness and fail to bring any of these behaviors to the threshold of reliable and consistent communication. Patients who cross this threshold are considered to have emerged from the MCS.

Diagnostic discernment is important when evaluating the MCS because prognosis relates to the degree of functional impairment. Patients in the MCS at the lower end of this functional continuum, although showing evidence of some degree of consciousness, might have little hope of additional cognitive recovery. Some of these patients have a degree of diffuse structural damage comparable to patients who are in the PVS.[18] To build on Beecher's framework, these patients are hopelessly damaged but without permanent loss of consciousness. Therapeutic efforts to restore cognitive function to some of these people will probably be as futile as in the case of patients in the PVS. However, it is wrong to presume that low-level functioning in a patient in the MCS can be equated with overwhelming structural damage.

Even within the PVS, widely varying residual cerebral activity has been shown.[19] In a collaborative study, my colleagues and I reported combined functional imaging studies from five patients in a chronic PVS, including three patients who showed unusual fragments of behavior. These studies showed the widely varying patterns of resting metabolic activity observed in these patients. Although these data showed that the patients had roughly half the normal rate of resting brain metabolism, all patients showed wide regional variation. One particular patient is worth special mention because of the unique pattern of injury that produced his PVS. Although the patient had widely preserved cerebral metabolism—indicating the potential for a higher level of functioning—overwhelming damage to the central mesodiencephalic region led to permanent loss of consciousness and onset of the PVS.

The role of the mesodiencephalic region in producing the PVS might be relevant to a pathophysiologic understanding of selective cortical responses in a celebrated single patient either undergoing the transition from the PVS to the MCS, or just entering the MCS.[20-22] This patient had suffered encephalitis that produced partial injury to deep structures of the mesodiencephalic region. Long-term recoveries from the MCS resulting from similar injuries have been described,[23] but are not comparable to, the overwhelming structural injuries that typically produce a chronic PVS.[24] In a series of studies functional integration in patients in the PVS, Laureys et al.[25, 26] provide quantitative evidence that distributed brain networks are typically widely disconnected in the PVS. Such cases are a powerful illustration of how different underlying structural injuries can lead to similar functional states with a different potential for recovery.

In contrast to the study of Laureys et al., recent functional magnetic resonance imaging studies in patients in the MCS near the borderline of emergence show a wide preservation of distributed networks that selectively activate in response to a spoken language.[27] These MCS patients, who intermittently follow commands, communicate or have awareness of self or loved ones, should raise crucial concerns about prognosis and the underlying mechanisms of functional impairment. The identification of retained, distributed activity points to the presence of a neuronal substrate that might support and sustain additional cognitive recovery. Such patients do not satisfy Beecher's clauses; they are neither permanently unconscious, nor are they necessarily hopelessly damaged. As such, these patients warrant additional evaluation.

This evolving 'typology' of brain states has profound ethical significance. It promotes diagnostic and normative clarity and indicts the common practice of conflating patients in disparate brain states. Refinement of our thinking promotes better stratification and diagnostic refinement, and ultimately leads to the identification of potential therapeutic modalities. For individual patients, it allows for the appropriate recognition of futility when such is justified, and for informed discussions with families about palliative care options. On the other hand, diagnostic clarity can identify the subset of patients in the MCS[28] (in the aggregate, there are between 100,000 and 300,000 people in the MCS in the United States alone) who might one day be helped and should be the object of serious study and therapeutic engagement. In this way, we can meaningfully distinguish utility from futility.[29]

Such a pursuit of diagnostic accuracy would seem to be an unimpeachable good and to be expected, given the endorsement of evidence-based medicine in clinical practice. Despite perspectives that have equated evidence-based practice with ethical obligation,[30] advocacy for evidence-based practice for impaired consciousness[31] has been met with some criticism. Critics have questioned the scientific merits of the categorization and the need for diagnostic refinement.[32, 33] A subtext to some of these concerns might be an unarticulated fear that if we even entertain the possibility of intervening with some patients with impaired consciousness, we might erode the hard-won right to die. Conversely, some people fear that these new categories could equate the attitudes of higher-functioning people in the MCS with the attitudes toward people in the PVS,[34] therefore minimizing the value of their lives.[35]

Having acknowledged these concerns, it is crucial to assert that the pursuit of diagnostic clarity should be an initial goal of research in this area. It should not become an ideological battleground that undermines the prerogative of informed refusal, or that threatens the rights of the disabled to advances in medical care and clinical research. This is important if we hope to overcome the neglect syndrome. Because futility was central to the justification of the right to die in patients in the PVS, many remain nihilistic about potential interventions in all patients with impaired consciousness. This therapeutic nihilism can only be addressed if we distinguish clinically between disparate brain states. If we fail to do so, we are left with the erroneous conclusion that nothing should be done for all patients with impaired consciousness.

## Therapeutic Hypothesis

Once we overcome the neglect syndrome and acknowledge that some people in the MCS deserve intervention, we are still faced with the question of whether anything might be done in addition to currently used cognitive rehabilitation strategies.[36] Although there is no interventional therapy to restore consciousness, Schiff and colleagues have marshaled clinical and experimental data to support the hypothesis of using emerging neuromodulation techniques to remediate chronically impaired cognitive function.[37–39]

These efforts follow a line of inquiry that began with studies of deep-brain stimulation in patients in the PVS that were carried out in the early 1990s in Japan. Although significant clinical improvements were not reported in these patients, wide activation of the cerebrum was shown by marked elevation in cerebral metabolic rates during stimulation.[40] This provocative finding did not result in cognitive benefit, but indicated a physiological effect.[41, 42] This did not result in a meaningful benefit, most probably because patients in the PVS are typically characterized by an overwhelming loss of functional integration.[37–39] However, such activation after deep-brain stimulation might actually prove useful in the context of patients in the MCS,[43, 44] and recent evidence from imaging studies showing that patients in the MCS near the borderline of communication (who also show marked decreases in resting cerebral metabolism) might still have preserved, if widely dispersed, integrative network responses.[27]

This developing work in neuromodulation forms an integral part of our arsenal for the treatment of Parkinson's disease,[45-47] chronic pain[48] and epilepsy. There are also investigational efforts exploring the use of deep-brain stimulation for psychiatric illnesses such as obsessive-compulsive disorder and depression.[49-52]

## Psychosurgery

Our ability to appropriately assess and study neuromodulation is further confounded by the perceived resemblance of modern neuromodulation techniques to psychosurgery.[53] Whereas pharmalogical, genetic or cellular therapeutic strategies have evaded this analogy so far, recent accounts have linked current efforts in neuromodulation to the debate over psychosurgery and the risks of 'mind control.'[54, 55] Allusion is often made to the crude surgeries and therapeutic adventurism of W. Freeman, a well-known supporter of the therapeutic use of lobotomy.[56] Similarly, reports in the lay press have linked neuromodulation with psychosurgery. These articles generally have been cautionary,[57-61] and often disregard the salient scientific differences between psychosurgical operations of the past and current neuromodulation techniques.[53] These omissions threaten the fair assessment of neuromodulation.[62]

Because historical biases threaten the legitimization of neuromodulation as both potential therapy and credible science, it is important that historical analogies are not to be shaped by unstudied recollections of the past. Allusions to the past need to be informed by the historical record. To this end, we need to recall that the United States Congress, under the National Research Act of 1974,[60] ordered the National Commission for the Protection of Human Subjects of Biomedical and Behavioral Research to issue a report on psychosurgery.[54]

The National Commission Report defined psychosurgery as ". . . implantation of electrodes, destruction or direct stimulation of the brain by any means . . ." when its primary purpose was to "control, change or affect any behavioral or emotional disturbance." This definition included both classical psychosurgeries (ablations) and electrical stimulation, but excluded the treatment of disorders like Parkinson's disease, epilepsy and pain management. It is interesting to note that this distinction illustrates a Cartesian dualism that continues to distinguish invasive procedures aimed at the ameliora-tion of movement disorders from those that address cognitive or psychiatric disabilities.[65]

The National Commission, the American Psychiatric Association Task Force on Psychosurgery[66] and the Behavior Control Research Group of the Hastings Center[67] did not find evidence that psychosurgery had been used for social control, political purposes or as an instrument for racist repression, as has been alleged. Contrary to popular expectations, the National Com-mission did not ban psychological procedures. Instead, it found sufficient evidence of efficacy of some psychosurgical procedures to endorse contin-ued experimental efforts as long as strict regulatory guidelines and limita-tions were in place.

The National Commission's conclusions were a surprise in their day[68] and have become a forgotten footnote in the annals of history. They remain a detail from the past that needs to be recalled, lest we allow distortion of sci-ence policy by erroneous historical analogy.

## Regulating Therapeutic Nihilism

The neglect syndrome and the debate on the legacy of psychosurgery have coalesced in a particularly malignant manner. If people with brain injury cannot be helped and there is not much to be done, society must protect them from the risks of therapeutic adventurism. This becomes especially important because these people constitute a vulnerable population owing to their inability to provide autonomous consent. These concerns transcend the parochial considerations of new therapies and threaten access to all peo-ple with delusional incapacity.

Such is the protectionist stance of The United States National Bioeth-ics Advisory Commission.[69] Although the recommendations of the Commis-sion have yet to be enacted into law, they would constrain neuromodulation research in people with impaired decision-making capacity when there was a more than minimal risk and no demonstrated prospect of direct medical benefit.[70-72]

No one should discount the importance of protecting human subjects. At the same time, it seems appropriate to appreciate how the neglect syn-drome and unfounded fears about neuromodulation have led to an ethical taxonomy in which risk-aversion dominates the ethical landscape when it comes to research on those with impaired decision-making abilities. This protectionist regulatory stance has had the unintended consequence of imperiling advances in the neurosciences that could benefit the very population that regulatory ethics has sought to protect from harm. This seems unfair when the inability to provide consent stems from the cognitive disability that would be the object of amelioration.

Given the emerging potential of neuromodulation and of other potential therapeutic approaches aimed at cognitive improvement, we must understand why regulatory research ethics has not been more motivated by a fiduciary ethic to enhance access to new interventions that might prove efficacious. Or to put it in the language of ethical principles, why have the obligations of distributive justice been so subsumed by an ethic of non-maleficence? Ultimately, it seems that the burdens of the neglect syndrome, coupled with fears of therapeutic excess, have led to an underappreciation of potential benefits and an overstatement of risks. Curiously, this is an inversion of the therapeutic misconception that often marks clinical research.[73] Recognizing this distortion of the ethical doctrine of proportionality is particularly important if we hope to calculate the right balance between access to potentially restorative clinical trials and vital human subjects protections, while still avoiding therapeutic nihilism.[6]

## Achieving Ethical Stereotaxy

Whereas the ethical principles of respect for persons, beneficence and justice require that subjects with impaired decision-making capacity be protected from harm, achieving an ethical stereotaxy among these principles requires us to ask whether current ethical norms deprive these subjects of interventions that might promote self-determination by restoring cognitive function.[6] If these norms are restrictive and abridge access to potentially restorative interventions, we should feel compelled to articulate a fiduciary obligation to promote well-designed and potentially valuable research. This is an ethical imperative for this historically marginalized population that has been unintentionally sequestered from research under the guise of nonmaleficence.[74]

Nonmaleficence is just one dimension of a proper ethical stereotaxy that should include the added dimensions of beneficence and distributive justice. This claim to the fruits of neuroscience research becomes particularly compelling as developments in neuromodulation show growing clinical potential.

The neuroscience community is well-positioned to make this claim, and must assume a greater role in responsibly representing their work to the public and engaging in deliberations about the ethical dimensions of their efforts. Their expertise will be essential if science is ever able to effect an 'awakening' of the patient in the MCS. When that occurs, as it inevitably will, society will be faced with a moment in the history of medicine that has deep philosophical implications. It will raise fundamental questions about the injured brain, and the nature and preservation of the self. It will also force us to ask the difficult question of whether regaining partial self aware-ness is always a benefit.[75] As we cannot know if a partial response to neuro-modulation portends recovery or additional suffering, decisions of this sort will be ethically complex.[6] Each deliberation will probably be made on an individual basis and will take into account both the patient's underlying physiology and psychology.[76]

But if patients who are near the frontier of functional communication re-gain the capacity to express their own preferences, we will need to contem-plate the meaning of a rediscovery of the self. We will be challenged to con-sider the moral significance of a reassertion of self-determination. When this happens, we will need to expand our idea of the human community, and at-tend better to the needs of those with impaired consciousness who are now often ignored and neglected.

If past is prologue, then it is fitting to end where we began and recall the wisdom of Beecher. In 1969, he cautioned his readers that the acceptance of brain death criteria was ". . . no mere academic matter but one that entails enormous consequences in terms of changes in medical philosophy and in medical practices." Then, as now, even as we shed light on the categoriza-tion and management of disparate cognitive states, the philosophical impli-cations of neuromodulation remains unforeseen. As Beecher reminds us, "Many of these consequences are clearly apparent; others can be as yet only dimly perceived."[77]

REFERENCES

1. W.J. Winslade, Confronting traumatic brain injury: devastation, hope and healing (New Haven, Connecticut: Yale University Press, 1998).

2. N.D. Schiff, Persistent vegetative state: a site map of the debate, in *HMS Beagle: The BioMedNet Magazine* 72 (February 18, 2000).

3. M. Bullinger et al., Quality of life in patients with traumatic brain injury—basic issues, assessment and recommendations, *Restorative Neurology and Neuroscience*, 20 (2002): 111–124.

4. E. Phipps and J. Whyte, Medical decision-making with persons who are minimally conscious, *American Journal of Physical and Medical Rehabilitation* 78 (1999): 78–82.

5. E. Phipps, M. DiPasquale, C.L. Blitz and J. Whyte, Interpreting responsiveness in persons with severe traumatic brain injury: beliefs in families and quantitative evaluations, *Journal of Head Trauma Rehabilitation*, 12 (1997): 52–69.

6. J.J. Fins, A proposed ethical framework for interventional cognitive neuroscience: a consideration of deep-brain stimulation in impaired consciousness, *Neurological Restoration*, 22 (2000): 273–278.

7. Consensus conference: Rehabilitation of persons with traumatic brain injury. NIH Consensus Development Panel on Rehabilitation of Persons with Traumatic Brain Injury, *Journal of the American Medical Association*, 282 (1999): 974–983.

8. N.L. Cantor, Twenty-five years after Quinlan: a review of the jurisprudence of death and dying, *Journal of Law and Medical Ethics* 29 (2001): 182–196.

9. R.E. Cranford, Medical futility: transforming a clinical concept into legal and social policies, *Journal of the American Geriatric Society* 42 (1994): 894–898.

10. G.J. Annas, The 'right to die' in America: sloganeering from Quinlan and Cruzan to Quil and Kevorkian, *Duquesne Law Review* 34 (1996): 875–897.

11. Landmark article August 5, 1968: a definition of irreversible coma. Report of the ad hoc committee of the Harvard Medical School to examine the definition of brain death, *Journal of the American Medical Association* 205 (1968): 337–340.

12. M.L. Stevents, Redefining death in America, 1968, *Caduceus* 11 (1995): 207–219.

13. H.K. Beecher, Ethical problems created by the hopelessly unconscious patient, *New England Journal of Medicine* 278 (1968): 1425–1430.

14. B. Jennett and F. Plum, Persistent vegetative state after brain damage, A syndrome in search of a name, *Lancet* 1 (1972): 734–737.

15. B. Jennett, *The Vegetative State*, (Cambridge, United Kingdom: Cambridge University Press, 2002).

16. J.L. Bernat, A defense of the whole-brain concept of death, *Hastings Center Report* 28 (1998): 14–28.

17. J.T. Giacino et al., The minimally conscious state: definition and diagnostic criteria, *Neurology* 58 (2002): 349–353.

18. B. Jennett, J.H. Adams, L.S. Murray and D.I. Graham, Neuropathology in vegetative and severely disabled patients after head injury, *Neurology* 56 (2001): 486–490.

19. N.D. Schiff et al., Residual cerebral activity and behavioral fragments can remain in the persistently vegetative brain, *Brain* 125 (2002): 1210–1234.

20. D.K. Menon et al., Cortical processing in persistent vegetative state. Wolfson Brain Imaging Centre Team, *Lancet* 352 (1998): 200.

21. B.A. Wilson, F. Gracey and K. Bainbridge, Cognitive recovery from 'persistent vegetative state': psychological and personal perspectives, *Brain Injury* 15 (2001): 1083–1092.

22. N.D. Schiff and F. Plum, Cortical function in the persistent vegetative state, *Trends in Cognitive Science* 3 (1999): 43–44.

23. P.H. van Domburg, H.J. ten Donelaar and S. Notermans, Akinetic mutism with Bithalaamic infarction: neurophysiological correlates, *Journal of Neurological Science* 139 (1996): 58–65.

24. A. Kampfl et al., Prediction of recovery from post-traumatic vegetative state with Cerebral magnetic-resonance imaging, *Lancet* 351 (1998): 1763–1767.

25. S. Laureys, M.E. Faymonville, G. Moonen, A. Luxen and P. Maquet, PET scanning and neuronal loss in acute vegetative state, *Lancet* 355 (2000): 1825–1826.

26. S. Laureys et al., Brain function in the vegetative state, *Acta Neurologica Belgica* 102 (2002): 177–185.

27. J. Hirsch et al., fMRI reveals intact cognitive systems in minimally conscious patients, *Society for Neuroscience Abstracts* 27 (2001): 529.14.

28. D.J. Strauss, S. Ashwal, S.M. Day and R.M. Shavelle, Life expectancy of children in vegetative minimally conscious patients, *Pediatric Neurology* 23 (2000): 312–319.

29. D. Callahan, Necessity, futility and the good society, *Journal of the American Geriatric Society* 42 (1994): 866–867.

30. E. Zarkovich and R.E. Upshur, The virtues of evidence, *Theoretical Medical Bioethics* 23 (2002): 403–412.

31. J.T. Giacino et al., Development of practice guidelines for assessment and mangagement of the vegetative and minimally conscious states, *Journal of Head Trauma Rehabilitation* 12 (1997): 79–89.

32. R.E. Cranford, The vegetative and minimally conscious states: ethical implications, *Geriatrics* 53 (1998): S70–S73.

33. D.A. Shewmon, The minimally conscious state: definition and diagnostic criteria, *Neurology* 58 (2002): 506–507.

34. K. Payne, R.M. Taylor, C. Stocking and G.A. Sachs, Physician's attitudes about the care of patients in the persistent vegetative state: a national survey, *Annals of Internal Medicine* 125 (1996): 104–110.

35. D. Coleman, The minimally conscious state: definition and diagnostic criteria, *Neurology* 58 (2002): 506–507.

36. K.D. Cicerone et al., Evidence-based cognitive rehabilitation: recommendations for clinical practice, *Archives of Psychological and Medical Rehabilitation* 81 (2000): 1596–1615.

37. N.D. Schiff and M. Pulver, Does vestibular stimulation activate thalamocortical mechanisms that reintegrate impaired cortical regions? *Proceedings of the Royal Society B/ Biological Sciences* 266 (1999): 421–423.

38. N.D. Schiff, M. Plum and A.R. Rezai, Developing prosthetics to treat cognitive disabilities resulting from acquired brain injuries, *Neurological Research* 24 (2002): 116–124.

39. N.D. Schiff and K.P. Purpura, Towards a neuropsychologic foundation for cognitive neuromodulation through deep brain stimulation, *Thalamus and Related Systems* 2 (2002): 55–69.

40. T. Tsubokawa and T. Yamamoto, in *Textbook of Stereotactic and Functional Neurosurgery* (eds. P.L. Gildenberg and R.R. Tasker 1979–1986). (New York: McGraw-Hill Professional, 1998.)

41. L.J. Schneiderman, N.S. Jecker and A.R. Jonsen, Medical futility: its meaning and ethical implications, *Annals of Internal Medicine* 112 (1990): 949–954.

42. L.J. Schneiderman, The futility debate: effective versus beneficial intervention, *Journal of the American Geriatric Society* 42 (1994): 883–886.

43. R. Smothers, Injury in '88, officer awakes in '96, *New York Times*, February 16, 1995.

44. J.W. Buruss and R.C. Chacko, Episodically remitting akinetic mutism following subarachnoid hemmorage, *Journal of Neuropsychiatry and Clin. Neuroscience* 11 (1999): 100–102.

45. J.D. Speelman and D.A. Bosch, Resurgence of functional neurosurgery for Parkinson's disease: a historical perspective, *Movement Disorders* 13 (1998): 582–588.

46. Deep-brain stimulation of the subthalamic nucleus or the pars interna of the globus pallidus in Parkinson's disease, *New England Journal of Medicine* 345 (2001): 956–963.

47. R. Kumar et al., Double-blind evaluation of subthalamic nucleus deep brain stimulation in advanced Parkinson's disease, *Neurology* 51 (1998): 850–855.

48. R.F. Young, Brain stimulation, *Neurosurgery Clinics of North America* 1 (1990): 865–879.

49. B.H. Kopell and A. Rezai, The continuing evolution of psychiatric neurosurgery, *CNS Spectrums* 5 (2000): 10:20–31.

50. R.M. Roth, L.A. Flashman, A.J. Saykin and D.W. Roberts, Deep brain stimulation in neuropsychiatric disorders, *Current Psychiatry Reports* 3 (2001): 366–372.

51. J.L. Rapoport and G. Inoff-Germain, Medical and surgical treatment of obsessive-compulsive disorder, *Neurologic Clinics* 15 (1997): 421–428.

52. B.D. Greenberg, Update on deep brain stimulation, *Journal of ECT.* 18 (2002): 1997.

53. J.J. Fins, From psychosurgery to neuromodulation and palliation: history's lessons for the ethical conduct and regulation of neuropsychiatric research, *Neurosurgery Clinics of North America* (in press).

54. J.M. Delgado and R.N. Anshen (eds.) *Physical Control of the Mind: Toward a Psychocivilized Society* (New York: Harper and Row, 1969).

55. W.M. Gaylin, J.S. Meister and R.C. Neville (eds.), *Operating on the Mind: the Psychosurgery Conflict* (New York: Basic Books, 1975).

56. E.S. Valenstein, *Great and Desperate Cures: The Rise and Decline of Psychosurgery and Other Radical Treatments for Mental Illness* (New York: Basic Books, 1986).

57. Editorial. The future of mind control, *The Economist* (London, May 25, 2002).

58. W. Safire, The but-what-if-factor, *The New York Times* (New York: May 16, 2002, A25).

59. J. El-Hai, The lobotomist, *The Washington Post Magazine* (Washington: February 4, 2001, pp. 16–31).

60. W. Herbert, Psychosurgery redux, *U.S. News and World Report* (Washington, November 3, 1997, 63).

61. M. Carmichael, Healthy shocks to the head, *Newsweek* (Washington, June 24, 2002, 56–58).

62. J.J. Fins, The ethical limits of neuroscience, *Lancet Neurology* 1 (2002): 213.

63. The National Research Act, Public Law 93-348 (July 12, 1974) (http://www.fas.harvard.edu/~research/PL93-348.html)

64. The National Commission for the Protection of Human Subjects of Biomedical and Behavioral Research. Use of psychosurgery in practice and research: report and recommendations of the National Commission for the Protection of Human Subjects of Biomedical and Behavioral Research, *Federal Register* 42 (1977): 26318–26332.

65. S.J. Matthew, S.C. Yudofsky, L.B. McCullough, T.A. Teasdale and J. Jankovic, Attitudes toward neurosurgical procedures for Parkinson's disease and obsessive-compulsive disorder, *Journal of Neuropsychiatry and Clinical Neuroscience* 11 (1999): 259–267.

66. J. Donnelly, The incidence of psychosurgery in the United States: 1971–1973, *American Journal of Psychiatry* 135 (1978): 1476–1480.

67. H. Blatte, State prisons and the use of behavior control, *Hastings Center Report* 4 (4) (1974): 11.

68. B.J. Culliton, Pyscosurgery: National Commission issues surprisingly favorable report, *Science* 194 (1976): 299–301.

69. National Bioethics Advisory Commission, *Research involving persons with mental disorders that may affect decision-making capacity* (Rockville, Maryland: December, 1998) (http://www.georgetown.edu/research/nrcbl/nbac/capacity/TOC.htm).

70. R. Michels, Are research ethics bad for our mental health? *New England Journal of Medicine* 340 (1999): 1427–1430.

71. J.M. Oldham, S. Haimowitz and S. Delano, Protection of persons with mental disorders from research risk, *Archives of General Psychiatry* 56 (1999): 688–693.

72. F.G. Miller and J.J. Fins, Protecting vulnerable research subjects without unduly constraining neuropsychiatric research, *Archive of General Psychiatry* 56 (1999): 701–702.

73. H.Y. Vanderpool and G.B. Weiss, False data and the therapeutic misconception, *Hastings Center Report* 17 (2) (1987): 16–19.

74. J.J. Fins and N.D. Schiff, Diagnosis and treatment of traumatic brain injury, *Journal of the American Medical Association* 283 (2000): 2392.

75. F. Cohadon, R. Richer, A. Bougiera, P. Deliack, and H. Loiseau, in *Neurostimulation: An Overview* (eds. Y. Lazrthesy and A.R.M. Upton) 247–250 (Mount Kisco, New York: Futura Publishing, 1985).

76. C.L. Osborn, *Over My Head* (Kansas City, Kansas: Andrews McNeal Publishing, 1998).

77. H.K. Beecher, After the "definition of irreversible coma," *New England Journal of Medicine* 281 (1969): 1070–1071.

JOSEPH J. FINS, M.D., is chief of the Division of Medical Ethics and professor of medicine, public health, and medicine in psychiatry at Weill Medical College of Cornell Medical Center. He is director of medical ethics at the New York-Presbyterian Weill Cornell Medical Center, associate for medicine at The Hastings Center, and a member of the adjunct faculty of Rockefeller University.

# Chapter 28

# Hope for "Comatose" Patients*

## *Nicholas D. Schiff* and *Joseph J. Fins*

On July 11, 2003, newspaper headlines proclaimed the dramatic awakening of Terry Wallis, a thirty-nine-year-old Arkansas father who had been in a "coma" after suffering a head injury in a July 1984 car accident. He had been riding with a friend when their car plunged into a creek. When they were found under a bridge the next day, his friend was dead and Wallis was comatose. But now, nineteen years later, he was talking. His first words were "Mom" and then "Pepsi," and, over the ensuing weeks, he began to speak with greater fluency. He apparently had no memories of the intervening time. In his world, Ronald Reagan was still president. The media described his recovery as a "miracle," and his doctors were stunned. What occurred seemed scientifically beyond the realm of possibility.

Despite the very unexpected (and as yet unexplained) nature of what happened to Wallis, we were, perhaps, less surprised than many people. For a decade we have been conducting research at the frontiers of understanding impaired consciousness and the ethical challenges posed by devastating brain injury. We have seen other patients like Wallis, whose improvements, although less heralded, also defy our understanding of impaired consciousness that follows brain injury. Our goal has been to understand both the mechanisms of recovery and biological differences between those patients who remain forever unconscious after catastrophic injury and those who regain at least limited awareness.

*From *Cerebrum* 5 (2003): 7–24.

From our own research and that of a handful of other cognitive neuro-scientists, we knew that the media's portrayal of Wallis's condition before he recovered was inaccurate, at best, and, at worst, seriously misleading. Although he was portrayed as being in an irreversible coma, or in a vegetative state, Wallis was in neither. He had not been in a coma immediately before his recovery, because "coma" describes a state of unconsciousness typically lasting only weeks from the time of injury. Comatose patients usually either recover or slip into various longer-term states of impaired consciousness. A review of literature about Wallis indicates that his behavior during the nineteen years after the accident was also inconsistent with being in a vegetative state. He had been able to respond to simple questions with a nod of the head or with grunting sounds, indicating some level of awareness and inter-action with his environment—neither is seen in the vegetative state. These behaviors, noted before his dramatic recovery, suggest a state that scientists are only beginning to characterize: the minimally conscious state.[1]

The importance of these distinctions cannot be overstated. Identifying those patients with severe brain injuries who have a chance of recovering is the first step in deciding who may benefit from the therapeutic approaches now being developed that might help them to regain function and inde-pendence. Having said that, we come face-to-face with a puzzling paradox: Why has this progress in understanding impaired states of consciousness been met by a surprising lack of interest—not to say an attitude of dismiss-al—on the part of the scientific community and society at large?

## Our Hidden Epidemic of Traumatic Brain Injury

At the outset, we should appreciate that what happened to Wallis when he was twenty years old is an all-too-common story. Although most of us do not think about traumatic brain injury (TBI) until a family member is touched by its tragedy, its incidence is staggering. TBI is the leading cause of long-term disability in children and young adults and, in the United States alone, has 1.5 to 2.0 million victims a year. Motorcycle, automobile, and sporting accidents are among the most frequent causes. The toll of TBI is still more graphically demonstrated when we consider that head trauma has left be-tween 2.5 and 6.5 million people in the United States with some degree of permanent impairment. The yearly cost for new cases of TBI is between $9 and $10 billion, and lifetime costs per individual have been estimated to be between $600,000 and $1,875,000.

Even these numbers, however, fail to do justice to the burden of TBI. Lives are suddenly and irrevocably altered by severe head trauma. If a patient is lucky enough to survive the acute phase of injury and intensive care, but remains severely impaired, he may face years of rehabilitation on the road to recovery. Unfortunately, rehabilitation services are often limited, and third-party payment depends on evidence of the patient's continual improvement and on demonstration of what is called "medical necessity." Some patients move from hospital to rehabilitation facility and then return to a life that is markedly different from their former existence. Such was the experience of Trisha Mellie, whose autobiography, *I Am the Central Park Jogger*, tells the story of a young investment banker attacked while jogging in 1988 and her recovery from the near fatal attack.

Mellie describes how she had to relearn all the tasks she had once mastered as a child and to develop new strategies to compensate for her loss of cognitive function. Even more daunting, she had to grapple not only with memory loss but with the realization that her injury had changed who she was as a person.

### The Hidden Epidemic of Traumatic Brain Injury in the United States

- 1.5 million TBI injuries occur every year.
- 50,000 people die from their injuries.
- 80,000 to 90,000 people experience long-term or lifelong disability as a result of their injuries.
- 2,000 people enter a persistent vegetative state following their injuries.
- One-third of all injury deaths are the result of a TBI.
- The cost of TBI is estimated to be more than $48 billion each year, including $10 billion for new cases.

### The Most Frequent Causes of Traumatic Brain Injury in the United States

- Vehicle crashes—This includes motor vehicles, bicycles, recreational vehicles, and pedestrians.
- Firearms—Firearm use is the leading cause of death relating to a TBI, and 90% of people with a firearm-related TBI die.
- Falls—Falls are the leading cause of TBI among the elderly, and 60% of fall-related TBI deaths involve people over seventy-five.

As monumental a challenge as Mellie faced, another class of patients with TBI has sustained even greater impairment. This class includes people such as Wallis, who recover from their acute injuries and have (perhaps minimal) evidence of cognitive awareness but fail to meet the criteria for medical necessity required to qualify for intensive ongoing rehabilitation. Our health care system fails them and their families. After a brief period of coma rehabilitation—or none at all—they are exiled to nursing homes for what is often impolitely called "custodial care." The experience of Wallis and his family is not unusual. His parents report that, after the accident, Terry was never seen by a neurologist. He was placed in a nursing home, and his family was told that an evaluation would be expensive. His father said to reporters, "They told us it would cost $120,000 just to evaluate him to see if he could be helped, and we didn't have that kind of money." The Wallis family applied for Medicaid to cover the cost of evaluation but was turned down. "They said the government will not put out that kind of money on no more chance than he's got to reenter the workforce," reported his father.

Even if the government was correct that the cost of evaluation would be more than Wallis's potential wages if he recovered, what is our ethical obligation to patients like him? We think society owes the many people with TBI, and their caregivers, some intellectual curiosity about severely impaired consciousness, as well as the potential fruits of scientific investigation. In the light of new understanding of various brain states after injury, clinicians, patients, and their families need accurate information to make informed choices about care. But, sadly, many clinicians are themselves ill informed, so they are unable to discuss the options for treatment—or refusal of it—in a meaningful way. If a society must ration scarce resources, the decision about whom to treat should be based on the best available science and an accurate assessment of diagnosis and prognosis. Such clinical precision seems to be the least that physicians should demand of themselves when caring for these patients.

## States of Disordered Consciousness: A Primer

What brain states can follow head injury? The immediate consequence of a severe brain injury, like the one that Wallis sustained, is a loss of consciousness that results in a brain state known as coma, an "unarousable unresponsiveness." The person does not respond to vigorous efforts to elicit a response of any kind—sound, movement, or eye-opening—and shows no variation in behavior, simply a sleeplike state with eyes closed.

The prognosis for someone in a coma very much depends on the person's age, the amount of structural damage (as identified by brain imaging), and whether there is evidence of direct injury to the brainstem. From coma, very severe brain injuries can progress to brain death, a total loss of whole brain function, including brainstem activity. In other cases, the comatose state, if uncomplicated by other factors, is typically followed within seven to fourteen days by an indeterminate period during which an eyes-open, "wakeful" appearance alternates with an eyes-closed, "sleep" state. These alternating periods represent a limited recovery of cyclical change in arousal pattern and characterize the vegetative state (VS), as originally defined by Bryan Jennett, M.D., and Fred Plum, M.D., in 1972.[2]

In all other respects, the vegetative state is similar to coma.

Patients in vegetative states demonstrate no evidence of awareness of self or response to their surroundings. If the patient remains in a vegetative state for more than thirty days, he is deemed to be in a persistent vegetative state (PVS). Prospects for the recovery of consciousness become grim when the vegetative state becomes chronic or permanent, after three months in the

Coma, Persistent Vegetative State, Minimally Conscious State

| Coma | Persistent Vegetative State (PVS) | Minimally Conscious State (MCS) |
| --- | --- | --- |
| Does not respond to vigorous efforts to elicit a response of any kind, shows no variation in behavior. | Demonstrates no awareness of self or response to surroundings. | Demonstrates directed behaviors and evidence of awareness of self or the environment, cannot demonstrate consistent functional communication. |
| Appears to be in a sleeplike state with eyes closed. | May demonstrate reflexive behaviors, including occasionally smiling or appearing momentarily to focus his/her gaze on something. | May demonstrate fluctuating behavior, including basic verbalization, gestures, memory, attention, intention, and awareness of self and environment. |
| Typically lasts only weeks from the time of injury. Can recover, die, or evolve into another state of impaired consciousness. | Can live for years in a PVS and never recover, or could progress to a minimally conscious state or recover. | Can emerge from a MCS and regain full consciousness. |

case of anoxic injury, caused by oxygen deprivation, and a year following traumatic injuries.

In other cases, a patient may recover to the point of very limited but definitely observable responses to his environment. Such a patient is classified as in a minimally conscious state (MCS). In this condition, a patient exhibits bits of directed behaviors that are different from the reflexive behaviors seen in PVS patients. The difference is that MCS patients demonstrate unequivocal—albeit fluctuating—evidence of awareness of self or the environment. Limited behavior exhibited by MCS patients can include basic verbalization, gestures, memory, attention, intention, and awareness of self and environment.

To know that a patient has emerged from MCS, we must observe consistent functional communication. Crossing this threshold requires more than the ability simply to follow commands. For example, a patient may be able to correctly identify a printed "yes" or "no" on a card held up by an examiner but not be able to answer questions reliably using such signaling. The patient would be considered only at the borderline of emergence from MCS. In Terry Wallis's case, we see him emerging from MCS after he passed through an initial coma and a period in the vegetative state. Although Wallis has often been described as vegetative right up until he began to speak, in fact he had been able (possibly within the first year after his accident) to respond to simple questions with a nod of his head or grunting—hallmarks of MCS. If so, his pattern of recovery is wholly consistent with scientific understanding of the recovery of consciousness from a vegetative state resulting from traumatic brain injury.

Because the progression from PVS to MCS may take months following a traumatic brain injury, physicians who do not fully understand what is happening, and who rely solely on their observations of a patient, can have unnecessarily negative expectations for the patient's recovery. Information about the patient's underlying brain function, gained from neuroimaging, may change those expectations. New neuroimaging techniques may eventually become an important adjunct to careful neurological examination and lead to much earlier identification of patients like Wallis who may be able to emerge from MCS. Just as surely, though, they might tell us that no further recovery can be expected, even if a patient exhibits some limited behavior above a vegetative level. Having more knowledge, sooner, will bring hope to some and despair to others.

## Seeing into the PVS Brain

The first brain-imaging studies of PVS patients were done by Fred Plum, M.D., David Levy, M.D., and their colleagues in the early 1980s using a technique called fluorodeoxyglucose positron emission tomography (FDG-PET). This imaging technique measures how much energy the brain is consuming. Plum and Levy discovered that overall brain-tissue metabolism in PVS patients was half or less than half of normal.

On the basis of their work, we can now pose a critical question: What functional activity might remain in severely injured brains? Seeking an answer, together with Plum, we have collaborated with research groups at the New York University Center for Neuromagnetism (directed by Urs Ribary, Ph.D., and Rodolfo Llinas, M.D., Ph.D.) and the Memorial Sloan-Kettering Center (directed by Brad Beattie, Ph.D., and Ron Blasberg, M.D.). To try to size up the remaining functional activity in several PVS patients, we used three neuroimaging techniques: magnetic resonance imaging (MRI), magnetoencephalography, and quantitative PET analysis.[3] What we have discovered is new evidence that the persistently vegetative brain can harbor still functional, but isolated, networks and that these networks, at times, can generate recognizable fragments of behavior.

We became interested in the possibility of this residual function through the case of a forty-nine-year-old woman who had suffered a series of cerebral hemorrhages as a result of malformed blood vessels in her brain. Despite some two decades in PVS, this woman occasionally uttered single words (typically expletives) without any external stimulus. MRI imaging showed that her right basal ganglia and thalamus were destroyed, and FDG-PET measurements confirmed a marked overall reduction of more than 50% in overall brain tissue metabolism, which is consistent with what we know of cerebral metabolism in PVS. What was intriguing, though, was that several isolated, relatively small regions in her left hemisphere showed higher levels of metabolism. These regions, in the normal adult brain, are associated with language functions. Two other patients we studied also revealed isolated metabolic activity in the brain that could be correlated with other unusual patterns of behavior. It seems, then, that residual cerebral metabolic activity that remains after severe brain injuries is not random; it is tied to local cerebral networks that have been preserved and to patterns of neuronal activity.

The work of other scientists helps to fill in the picture. Using a different PET technique, David Menon, M.D., and his colleagues in Cambridge,

## The Terri Schiavo Case

The plight of Terri Schiavo, the young woman at the center of the much-publicized legal battle in Florida, illustrates the devastation of the chronic vegetative state following anoxic injury (one that deprives the brain of oxygen). Adults who suffer cardiac arrest and oxygen deprivation to the brain leading to a persistent vegetative state that lasts beyond three months have essentially no statistical chance of further recovery. This grim prognosis rests on a convergence of evidence from studying outcomes in large numbers of patients; in addition, it is supported by evidence of diffuse neuronal damage in the cerebral cortex and other higher brain regions in such an injury. Measurement of patients' brain activity further demonstrates the loss of cerebral function, with the resting metabolic activity in chronic vegetative states following anoxic injuries averaging less than half the level in normal brains. Structural imaging studies in such cases typically reveal widespread neuron loss and cerebral atrophy similar in extent to that observed in end-stage Alzheimer's disease. Brain electrical activity is grossly disturbed, if evident at all. Patients in this state may live for years, occasionally smiling, shedding a tear, or briefly appearing to fix their gaze on something. These reflex behaviors in the chronic phase of PVS do not reflect awareness or the potential for further recovery. Many of these facial displays are organized by intrinsic circuits of the brainstem and do not depend on the integrity of higher centers of the brain, including the cortex and thalamus, which are overwhelmingly damaged in patients who chronically remain in a vegetative state. In rare instances, islands of cerebral activity do remain on levels higher than the brain stem, producing fragments of behavior that are not responses to anything in the environment. The presence of these fragmentary behaviors, unfortunately, does not improve the prognosis or suggest greater potential for recovery in patients remaining in vegetative states beyond the three-month period following an anoxic injury. Thus, each patient's examination should be considered in the overall context of the history of their illness, along with the results of structural brain imaging and studies of function. To the untrained observer, the simple appearance of wakefulness is difficult to dissociate from an inference of awareness, especially if this appearance is accompanied by brief, out-of-context, reflexive behaviors that also can be misinterpreted. This emotionally charged situation dictates extraordinarily careful and repeated efforts to reduce the uncertainty in making the diagnosis. In the Schiavo case, many qualified experts testified that repeated examinations of the patient revealed a vegetative state, that structural imaging confirmed the neuron loss and widespread atrophy, and that repeated testing documented the absence of brain electrical activity. In the aggregate, this evidence is as unequivocal, and lacking in reasons for hope, as any obtainable in these circumstances.

—N.D.S. and J.J.F.

England, found that a patient who was recovering from PVS into MCS had isolated neural networks that responded to human faces. Steven Laureys, M.D., and his colleagues in Belgium have examined functioning in the PVS brain by comparing its responses to simple auditory and other stimuli with its baseline resting state. For both types of stimuli, these PVS patients demonstrated a loss of brain activation in so-called higher-order regions— regions outside of their primary sensory cortices.[4] This seems to indicate that there is a wholesale disconnection between functions in the PVS brain that prevents basic sensory input from being processed anywhere but the earliest cortical levels. This evidence is consistent with our own results and supports the conclusion that residual cortical activity seen in PVS patients does not signify any awareness.

One additional observation may shed light on the significance of residual islands of activity in the PVS brain. The observation involves one patient who had exceptionally widely preserved metabolism in the cerebral cortex, despite six years in a vegetative state after a traffic accident. The patient's behavior had been completely unremarkable. The unusual observation was that this patient's cortical metabolism was near normal, except for marked reductions in the severely damaged region of the upper brainstem and central thalamus. We conjectured that this well-preserved cortical metabolism probably meant that there were many partially functioning brain networks. In other words—and this is crucial—nothing linked these islands of activity as before the injury. It is relevant that the upper brain stem and central thalamus regions were damaged in this patient because those regions have a critical role in the functional integration of parallel neural networks. In a different patient, who had recovered from a vegetative state, Laureys had identified a return of activity in these regions.

## Inside the MCS Brain

Everything we had learned about the brains of PVS patients made us want to study what remaining cerebral activity might be found in MCS patients, particularly those recovering almost to the level of emergence from MCS. Patients who remain near this borderline raise different questions. What underlying mechanisms could be limiting their recovery of communication? In collaboration with Joy Hirsch, Ph.D., and her colleagues, we studied two such patients and compared what we found with our findings in PVS patients.[5] Both MCS patients could intermittently follow simple commands with eye movements, occasionally made attempts to vocalize, and showed significant fluctuations in their responses. Would their brains respond to language? To test this, we played them taped narratives, spoken by a familiar relative. Tapes were played both as normal speech and backward.

We found that when the story was played forward, as normal speech, the two MCS patients showed activation of cerebral networks underlying language comprehension. The activation was similar to activation in normal subjects. Not so for the tape played backward. Normal subjects showed similar activation patterns for both, but the MCS patients failed to activate the language comprehension networks when they heard the tape reversed. This failure indicates to us that in some MCS patients there are forebrain networks that might be potentially functional, yet fail to establish the patterns of activity needed for consistent communication. The preservation of forebrain networks associated with higher cognitive functions, such as language, could provide a neurobiological basis for wide fluctuations in behavior, such as was observed in Terry Wallis.

Obviously, these studies suggest the crucial importance of functional integration. Although these MCS patients demonstrated functioning forebrain networks and could respond to forward language, their overall resting cerebral metabolism was low, near PVS levels. To our surprise, we also found that these patients seemed to have intact integrative responses in both cerebral hemispheres. This leads us to believe that differences in the integration of functions, more than levels of resting brain activity, are what separate PVS from MCS.

## Emerging from MCS

With this insight about what may be happening in the brains of some patients with severe brain injury, we can revisit the question, How could someone like Terry Wallis harbor residual cognitive capacities that lay dormant for so many years? One possibility is that, over time, as patterns of activation come and go in intact regional networks, one result may be improved awareness and cognition; but another result, exactly the contrary, may be the inhibition of recovery.

From a physiologic standpoint, several mechanisms may possibly be at work here, yielding changes in the capacities of patients who have complex brain injuries. How and to what extent these mechanisms may limit recovery, we do not know, as yet, but several observations are suggestive. It is relatively common, even after a localized stroke or brain injury, to have reduced cerebral metabolism in brain regions that are remote from the site of injury. The cause seems to be a loss of excitatory inputs from nerves at the site of the original injury. This process, which is reversible, results from a strong inhibition of the distant neurons brought about by a lack of incoming synaptic activity.

Another mechanism that conceivably could affect the delicate balance of excitation and inhibition is abnormally increased synchronization of populations of neurons (such as is seen in epilepsy and other brain disorders). It is possible that, following structural brain injuries, such changes may arise in specific brain networks that play an important role in the functional integration of networks in the normal brain. An alteration of this kind could have played a role for Wallis, limiting his capacity for producing speech through active inhibition of language networks. While this explanation is, of course, speculation, some kind of functionally reversible process must play a role in such cases. Wallis's doctors speculated that adding the antidepressant Paxil to his medications could have been connected in some way with his later recovery of speech, although he had taken Paxil for eighteen months before his recovery. Did this antidepressant have a role in slowly changing his patterns of cerebral integration, and eventually unmasking residual function in his brain?

As neuroimaging is used to study additional patients with severe brain injuries, more questions will arise about the mechanisms underlying their functional disabilities. For patients like Wallis, neuroimaging could also reveal previously unrecognized residual capacities that can be at least partially restored by new kinds of therapy.

## Exploring Deep Brain Stimulation

If some patients with severe cognitive impairment could be limited, in part, by a lack of functional integration among intact regions of their brains, we should look for ways to foster reintegration. Patients like Wallis who have recovered to functional levels that are near the threshold of emergence from MCS would be the first likely candidates for new therapies to improve consistent communication. Of many new medical technologies, the most promise may lie in emerging techniques for deep brain stimulation.

Over the past fifteen years, deep brain stimulation has advanced the treatment of drug-resistant Parkinson's disease, sometimes dramatically, and is approved by the Food and Drug Administration for that use. More than 15,000 patients with Parkinson's have been treated, to date, and new uses of deep brain stimulation are being investigated to help patients with chronic pain, epilepsy, and psychiatric disorders such as depression and obsessive-compulsive disorder.

In an interdisciplinary project with the Cleveland Clinic Foundation, the JFK Johnson Rehabilitation Institute, and the Columbia University Functional MRI Research Center, we have been planning how to use deep brain stimulation to raise the functional level of MCS patients. Our efforts follow some provocative work over the past two decades that studied this technique with patients in a vegetative state (including, in fact, Terri Schiavo). Specifically, there have been several attempts to use deep brain stimulation in regions of the central thalamus, an area with many connections to the cerebral cortex. Activating these brain regions with an electrical current induces many of the standard signs of arousal, confirming experiments with animals in which electrical stimulation induced wakeful arousal. Unfortunately, in some fifty PVS patients studied worldwide, the stimulation evoked no evidence of sustained recovery of interactive awareness.

In contrast, deep brain stimulation did succeed in bringing about significant physiologic responses in PVS patients who had large increases in global and regional cerebral metabolism and changes in brain-wave activity toward a more normal profile for a wakeful state. The behavioral and physiologic arousal seen in all the patients demonstrated that, despite overwhelming brain damage, it was still possible to activate the cortex. It may be that, given the overwhelming brain injury in PVS, this increased activation was not enough to restore interactive awareness.

The areas electrically stimulated in these PVS patients are parts of the thalamus that are known to link a state of arousal with some aspects of moment-to-moment behavior. Here, then, is a new rationale for using deep brain stimulation in MCS patients who demonstrate limited integrative forebrain activity. Unlike the PVS patients—who initiate no behavior, follow no commands, and attempt no communication—MCS patients near the borderline of emergence typically have changes in cognitive functioning that come and go over hours, days, weeks, or even longer. This fluctuation might be the result of unstable interactions of the arousal state with the organization and maintenance of behaviors, which these patients can initiate, but not sustain. If so, then deep brain stimulation of the central thalamus might improve integration in the damaged networks that underlie these limited behaviors. By contrast, functional and structural neuroimaging studies demonstrate that these networks in patients with chronic PVS have been overwhelmingly damaged.

Before we can go beyond these planned pilot studies of deep brain stimulation in MCS patients, the criteria for selecting patients must be worked out, and important ethical questions considered. At present, it seems that those who have recovered to functional levels near the threshold of emergence from MCS could be the first candidates for new therapies to improve consistent communication. But new therapies for patients with severely impaired consciousness encounter challenges in the form of attitudes and preconceptions that could pose greater difficulties than the science. In particular, as we work with MCS patients, we will have to address two sources of skepticism: the right-to-die movement and the troubled record of psychosurgery.

## Seeking a New Moral Warrant

The first hurdle will be for our society to reexamine its attitudes toward patients with severe brain injury, attitudes shaped by the right-to-die movement. By taking to heart the possibility of new hope for these patients, we are asking for a moral warrant to intervene in patients who resemble those patients for whom the right to die was first established in the 1960s and 1970s. This hard-won—and important—right was vouchsafed to patients closest to death and for whom, therefore, the withdrawal of life-sustaining therapy seemed justifiable: patients with permanent and irreversible loss of cortical function.

Because the futility of any potential treatment was pivotal in justifying the right to die for PVS patients, many physicians remain nihilistic about potential interventions in these patients with severely impaired consciousness. Why bother, they wonder, because these people are essentially dead? In ruminations like these, which underlie judgments but are seldom explicitly voiced, people echo perceptions that were critical in establishing the rights of patients to refuse life-sustaining therapies. In the seminal 1976 Karen Quinlan case, the New Jersey Supreme Court allowed the removal of life-sustaining therapy because Quinlan was in a vegetative state without, they held, any possibility of return to a "cognitive sapient state." In 1968, a similar justification was urged by Harvard anesthesiologist Henry K. Beecher, M.D. when he advanced the concept of brain death, although that was in the context of seeking organs for transplantation. In both cases, however, the moral value placed on life and death hinged, in large part, on a person's cognitive state.

We hope, of course, to intervene for patients who are in the minimally conscious state, not the hopeless condition of chronic PVS patients, but the various states are often conflated or simply confused or the crucial differences among them are considered unimportant. The sense of nihilism is so pervasive that even the delineation of MCS in the scientific literature has come under attack from some medical quarters. We believe that refining the definitions of brain states is value-neutral, but many physicians have resisted this diagnostic clarification. Some proponents of the right to die have been concerned that this newly identified brain state might erode the hard-won right to forgo life-sustaining therapy. Disability advocates have also voiced their concern, worrying that adding MCS to the categories of impaired brain states could be used nefariously to equate higher-functioning individuals with those in PVS, thus minimizing the value of their lives.

For the record: We support both the right to die and the right to appropriate medical care. We do not see these rights as mutually exclusive, and we view decisions to pursue or refuse care as a matter of ethically balancing the potential benefits and burdens. What is interesting about the discord over MCS is not that the designation could justify either more treatment or less in any particular case but, rather, how emotionally charged the entire issue has become. As a response to discussion of the designation's scientific basis, the reactions seem way out of proportion. If nothing else, they are a cultural marker for our implicit assumptions about severe cognitive impairment.

## The Specter of Psychosurgery

Progress in developing therapies for severe cognitive impairment will also be hampered by the association of deep brain stimulation with psychosurgery and the abuses that have marred its history. News stories about deep brain stimulation often allude to the crude and unjustifiable lobotomies performed by Walter Freeman, a neurologist who performed more than 3,000 procedures on mentally incapacitated patients, and to the specter of mind control. These fears hark back to the debate over psychosurgery in the late 1960s and early 1970s, when Spanish neurophysiologist José M. R. Delgado, M.D., advanced the use of an implantable electrode operated by remote control as a way to "psychocivilize society" and cope with the social unrest of the day. Delgado had already won some notoriety by using his "Stimociever" to halt a charging bull in a bullring in 1965. His work entered popular culture through novels and films like Michael Crichton's *Terminal Man* and Stanley Kubrick's *A Clockwork Orange*. Concerns about the ethics of psychosurgery moved the U.S. Congress to direct the National Commission for the Protection of Human Subjects of Biomedical and Behavioral Research to report on psychosurgery as part of the landmark National Research Act of 1974. The commission did not find that psychosurgery had been used for social control; in fact, it found enough evidence of potential efficacy to recommend that the investigational use of some psychosurgical procedures proceed with appropriate regulation and oversight. This conclusion ran against popular opinion at the time, and popular opinion has never really changed.[6]

## Moving Beyond Scientific Stasis

Whatever the merits of deep brain stimulation in neurology and psychiatry (and we believe the merits are potentially great), public perception of its value and even its ethical standing is colored by the legacies of the right-to-die movement and psychosurgery. We believe that this perception does a profound disservice to some of our society's most desperately burdened patients by contributing to their being marginalized and even abandoned. When patients with severe cognitive impairment across the whole spectrum of such conditions are perceived as beyond hope, any potential interventions that might be developed are automatically deemed ethically disproportionate.[7] When this perception is combined with the problem that individuals with severe head trauma often lack the capacity to make decisions and, therefore, cannot give their own consent to enroll in clinical trials, you have a recipe for scientific deadlock.

Future Terry Wallises deserve better. They deserve protection from errors of commission, but equally of omission. They desperately need access to the fruits of science and all the assistance we can provide as they return from the limbo of impaired consciousness and try to reenter the world of human interaction.

REFERENCES

1. J. T. Giacino, S. Ashwal, N. Childs, et al., The minimally conscious state: Definition and diagnostic criteria, *Neurology* 58, no. 3 (2002): 349–353.

2. B. Jennett, *The vegetative state: Medical facts, ethical and legal dilemmas.* (Cambridge, UK: Cambridge University Press, 2002); B. Jennett and F. Plum, "Persistent vegetative state after brain damage. A syndrome in search of a name, *Lancet* 1 (1972): 734–737.

3. N. D. Schiff, U. Ribary, D. R. Moreno, et al., Residual cerebral activity and behavioural fragments can remain in the persistently vegetative brain, *Brain* 125 (2002): 1210–1234.

4. S. Laureys, M. E. Faymonville, P. Peigneux, et al., Cortical processing of noxious somatosensory stimuli in the persistent vegetative state, *Neuroimage* 17, no. 2 (2002): 732–741.

5. J. Hirsch, A. Kamal, D. Moreno, et al., fMRI reveals intact cognitive systems for two minimally conscious patients, *Society for Neuroscience, Abstracts* 271, no. 1 (2001): 1397.

6. J. J. Fins, From psychosurgery to neuromodulation and palliation: History's lessons for the ethical conduct and regulation of neuropsychiatric research, *Neurosurgery Clinics of North America* 14, no. 2 (2003): 303–319.

7. J. J. Fins, Constructing an ethical stereotaxy for severe brain injury: Balancing risks, benefits and access, *Nature Reviews Neuroscience* 4 (2003): 323–327.

NICHOLAS D. SCHIFF, M.D., is assistant professor of neurology and neuroscience at the Weill Medical College of Cornell University and assistant attending neurologist at the New York Presbyterian Hospital. He is director of the Laboratory of Cognitive Neuromodulation at New York-Presbyterian Weill Cornell Medical Center.

JOSEPH J. FINS, M.D., is chief of the Division of Medical Ethics and professor of medicine, public health, and medicine in psychiatry at Weill Medical College of Cornell Medical Center. He is director of medical ethics at the New York-Presbyterian Weill Cornell Medical Center, associate for medicine at The Hastings Center, and a member of the adjunct faculty of Rockefeller University.

*Chapter 29*

# Rethinking Disorders of Consciousness: *New Research and Its Implications**

*Joseph J. Fins*

Over the past several years, deciding whether to withdraw life-sustaining therapy from patients who have sustained severe brain injuries has become much more difficult. The problem is not the religious fundamentalism that infused the debate over the care of Terri Schiavo, the Florida woman in a permanent vegetative state whose case has drawn national attention. Rather, the difficulty stems from emerging knowledge about the diagnosis and physiology of brain injury and recovery. The advent of more sophisticated neuroimaging techniques like MRI and PET scans, in tandem with electrophysiologic and observational studies of brain-injured patients, has led to an effort to differentiate disorders of consciousness more precisely. The crude categories that have informed clinical practice for a quarter century are becoming obsolete.

It used to be enough for a neurologist or neurosurgeon to write a note in the chart grimly recording the patient's neurological exam and then concluding with the global statement "no hope for meaningful recovery." It can no longer be so simple. With a better understanding of brain injury and mechanisms of recovery, we should be suspicious of blanket statements that

---

*From *Hastings Center Report* 35, no. 2 (2005): 22–24.

might, we now believe, obscure important differences among different patients' prospects for recovery, although even those patients we now think
may recover may still be left with profound and perhaps intolerable burdens
of disability.

Recovery from coma depends on a patient's age, the site of injury, and
whether the damage was done by trauma, anoxia (oxygen deprivation), or
other processes. The most severe brain injuries may lead to brain death. If
patients survive and begin to recover from coma, they often first enter into
the vegetative state, first described by Bryan Jennett and my teacher, Fred
Plum, in 1972. The vegetative state is a paradoxical state of "wakeful unresponsiveness" in which the eyes are open but there is no awareness of self or
environment. When a vegetative state continues beyond thirty days, it is described as "persistent." A vegetative state is generally considered permanent
three months after anoxic injury and twelve months after trauma.

All of this is news since I went to medical school. I was taught that the
vegetative state was immutable and fixed. Vegetative brains were, if I recall the phrase correctly, "gelatinous gels." The futility of this brain state
was the basis for the establishment of the right to die in cases like Quinlan's
and Cruzan's. Recent studies have shown, however, that patients can regain
some evidence of consciousness before the vegetative state becomes permanent. In the window between the persistent and permanent vegetative state,
patients can progress to what has been described as the "minimally conscious state" (MCS). Unlike vegetative patients, the minimally conscious
demonstrate unequivocal, but fluctuating, evidence of awareness of self and
the environment. The natural history of MCS patients is not yet known.
Near the upper boundary of this category, patients may say words or phrases
and gesture. They also may show evidence of memory, attention, and intention. Patients are considered to have "emerged" from MCS only when they
can reliably and consistently communicate.

Unfortunately, all of this is easier to explain in theory than to observe in
practice. First and foremost is the challenge of diagnosis. To the untrained
eye, MCS patients may appear very similar to those who are vegetative.
These diagnoses can be confused and conflated and in the earlier phases
of illness need to be considered very carefully in the context of the mechanism of injury. In a patient with non-anoxic injury, even small gains beyond
the vegetative level may herald the potential for significant further recovery.
Some recent studies suggest that the diagnostic distinction between MCS
and PVS is missed by neurologists at rates that would be intolerable in oth-

er clinical domains. To be fair, however, a neurologist acting in good faith might examine an MCS patient when his level of arousal was low and elicit an exam that is indistinguishable from that of a vegetative patient.

But there is another sort of diagnostic error that occurs when the objectivity of diagnosis is infiltrated by value judgments. Instead of dealing with the moral ambiguity associated with balancing the burdens and possible benefits of continuing care, there is a tendency among some practitioners to act paternalistically and label some who might be minimally conscious as vegetative. By being categorical, they sidestep the more difficult choices and avoid the morass of the minimally conscious state. With "no hope for meaningful recovery," care can be withdrawn. But even if this is true—and it may be, since a patient in a minimally conscious state may indeed have no hope for meaningful recovery—our greater level of knowledge about these conditions calls for more diagnostic clarity.

Diagnostic distortion has also been used to undermine the right to die. In Schiavo, right-to-life advocates asserted that she was not vegetative. By suggesting consciousness where there was none, these opponents of choice at the end of life cast doubt on the ethical propriety of removing life-sustaining therapy. They persisted even though court-appointed physicians found that she was vegetative, and even when the Florida Supreme Court determined that there was clear and convincing evidence for this diagnosis.

A third sort of diagnostic distortion is journalistic. Differing brain states can be conflated either through ignorance of the facts or deliberately—to hype a case or a new scientific development. The latter occurred in a *New York Times Magazine* article that discussed our work with minimally conscious patients and its implications for the centrality of diagnostic discernment and the use of neuroimaging techniques. Although the text was for the most part accurate, the headlines and pull quotes mistakenly labeled the patients as vegetative. One notable header: "New research suggests that many vegetative patients are more conscious than previously supposed—and might eventually be curable. A whole new way of thinking about pulling the plug."

When we learned about such errors prior to publication, Fred Plum and my colleague Nicholas Schiff and I contacted the magazine's editor to request a change. She told us the distinction was unimportant. It was, in her view, merely a matter of semantics. The article ran without the changes we had requested. The magazine published a correction weeks later along with our letter to the editor, but by then it was too late. Few people read corrections. Just recently a family we were counseling discovered the article in an Internet search and brought it up during an ethics consultation. If there was hope for vegetative patients, could there not be hope for their loved one? We sought to explain how the *Times* had gotten it wrong and to provide as much diagnostic and prognostic information as we could.

Each of these distortions is troubling. If a distortion is a physician's it undermines the integrity of the clinical transaction. If it is inspired by ideology it politicizes a process that is better left to scientific judgment. And if it occurs through journalistic hubris, it perpetuates misunderstanding in the popular culture.

Families will have more than enough difficulty contending with disorders of consciousness even when they are properly diagnosed. Assuming that families can ascertain a credible diagnosis and prognosis, how should they make decisions about care? How should a slim prospect of recovering consciousness be balanced against the burdens associated with enduring disability? The protracted time frame during which recovery might occur could require a vigil that lasts for months and still might lead only to disappointment.

A long vigil may also preclude options to withdraw life-sustaining therapy. Consider the implications of the recent papal statement on the ethical mandate to provide artificial nutrition and hydration to vegetative patients. If an observant Catholic family were to follow Church teachings, they might be able to discontinue "extraordinary" measures early in the patient's course when the prognosis was still unknown, but they might not be able to discontinue artificial nutrition and hydration later on, once it was clear that the patient would not make any progress from the vegetative state. This might cause some families to be more risk-aversive and withdraw extraordinary measures earlier in the course of illness while treatments like ventilators were still in place. The paradox is striking: a papal statement intended to promote life might have the unintended consequence of limiting the chance of recovery for some.

To make matters even more complicated, these decisions will likely take place beyond the reach of the hospital and the expertise that is available in clinical ethics and neurology. Transfers out of the acute-care setting can lead to errors of diagnostic omission and a failure to follow patients longitudinally as their condition evolves.

Such was the fate of Terry Wallis, an Arkansan who suffered traumatic brain injury in 1984 after a car accident. After he was diagnosed as being in a vegetative state, he was discharged to a nursing home, where he lingered for nineteen years. Although his family saw evidence of awareness, he did not receive an examination by a neurologist and never underwent an imaging study. His family was told that a workup would be too expensive. The implication was that it was also pointless.

His case gained national press coverage in July 2003 when he began to speak. Headlines suggested that he had miraculously emerged from a coma. A closer examination of the record reveals that he had probably moved from the vegetative state to the minimally conscious state within the first months after his injury and then remained improperly diagnosed for years.

Stories like these send a chill down my spine. Some patients diagnosed as vegetative are probably in fact intermittently sentient but unable to communicate. The isolation, abandonment, and neglect they experience is unimaginable. Though their numbers may be small—there are no reliable data on how common this phenomenon is—they still make a claim of justice on all of us who know that some conscious but noncommunicative individuals may have been relegated to the margins of the human community. And they are but a small segment of a larger group of institutionalized patients with severe brain injuries who are receiving what has been described as merely "custodial care."

All of these patients deserve better. The small community of neuroscientists who have taken an interest in mechanisms of brain injury and recovery needs to be expanded, and bioethicists need to grapple with the imponderables, both theoretical and practical, that attend to disorders of consciousness. There is no shortage of questions about the nature of the self, personal identity, and autonomy to occupy us. Colloquially put, how much of yourself do you have to lose to cease to be you? The implications for an ethic grounded in self-determination are obvious and ripe for engagement by both theoretical and practical ethicists.

The lesson from narratives of individuals who have suffered from brain injury is that the physiologic is only part of the story. Although injury to the same brain substrate might produce memory loss or language difficulties, these impairments are superimposed on each patient's personal psychology and past, producing highly individual losses rather than generic deficits. Likewise the recovery will be highly individual. Consider the physician who sustains an injury to her frontal lobe, spends the next ten years learning how to sequence daily tasks that previously were second nature, and along the way becomes an accomplished abstract artist, her creative impulses disinhibited perhaps by her injury. Or the athlete who worked in the financial sector and struggles to relearn the arithmetic skills that she once possessed, accommodating her expectations to her altered abilities. Each of these stories is about rediscovering a new self while recalling a lost identity.

If we hope to help patients and families make the tough choices following brain injury, we will need to embrace the ambiguity that goes along with long courses of recovery and questions about altered selves. These decisions will be more challenging than decisions to remove life support in the face of overwhelming sepsis or to pursue treatment in the face of widely metastatic cancer. We will also need to demand diagnostic honesty and precision. In discussing diagnoses with families we will need to strike a balance between realism and hope. The objective must be to bring greater attention to the minimally conscious patient without engendering expectations for the permanently unconscious. If we are successful, we will protect both the right to die and the right to care, as paradoxical as that may seem in today's clinical and political climate.

## Acknowledgments

The writing of this essay was funded in part by grants from the Charles A. Dana Foundation and the Buster Foundation.

JOSEPH J. FINS, M.D., is chief of the Division of Medical Ethics and professor of medicine, public health, and medicine in psychiatry at Weill Medical College of Cornell Medical Center. He is director of medical ethics at the New York-Presbyterian Weill Cornell Medical Center, associate for medicine at The Hastings Center, and a member of the adjunct faculty of Rockefeller University.

# Epilogue

*Chapter 30*

# Ethics in a Neurocentric World[*]

*Steven Rose*

The concerns that have occupied the preceding eleven chapters [of the book from which this chapter is excerpted—Ed.] have nagged away at me for the best part of the past decade, as the neuroscience whose excitement has shaped my researching life seemed to be segueing seamlessly into neurotechnology. The claims of some of my colleagues have become ever more ambitious and comprehensive. Human agency is reduced to an alphabet soup of As, Cs, Gs, and Ts in sequences patterned by the selective force of evolution, while consciousness becomes some sort of dimmer switch controlling the flickering lights of neuronal activity. Humans are simply somewhat more complex thermostats, fabricated out of carbon chemistry. In the meantime, I have been becoming uncomfortably aware that the issues raised in the shift from neuroscience to neurotechnology have close and obvious precedents.

By the time that the National Institutes of Health in the United States had titled the 1990s "The Decade of the Brain," advances in genetics and the new reproductive technologies were already beginning to generate ethical concerns. Back in the 1970s, when the prospects of genetic manipulation of microorganisms—though not yet of mammals—emerged, concerned geneticists called a conference at Asilomar, in California, to consider the implications of the new technology and draw up guidelines as to its use. They

*From *The Future of the Brain: The Promise and Perils of Tomorrow's Neuroscience* by Steven Rose (Oxford: Oxford University Press, 2005), Chapter 12, pp. 297–305.

called a temporary moratorium on research while the potential hazards were investigated. However, the moratorium didn't last—before long the prospects of fame and riches to be made from the new technologies became too tempting, the hazards seemed to have been exaggerated, and the modern age of the commercialization of biology began. By the 1990s the biotech boom was well under way, genes were being patented, the ownership of genetic information was under debate, and speculations about the prospects of genetically engineering humans were rife. As social scientists, philosophers, and theologians began to consider the implications of these new developments, a new profession was born, that of bioethicist, charged, like any preceding priesthood, with condoning, and if possible blessing, the enterprise on which the new biology was embarked. In all the major researching nations, bioethics committees, either institutionalized by the government as in the United States or France, or by charitable foundations, as in the UK, began to be established, with the task of advising on what might or might not be acceptable. Thus somatic genetic therapy was deemed acceptable, germ line therapy unacceptable. Therapeutic cloning is okay, reproductive cloning is not. Human stem cell research is permissible in Britain, but not (with certain exceptions) under federal regulations in the United States.

This experience helps to explain why it was that in 1990, when the Human Genome Project (HGP) was launched, U. S. and international funding agencies—unlike the private funders against whom they were competing—took the unprecedented step of reserving some 3% of the HGP's budget for research on what they defined as the ethical, legal, and social implications of the project. The acronym that this has generated, ELSI, has entered the bioethicists' dictionary (although Europeans prefer to substitute A, meaning "Aspects," for the I of "Implications"). Admittedly, some of those responsible for ring-fencing this funding had mixed motives in so doing. Central to the entire project was James Watson, co-discoverer of the structure of DNA and for half a century the éminence grise of molecular biology, who, with characteristic bluntness, referred to it as a way of encapsulating the social scientists and ethicists and hence deflecting potential criticism. Nonetheless, whatever the complex motivation that led to these programs being set up and funded, ELSI/ELSA is here to stay.

As a neuroscientist, I watched these developments with much interest; it seemed to me that my own discipline was beginning to throw up at least as serious ethical, legal, and social concerns as did the new genetics. But such was and is the pace of genetic research, and the lack of control over the work of the biotech companies, that the various national ethics councils were often at work closing the stable door after the horse—or in this case the sheep—had bolted. The successful cloning of Dolly, announced to great fanfares of publicity in 1996, took the world seemingly by surprise, prompting a flurry of urgent ethical consultations and hasty efforts to legislate. No one, it seemed, had thought the cloning of a mammal possible, although in fact the basic science on which the success was based had been done years before.

Comparable technological successes in the neurosciences are still in the future. Whereas in the case of genetics ethical considerations often seemed to occur post hoc, in the neurosciences it might be possible to be more proactive—to find ways of publicly debating possible scientific and technological advances before they have occurred, and engaging civil society in this debate. Over the last five years my own concerns have become more widely shared. A new term has appeared in the bioethical and philosophical literature: "neuroethics." In Britain the Nuffield Council on Bioethics has held consultations on the implications of behavior genetics. In the United States, the President's Bioethics Council has discussed topics as disparate as cognitive enhancers and brain stimulation. The Dana Foundation, a U.S.-based charity with a European branch, has sponsored discussions on neuroethics as a regular component of major neuroscience meetings.[1] The European Union has organized its own symposia.[2] Critiques of both the power and the pretensions of neuroscience have appeared.[3] As I was completing this very chapter, a Europe-wide citizens' consultation on neuroethics, scheduled to last for the next two years, was announced.

The issues raised in such discussions range from the very broadest to the most specific. Can we, for instance, infer some universal code of ethics from an understanding of evolutionary processes? What are the boundaries between therapy and enhancement—and does it matter? How far can biotechnology aid the pursuit of happiness? Could one—should one—attempt to prevent aging and even death? To what extent should neuroscientific evidence be relevant in considering legal responsibility for crimes of violence? Should there be controls on the use of cognitive enhancers by students sitting examinations? On what issues should governments and supranational bodies endeavour to legislate?[4]

As usual, novelists and moviemakers have got here before us. The daddy of them all is of course Aldous Huxley, but the cyborgian worlds of William Gibson's *Neuromancer* (published twenty years ago) and *Virtual Light* offer prospects that many will see as potential if unpalatable. Recent films have trodden similar paths. Thus Charlie Kauffman's *Eternal Sunshine of a Spotless Mind*, released in 2004, plays with the prospect of a neurotech company (Lacuna, Inc.) specializing in erasing unwanted memories via transcranial brain stimulation.

It is far from my intention to pronounce dogmatically on these quite deep questions. This isn't simply a cop-out . . .; it is that I don't think that I have the right or authority to do so. They are, after all, the domain of civil society and political debate. My task as an expert with special competence in some of these areas is to lay out as clearly as I can what seem to me to be the theoretical and technical possibilities, which is what [my] preceding eleven chapters have attempted to do, and then to participate, as a citizen, in discussing how we should try to respond to them. What is certain is that society needs to develop methods and contexts within which such debates can be held. The processes that have been attempted so far include the establishment of statutory bodies, such as, in the UK, the Human Genetics Commission (HGC), and various forms of public consultation in the form of Technology Assessment panels, Citizens' Juries, and so forth. None are yet entirely satisfactory. Engaging "the public" is not straightforward, as of course there are multiple publics, not one single entity. Furthermore, governments are seemingly quite willing to bypass their own agencies, or set aside public opinion, when it suits. Legislation permitting the use of human stem cells in Britain bypassed the HGC. When the public consultation on GM crops produced the "wrong" answer—an overwhelming public rejection—the government simply ignored it and authorized planting.

Nonetheless, the very recognition that public engagement is essential, that technologies cannot simply be introduced because they are possible and might be a source of profit for a biotech company, becomes important. In Europe we are becoming used to developing legislative frameworks within which both research and development can be regulated. The situation is different in the United States, where, although the uses of federal funds are restricted, there is much more freedom for private funders to operate without legal restriction—as the flurry of claims concerning human cloning by various maverick researchers and religious sects has demonstrated. This is perhaps why Francis Fukuyama, in contrast to other American writers concerned with the directions in which genetics and neuroscience might be leading, argues in favor of the European model.

The framework within which I approach these matters is that presaged in earlier chapters. It pivots on the question of the nature of human freedom,

I have tried to explain why, although I regard the debates about so-called "free will" and "determinism" as peculiar aberrations of the Western philosophical tradition, we as humans are radically undetermined—that is, living as we do at the interface of multiple determinisms we become free to construct our own futures, though in circumstances not of our own choosing.[5] This is what I was driving at in Chapter 6 [of *The Future of the Brain*—Ed.] with my account of the limited power of the imaginary cerebroscope.

We are both constrained and freed by our biosocial nature. One of these enabling constraints is evolutionary. Evolutionary psychologists, in particular, have claimed that it is possible to derive ethical principles from an understanding of evolutionary processes, although many philosophers have pointed out with asperity that one cannot derive an ought from an is.[6] I don't want to get hung up on that here; to me the point is that no immediately conceivable neurobiological or genetic tinkering is going to alter the facts that the human life span is around a hundred, plus or minus a few years, and that human babies take around nine months from conception to birth and come into the world at a stage of development that requires many years of postnatal maturation to arrive at adulthood. These facts, along with others such as our size relative to other living organisms and the natural world, our range of sensory perception and motor capacities, our biological vulnerabilities and strengths, shape the social worlds we create just as the social worlds in turn affect how these limits and capacities are expressed. This is the evolutionary and developmental context within which we live and which helps define our ethical values. There may be a remote cyborgian future in which other constraints and freedoms appear, but we may begin to worry about those if— and it is a disturbingly large if—humanity manages to survive the other self-inflicted perils that currently confront us. But we also live in social, cultural, and technological contexts, which equally help to define both our self-perceptions and our ethical values. This is what the sociologist Nikolas Rose is driving at when he refers to us, in the wake of the psychopharmacological invasion of our day-to-day lives, becoming "neurochemical selves," in which we define our states of mind and emotion in terms of medical categories and the ratios of serotonin to dopamine transmitters in our brains.

As I argued in the [my] two preceding chapters, the neurosciences are moving us toward a world in which the prospects both for authoritarian control of our lives and the range of "choices" available for the relatively prosperous two-thirds of our society are increasing. On the one hand, we have the prospect of profiling and prediction on the base of gene scans and brain imaging, followed by the direct electromagnetic manipulation of neural processes or the targeted administration of drugs to "correct" and "normalize" undesirable profiles. On the other, an increasing range of available Somas to mitigate misery and enhance performance, even to create happiness.

There's always a danger of overstating the threats of new technologies, just as their advocates can oversell them. The powers of surveillance and coercion available to an authoritarian state are enormous: ubiquitous CCTV cameras, hypersensitive listening devices and bugs, satellite surveillance; all create an environment of potential control inconceivable to George Orwell in 1948 when he created 1984's Big Brother, beaming out at everyone from TV cameras in each room. The neurotechnologies will add to these powers, but the real issue is probably not so much how to curb the technologies, but how to control the state. As for thought control, in a world whose media, TV, radio, and newspapers are in the hands of a few giant and ruthless global corporations, perhaps there isn't much more that transcranial brain stimulation can add.

The same applies to arguments about enhancement. In Britain today, as in most Western societies (with the possible exception of Scandinavia), enhancement is readily available by courtesy of wealth, class, gender, and race. Buying a more privileged, personalized education via private schools, and purchasing accomplishments for one's children—music teaching, sports coaching, and so forth—are sufficiently widespread as to be taken almost for granted except among those of us who long for a more egalitarian society. No major political party in power seems prepared to question such prerogatives of money and power. Compared with this, what difference will a few smart drugs make?

But of course, they will—if only because we seem to feel that they will. Incremental increases in the power of the state need to be monitored carefully: think of the widespread indignation at the new U. S. practice of fingerprinting, or using iris identification techniques, on foreign passport holders entering the country, or the hostility in the UK to the introduction of identity cards carrying unique biodata. Similarly with the cognition-enhancing drugs—as with the use of steroids by athletes—their use, at least in a competitive context, is seen as a form of cheating, of bypassing the need for hard work and study. Enthusiasts are too quick to dismiss these worries, deriding them as hangovers from a less technologically sophisticated past (as when slide rules used to be banned from math examinations). They speak of a "yuck factor," an initial response to such new technologies that soon disappears as they become familiar, just as computers and mobile phones are now taken-for-granted additions to our range of personal powers. But we should beware of enthusiasts bearing technological gift horses; they do need looking at rather carefully in the mouth, and it should be done long before we ride off on their backs, because they may carry us whither we do not wish to go. For, as I continue to emphasize, the dialectical nature of our existence as biosocial beings means that our technologies help shape who we are, reconstructing our very brains; as technology shifts, so do our concepts of personhood, of what it means to be human.

Which brings me, at last, to the issues of freedom and responsibility. It is in the context of the law that these concepts are most acutely tested. For a person to be convicted of a crime he or she must be shown to be responsible for his or her actions—that is, to be of sound mind. As Judge Stephen Sedley explains,

> When a defendant claims in an English Court that a killing was not
> murder but manslaughter by reason of diminished responsibility, the
> judge has to explain to the jury that "Where a person kills or is party
> to the killing of another, he shall not be convicted of murder if he
> was suffering from such abnormality of mind (whether arising from a
> condition of arrested or retarded development of mind or any inherent
> causes or induced by disease or injury) as substantially impaired his
> mental responsibility for his acts or omissions in doing or being a party
> to the killing."[7]

As Sedley goes on to point out, the questions of "abnormality of mind" and "mental responsibility" raise endless tangles. They are enshrined in English law by way of the McNaghten rules under which an insanity plea can be entered, laid down by the House of Lords in 1843 and refined, as in the quote above, in 1957. From a neuroscientific point of view the definitions make little sense—as Sedley would be the first to agree. If, for instance, presence of a particular abnormal variant of the MAOA gene predisposes a person to a violent temper and aggressive acts, can the possession of this gene be used as a plea in mitigation by a person convicted of murder? If Adrian Raine were right, and brain imaging can predict psychopathy, could a person showing such a brain pattern claim in defense that he was not responsible for his actions? Can a person committing homicide while under the influence of a legally prescribed drug like Prozac claim that it was not he but the drug that was responsible for his behavior? As I've pointed out, such defences have been tried in the United States, and at least admitted as arguable in court, though not yet in the UK.

Such genetic or biochemical arguments seem to go to the core of our understanding of human responsibility, but they raise profound problems. If we are neurochemical beings, if all our acts and intentions are inscribed in our neuronal connections and circulating neuromodulators, how can we be free? Where does our agency lie? Once again, it is instructive to turn the argument around. Back in the 1950s it became fashionable to argue that many criminal acts were the consequences of an impoverished and abused childhood. Is there a logical difference between arguing "It was not me, it was my genes" and "It was not me, it was my environment"? I would suggest not. If we feel that there is, it is because we have an unspoken commitment to the view that "biological" causes are more important, somehow more determining, in a sense, than "social" ones. This is the biological determinist trap. Yet the study by Caspi and his colleagues of the ways in which gene and environment "interact" during development in the context of any putative relationship between MAOA and "aggressive behavior" points to the mistake of giving primacy to a genocentric view.

Of course, it is probable both that many people with the gene variant are not defined as criminal, while most people who are defined as criminal do not carry the gene. At best such predictors will be weakly probabilistic even if one sets aside the problems raised in the previous chapter [of *The Future of the Brain*] about the impossibility of extracting definitions of either "aggression" or "criminality" from the social context in which specific acts are performed. It follows that within a very broadly defined range of "normality," claims of reduced responsibility on the basis of prior genetic or environmental factors will be hard to sustain. And there is a final twist to this argument. People found guilty of criminal acts may be sentenced to prison. So what should be done with persons who are found not guilty by virtue of reduced responsibility? The answer may well be psychiatric incarceration—treatment, or restraint if treatment is deemed impossible, rather than punishment. Assuming that prison is intended to be rehabilitative as well as retributive, defining the difference between the consequences of being found responsible rather than irresponsible may become a matter of semantics.[8] Despite the increasing explanatory powers of neuroscience, I suspect that many of these judgements are best left to the empirical good sense of the criminal justice system, imperfect, class-, race-, and gender-bound though it may be.

But, more fundamentally, the question of how we can be free if our acts and intentions are inscribed in our neurons is a classic category error. When Francis Crick tried to shock his readers by claiming that they were "nothing but a bunch of neurons,"[9] he splendidly missed the point. "We" are a bunch of neurons, and other cells. We are also, in part by virtue of possessing those neurons, humans with agency. It is precisely because we are biosocial organisms, because we have minds that are constituted through the evolutionary, developmental, and historical interaction of our bodies and brains (the bunch of neurons) with the social and natural worlds that surround us, that we retain responsibility for our actions, that we, as humans, possess the agency to create and re-create our worlds. Our ethical understandings may be enriched by neuroscientific knowledge, but they cannot be replaced, and it will be through agency, socially expressed, that we will be able, if at all, to manage the ethical, legal, and social aspects of the emerging neurotechnologies.

REFERENCES

1. S. J. Marcus, ed., *Neuroethics: Mapping the field* (New York: Dana Press, 2002.)

2. P. Busquin et al., *Modern biology and visions of humanity* (Brussels: Multi-Science Publishing, 2004).

3. J. Horgan, *The undiscovered mind: How the brain defies explanation* (London: Weidenfeld and Nicolson, 1999).

4. R. H. Blank, *Brain policy: How the new neuroscience will change our lives and our politics* (Washington, DC: Georgetown University Press, 1999).

5. S. Rose, *Lifelines: Biology, freedom, determinism* (New York: Penguin, 1997; 2nd ed., Vintage, forthcoming); D. Rees and S. Rose, eds., *The new brain sciences: Prospects and perils* (especially the chapters by Stephen Sedley, Peter Lipton, and Mary Midgley) (Cambridge, UK: Cambridge University Press, 2004).

6. A. Rosenberg, *Darwinism in philosophy, social science, and policy* (Cambridge, UK: Cambridge University Press, 2000).

7. S. Sedley, in D. Rees and S. Rose, eds., *The new brain sciences*, 123–180.

8. L. Reznek, *Evil or ill? Justifying the insanity defence* (London: Routledge, 1997).

9. F. H. C. Crick, *The astonishing hypothesis: The scientific search for the soul* (New York: Simon and Schuster, 1994).

STEVEN ROSE, PH.D., is professor and chair of the Department of Biology and director of the Brain and Behavior Research Group at the Open University.

# Further Reading

Ackerman, S. J. *Hard Science, Hard Choices: Facts, Ethics, and Policies Guiding Brain Science Today*. New York and Washington: Dana Press, 2006.

Blank, R. *Brain Policy: How the New Neuroscience Will Change Our Lives and Our Politics*. Washington: Georgetown University Press, 1999.

Cacioppo, T., et al. *Social Neuroscience: People Thinking about Thinking People*. Cambridge, MA: MIT Press, 2005.

Caplan, A. "No Brainer: Can We Cope with the Ethical Ramifications for New Knowledge of the Human Brain?" in S. J. Marcus, ed., *Neuroethics: Mapping the Field*, 95–106. New York and Washington: Dana Press.

Churchland, P. S. *Brain-Wise: Studies in Neurophilosophy*. Cambridge, MA: MIT Press, 2002.

Damasio, A. *Descartes' Error: Emotion, Reason, and the Human Brain*. New York: Putnam, 1994.

_____. *The Feeling of What Happens: Body and Emotion in the Making of Consciousness*. New York: Harcourt Brace, 1999.

_____. *Looking for Spinoza: Joy, Sorrow, and the Feeling Brain*. Orlando: Harcourt, 2003.

_____. "Neuroscience and Ethics: Intersections," *The American Journal of Bioethics–Neuroscience*. 7, (2007): 3–7.

Dennett, D. *Freedom Evolves*. New York: Viking, 2003.

Elliott, Carl. *Better than Well: American Medicine Meets the American Dream*. New York: Norton, 2003.

Fins, J., and Schiff, N. "Shades of Gray: New Insights into the Vegetative State," *Hastings Center Report* 36 (November-December, 2006): 8.

Fuchs, T. "Ethical Issues in Neuroscience," *Current Opinion in Psychiatry* 19 (2006): 600–607.

Garland, B., ed. *Neuroscience and the Law: Brain, Mind, and the Scales of Justice*. New York and Washington: Dana Press, 2004.

Gazzaniga, M. *The Ethical Brain*. New York Dana Press, 2005.

Gillett, G. "Cyborgs and Moral Identity." *Journal of Medical Ethics* 32 (2006): 79–83.

Glannon, W. *Bioethics and the Brain*. New York: Oxford University Press, 2006.

_____. "Free Will and Moral Responsibility in the Age of Neuroscience." *Lahey Clinic Medical Ethics* 13, 2 (2006): 1–2.

_____. "Neuroethics," *Bioethics* 20 (2006): 37–52.

Greene, J. D., et al. "An fMRI Investigation of Emotional Engagement in Moral Judgment," *Science* 293 (2001): 2105–2108.

Greene, J. D., et al. "The Neural Basis of Cognitive Conflict and Control in Moral Judgment." *Neuron* 44 (2004): 389–400.

Hanes, J.D., and G. Rees, "Neuroimaging: Decoding Mental States from Brain Activity in Humans." *Nature Reviews Neuroscience* 7 (2006): 523–524.

Hauser, M. *Moral Minds: How Nature Designed Our Universal Sense of Right and Wrong.* New York: HarperCollins, 2006.

Healy, D. *The Antidepressant Era.* Cambridge, MA: Harvard University Press, 1997.

_____. *The Creation of Psychopharmacology.* Cambridge, MA: Harvard University Press, 2002.

Hochberg, L. R., et al. "Neuronal Ensemble Control of Prosthetic Devices by a Human with Tetraplegia." *Nature* 442 (2006): 164–171.

Hyman, S., and W. Fenton. "What Are the Right Targets for Psychopharmacology?" *Science* 299 (2003): 350–351.

Illes, J. ed. *Neuroethics: Defining the Issues in Theory, Practice, and Policy.* New York and Oxford: Oxford University Press, 2006.

Laureys, S, ed. *The Boundaries of Consciousness: Neurobiology and Neuropathology.* London: Elsevier, 2006.

Laureys, S., et al. "Brain Function in Coma, Vegetative State, and Related Disorders." *Lancet Neurology* 3 (2004): 537–546.

LeDoux, J. *The Emotional Brain.* New York: Simon and Schuster, 1996.

_____. *The Synaptic Self: How Our Brains Become Who We Are.* New York: Viking, 2002.

Levy, N. *Neuroethics: Challenges for the 21st Century.* Cambridge, UK: Cambridge University Press, 2007.

Lisanby, S. H., ed. *Brain Stimulation in Psychiatric Treatment.* Washington: American Psychiatric Association, 2004.

Marcus, S. J., ed. *Neuroethics: Mapping the Field.* New York and Washington: Dana Press, 2002.

McGaugh, J. *Memory and Emotion: The Making of Lasting Memories.* New York: Columbia University Press, 2003.

_____. "Remembering and Forgetting: Physiological and Pharmacological Aspects." Testimony before the U.S. President's Council on Bioethics. Seventh meeting, session 3, October 17, 2002. Http://www.bioethics.gov/transcripts/oct02/session3.html

Mayberg, H., et al. "Deep-Brain Stimulation for Treatment-Resistant Depression." *Neuron* 45 (2005): 651–660.

McClure, S., et al. "Neural Correlates of Behavioral Preferences for Culturally Familiar Drinks." *Neuron* 44 (2004): 379–387.

Moreno, J. *Mind Wars: Brain Research and National Defense.* New York and Washington: Dana Press, 2006.

Nicolelis, M. "Brain-Machine Interfaces to Restore Motor Function and Probe Neural Circuits." *Nature Reviews Neuroscience* 4 (2003): 417–422.

Parens, E. ed. *Enhancing Human Traits: Ethical and Social Implications.* Washington: Georgetown University Press, 1998.

President's Council on Bioethics. *Beyond Therapy: Bioethics and the Pursuit of Happiness.* New York and Washington: Dana Press, 2003.

Racine, E., et al. "fMRI in the Public Eye," *Nature Reviews Neuroscience* 6 (2005): 159–164.

Rees, D., and S. Rose, eds. *The New Brain Sciences: Perils and Prospects.* Cambridge: Cambridge University Press, 2004.

Rose, S. *The Twenty-first Century Brain: Explaining, Mending, and Manipulating the Mind.* London: Cape, 2005.

Roskies, A. "Neuroscientific Challenges to Free Will and Responsibility," *Trends in Cognitive Sciences* 10 (2006): 419–423.

Schacter, D. "Remembering and Forgetting: Psychological Aspects." Testimony before the U.S President's Council on Bioethics. Seventh meeting, session 4, October 17, 2002. http://www.bioethics.gov/transcripts/oct02/session4.html.

_____. *Searching for Memory: The Brain, the Mind, and the Past.* New York: Basic Books, 1996.

Tancredi, L. *Hardwired Behavior: What Neuroscience Reveals About Morality.* New York: Cambridge University Press, 2005.

Wexler, B. E. *Brain and Culture: Neurobiology, Ideology, and Social Change.* Cambridge, MA: MIT Press, 2006.

Winslade, W. *Confronting Traumatic Brain Injury.* New Haven: Yale University Press, 1999.

Wolpe, P. R. "Emerging Neurotechnologies for Lie Detection: Promises and Perils." *American Journal of Bioethics* 5 (2005): 15–26.

Wolpe, P. R. "Neurotechnology, Cyborgs, and the Sense of Self." In S. J. Marcus, ed., *Neuroethics: Mapping the Field,* (2002): 159–167.

World Health Organization. *The World Health Report, 2001. Mental Health: New Understanding, New Hope.* Geneva: World Health Organization, 2001.

# Index

# OTHER DANA PRESS
# BOOKS AND PERIODICALS

www.dana.org/books/press

## Books For General Readers

### Brain and Mind:

### BEST OF THE BRAIN FROM SCIENTIFIC AMERICAN:
Mind, Matter, and Tomorrow's Brain

*Floyd E. Bloom, M.D., Editor*

Top neuroscientist Floyd E. Bloom has selected the most fascinating brain-related articles from *Scientific American* and *Scientific American Mind* since 1999 for this collection. Divided into three sections—Mind, Matter, and Tomorrow's Brain—this compilation offers the latest information from the front lines of brain research. 30 full-color illustrations.

Cloth, 300 pp. 978-1-932594-22-5 • $25.00

### CEREBRUM 2007: Emerging Ideas in Brain Science

*Cynthia A. Read, Editor*

*Foreword by Bruce McEwen, Ph.D.*

Prominent scientists and other thinkers explain, applaud, and protest new ideas arising from discoveries about the brain in this first yearly anthology from *Cerebrum*'s Web journal for inquisitive general readers. 10 black-and-white illustrations.

Paper 225 pp. 978-1-932594-24-9 • $14.95

### MIND WARS: Brain Research and National Defense

*Jonathan Moreno, Ph.D.*

A leading ethicist examines national security agencies' work on defense applications of brain science, and the ethical issues to consider.

Cloth 210 pp. 1-932594-16-7 • $23.95

.THE DANA GUIDE TO BRAIN HEALTH: A Practical Family Reference from Medical Experts (with CD-ROM)

*Floyd E. Bloom, M.D., M. Flint Beal, M.D., and David J. Kupfer, M.D., Editors*

*Foreword by William Safire*

The only complete, authoritative family-friendly guide to the brain's development, health, and disorders. *The Dana Guide to Brain Health* offers ready reference to our latest understanding of brain diseases as well as information to help you participate in your family's care. 16 full-color illustrations and more than 200 black-and-white drawings.

Paper (with CD-ROM) 733 pp. 1-932594-10-8 • $25.00

THE CREATING BRAIN: The Neuroscience of Genius

*Nancy C. Andreasen, M.D., Ph.D.*

A noted psychiatrist and best-selling author explores how the brain achieves creative break-throughs, including questions such as how creative people are different and the difference between genius and intelligence. She also describes how to develop our creative capacity. 33 illustrations/photos.

Cloth 197 pp. 1-932594-07-8 • $23.95

THE ETHICAL BRAIN

*Michael S. Gazzaniga, Ph.D.*

Explores how the lessons of neuroscience help resolve today's ethical dilemmas, ranging from when life begins to free will and criminal responsibility. The author, a pioneer in cognitive neuroscience, is a member of the President's Council on Bioethics.

Cloth 201 pp. 1-932594-01-9 • $25.00

A GOOD START IN LIFE: Understanding Your Child's Brain and Behavior from Birth to Age 6

*Norbert Herschkowitz, M.D. and Elinore Chapman Herschkowitz*

The authors show how brain development shapes a child's personality and behavior, discuss-ing appropriate rule-setting, the child's moral sense, temperament, language, playing, aggres-sion, impulse control, and empathy. 13 illustrations.

Cloth 283 pp. 0-309-07639-0 • $22.95
Paper (Updated with new material) 312 pp. 0-9723830-5-0 • $13.95

BACK FROM THE BRINK: How Crises Spur Doctors to New
Discoveries About the Brain

*Edward J. Sylvester*

In two academic medical centers, Columbia's New York Presbyterian and Johns Hopkins
Medical Institutions, a new breed of doctor, the neurointensivist, saves patients with life-
threatening brain injuries. 16 illustrations/photos.

Cloth 296 pp. 0-9723830-4-2 • $25.00

THE BARD ON THE BRAIN: Understanding the Mind Through the Art of
Shakespeare and the Science of Brain Imaging

*Paul Matthews, M.D. and Jeffrey McQuain, Ph.D. Foreword by Diane Ackerman*

Explores the beauty and mystery of the human mind and the workings of the brain, following
the path the Bard pointed out in 35 of the most famous speeches from his plays. 100 illustrations.

Cloth 248 pp. 0-9723830-2-6 • $35.00

STRIKING BACK AT STROKE: A Doctor-Patient Journal

*Cleo Hutton and Louis R. Caplan, M.D.*

A personal account with medical guidance from a leading neurologist for anyone enduring the
changes that a stroke can bring to a life, a family, and a sense of self. 15 illustrations.

Cloth 240 pp. 0-9723830-1-8 • $27.00

UNDERSTANDING DEPRESSION: What We Know and
What You Can Do About It

*J. Raymond DePaulo Jr., M.D. and Leslie Alan Horvitz*

*Foreword by Kay Redfield Jamison, Ph.D.*

What depression is, who gets it and why, what happens in the brain, troubles that come with
the illness, and the treatments that work.

Cloth 304 pp. 0-471-39552-8 • $24.95
Paper 296 pp. 0-471-43030-7 • $14.95

KEEP YOUR BRAIN YOUNG: The Complete Guide to Physical and Emotional Health and Longevity

*Guy McKhann, M.D. and Marilyn Albert, Ph.D.*

Every aspect of aging and the brain: changes in memory, nutrition, mood, sleep, and sex, as
well as the later problems in alcohol use, vision, hearing, movement, and balance.

Cloth 304 pp. 0-471-40792-5 • $24.95
Paper 304 pp. 0-471-43028-5 • $15.95

## THE END OF STRESS AS WE KNOW IT

*Bruce McEwen, Ph.D. with Elizabeth Norton Lasley*

*Foreword by Robert Sapolsky*

How brain and body work under stress and how it is possible to avoid its debilitating effects.

Cloth 239 pp. 0-309-07640-4 • $27.95
Paper 262 pp. 0-309-09121-7 • $19.95

## IN SEARCH OF THE LOST CORD: Solving the Mystery of Spinal Cord Regeneration

*Luba Vikhanski*

The story of the scientists and science involved in the international scientific race to find ways to repair the damaged spinal cord and restore movement. 21 photos; 12 illustrations.

Cloth 269 pp. 0-309-07437-1 • $27.95

## THE SECRET LIFE OF THE BRAIN

*Richard Restak, M.D.*

*Foreword by David Grubin*

Companion book to the PBS series of the same name, exploring recent discoveries about the brain from infancy through old age.

Cloth 201 pp. 0-309-07435-5 • $35.00

## THE LONGEVITY STRATEGY: How to Live to 100 Using the Brain-Body Connection

*David Mahoney and Richard Restak, M.D.*

*Foreword by William Safire*

Advice on the brain and aging well.

Cloth 250 pp. 0-471-24867-3 • $22.95
Paper 272 pp. 0-471-32794-8 • $14.95

## STATES OF MIND: New Discoveries About How Our Brains Make Us Who We Are

*Roberta Conlan, Editor*

Adapted from the Dana/Smithsonian Associates lecture series by eight of the country's top brain scientists, including the 2000 Nobel laureate in medicine, Eric Kandel.

Cloth 214 pp. 0-471-29963-4 • $24.95
Paper 224 pp. 0-471-39973-6 • $18.95

*Immunology:*

RESISTANCE: The Human Struggle Against Infection

*Norbert Gualde, M.D., translated by Steven Rendall*

Traces the histories of epidemics and the emergence or reemergence of diseases, illustrating how new global strategies and research of the body's own weapons of immunity can work together to fight tomorrow's inevitable infectious outbreaks.

Cloth 219 pp. 1-932594-00-0 $25.00

FATAL SEQUENCE: The Killer Within

*Kevin J. Tracey, M.D.*

An easily understood account of the spiral of sepsis, a sometimes fatal crisis that most often affects patients fighting off nonfatal illnesses or injury. Tracey puts the scientific and medical story of sepsis in the context of his battle to save a burned baby, a sensitive telling of cutting-edge science.

Cloth 225 pp. 1-932594-06-x • $23.95
Paper 225 pp. 1-932594-09-4 • $12.95

*Arts Education:*

A WELL-TEMPERED MIND: Using Music to Help Children
Listen and Learn

*Peter Perret and Janet Fox*

*Foreword by Maya Angelou*

Five musicians enter elementary school classrooms, helping children learn about music and contributing to both higher enthusiasm and improved academic performance. This charming story gives us a taste of things to come in one of the newest areas of brain research: the effect of music on the brain. 12 illustrations.

Cloth 231 pp. 1-932594-03-5 • $22.95
Paper 231 pp. 1-932594-08-6 • $12.00

## Free Educational Books

(Information about ordering and downloadable PDFs are available at www.dana.org.)

PARTNERING ARTS EDUCATION: A Working Model from
ArtsConnection

This publication describes how classroom teachers and artists learned to form partnerships as they built successful residencies in schools. *Partnering Arts Education* provides insight and concrete steps in the ArtsConnection model. 55 pp.

ACTS OF ACHIEVEMENT: The Role of Performing Arts Centers
in Education

Profiles of more than 60 programs, plus eight extended case studies, from urban and rural communities across the United States, illustrating different approaches to performing arts education programs in school settings. Black-and-white photos throughout. 164 pp.

PLANNING AN ARTS-CENTERED SCHOOL: A Handbook

A practical guide for those interested in creating, maintaining, or upgrading arts-centered schools. Includes curriculum and development, governance, funding, assessment, and community participation. Black-and-white photos throughout. 164 pp.

THE DANA SOURCEBOOK OF BRAIN SCIENCE: Resources for
Teachers and Students
Fourth Edition

A basic introduction to brain science, its history, current understanding of the brain, new developments, and future directions. 16 color photos; 29 black-and-white photos; 26 black-and-white illustrations. 160 pp.

THE DANA SOURCEBOOK OF IMMUNOLOGY: Resources for Secondary and Post-Secondary Teachers and Students

An introduction to how the immune system protects us, what happens when it breaks down, the diseases that threaten it, and the unique relationship between the immune system and the brain. 5 color photos; 36 black-and-white photos; 11 black-and-white illustrations. 116 pp.
ISSN: 1558-6758

PERIODICALS

Dana Press also offers several periodicals dealing with arts education, immunology, and brain science. These periodicals are available free to subscribers by mail. Please visit www.dana.org.